Biomarkers of Radiation in the Environment

NATO Science for Peace and Security Series

This Series presents the results of scientific meetings supported under the NATO Programme: Science for Peace and Security (SPS).

The NATO SPS Programme supports meetings in the following Key Priority areas: (1) Defence Against Terrorism; (2) Countering other Threats to Security and (3) NATO, Partner and Mediterranean Dialogue Country Priorities. The types of meeting supported are generally "Advanced Study Institutes" and "Advanced Research Workshops". The NATO SPS Series collects together the results of these meetings. The meetings are co-organized by scientists from NATO countries and scientists from NATO's "Partner" or "Mediterranean Dialogue" countries. The observations and recommendations made at the meetings, as well as the contents of the volumes in the Series, reflect those of participants and contributors only; they should not necessarily be regarded as reflecting NATO views or policy.

Advanced Study Institutes (ASI) are high-level tutorial courses to convey the latest developments in a subject to an advanced-level audience.

Advanced Research Workshops (ARW) are expert meetings where an intense but informal exchange of views at the frontiers of a subject aims at identifying directions for future action.

Following a transformation of the programme in 2006, the Series has been re-named and re-organised. Recent volumes on topics not related to security, which result from meetings supported under the programme earlier, may be found in the NATO Science Series.

The Series is published by IOS Press, Amsterdam, and Springer, Dordrecht, in conjunction with the NATO Emerging Security Challenges Division.

Sub-Series

A.	Chemistry and Biology	Springer
B.	Physics and Biophysics	Springer
C.	Environmental Security	Springer
D.	Information and Communication Security	IOS Press
E.	Human and Societal Dynamics	IOS Press

http://www.nato.int/science
http://www.springer.com
http://www.iospress.nl

Series A: Chemistry and Biology

Biomarkers of Radiation in the Environment

Robust Tools for Risk Assessment

edited by

Michael D. Wood
School of Science, Engineering and Environment
University of Salford
Salford, UK

Carmel E. Mothersill
McMaster University
Hamilton, ON, Canada

Gohar Tsakanova
Institute of Molecular Biology NAS RA
Yerevan, Armenia

Tom Cresswell
Australian Nuclear Science and Technology Organisation, ANSTO
Lucas Heights, NSW, Australia

and

Gayle E. Woloschak
Feinberg School of Medicine
Northwestern University
Chicago, IL, USA

Springer

Published in Cooperation with NATO Emerging Security Challenges Division

Proceedings of the NATO Advanced Research Workshop on Biomarkers of Radiation in the Environment: Robust Tools for Risk Assessment (BRITE)
Yerevan, Armenia
27–30 November 2017

ISBN 978-94-024-2103-3 (PB)
ISBN 978-94-024-2100-2 (HB)
ISBN 978-94-024-2101-9 (eBook)
https://doi.org/10.1007/978-94-024-2101-9

Published by Springer,
P.O. Box 17, 3300 AA Dordrecht, The Netherlands.

www.springer.com

Printed on acid-free paper

This Springer imprint is published by the registered company Springer Nature B.V.
The registered company address is: Van Godewijckstraat 30, 3311 GX Dordrecht, The Netherlands

To our friend and colleague Margarita Malakyan, whose warmth, kindness and dedication to science and the promotion of collaborative research enabled this workshop to occur in Armenia and gave us all an opportunity to visit this ancient country and learn about its history.
Always remembered.

Acknowledgements

The Editors would like to acknowledge Andrej Rusin, an M.Sc. candidate in the Mothersill/Seymour Laboratory at McMaster University, Canada, for his assistance in the final preparation of this manuscript.

About This Book

This book presents papers from an Advanced Research Workshop sponsored by the North Atlantic Treaty Organisation (NATO). The workshop, *'Biomarkers of Radiation In The Environment: Robust tools for risk assessment (BRITE)'*, took place in Yerevan, Armenia from 27th – 30th November 2017. It brought together more than 40 international scientists from a broad range of disciplines including environmental protection, radiation metrology, radiobiology and radioecology.

The aim of the workshop was to identify biomarkers that are practical for use in the field and are robust enough to aid in risk assessment. Biomarkers were loosely defined as biological indicators of change that can be used to identify exposure events and inform risk assessment.

For effective use of biomarkers in an environmental context, biomarker samples need to be collected non-lethally. Otherwise it is difficult to justify that the use of BRITE is helping to ensure protection of wildlife. In human biomarker research, non-lethal sampling is clearly the norm. Therefore, the BRITE workshop facilitated knowledge exchange between the radiobiology and radioecology communities. Insights from cancer research, epigenetics and risk assessment (both human and non-human) were discussed, recognising that state-of-the-art biomarkers being developed for humans deserve consideration for environmental applications and vice versa. Sessions were wide-ranging, covering methods, mechanisms, cross disciplinary application and regulation.

The chapters in this book have been grouped into five major themes that were covered by the BRITE workshop:

- Techniques for biomarker development
- Low-dose effect mechanisms
- Biomarkers for risk evaluation
- Biomarkers in wildlife
- Biomarker use and responses

Each chapter has been written independently and reflects the views of the chapter author(s), allowing the reader to form their own balanced view of the different perspectives on biomarkers of radiation in the environment. Given the breadth of topics covered and the state-of-the-art perspectives shared by leading experts in their respective fields, this book should form a valuable resource for anyone with an interest in how biomarkers can be used to improve our understanding of radiation in the environment and its potential impacts.

Contents

About the Editors

Michael D. Wood is Professor of Applied Ecology at the University of Salford, UK. A Chartered Radiation Protection Professional, his radioecological research focuses on the behaviour, fate and impacts of radionuclides in the environment. This research has led to various prestigious awards, including the Times Higher Education award for research and the Society for Radiological Protection's Founders' Medal. A regular keynote speaker at international conferences, Professor Wood has also acted as an invited expert for the International Atomic Energy Agency and the European Commission in the area of environmental radiation protection. Beyond his core scientific research, Professor Wood is internationally recognised for his expertise in communicating radiation science and risk, including the innovative use of virtual reality to engage audiences with his Chernobyl-based research.

Carmel Mothersill is a Professor and Canada Research Chair in Environmental Radiobiology at McMaster University in Canada. Her research is focused on low dose effects of radiation, in particular non-targeted effects where radiation effects are seen in cells and organisms that have not received a direct dose of radiation but receive signals from irradiated cells or organisms. Professor Mothersill has received many prestigious awards including the Nikolai Timofeev-Ressovsky award from the Russian Academy of Sciences, The Marie Curie medal, and the St Luke's Lecture and medal from the Irish Academy of Medicine. She also received an Honorary Doctorate from Heriot Watt University in Scotland. Professor Mothersill has organised a number of international conferences and workshops including the International Congress on Radiaoecology and Environmental Radiation in Canada and the International Congress of Radiation Research in Ireland. She is currently Treasurer of the International Union of Radioecologists.

Gohar Tsakanova is the deputy director and senior scientist at the Institute of Molecular Biology NAS RA, and head of Laboratory of Experimental Biology and senior scientist at CANDLE Synchrotron Research Institute, Yerevan, Armenia. Her primary work is related to the molecular and cellular pathomechanisms of cerebrovascular diseases, aging and radiation biology aiming to development of effective treatment strategies. She is member of editorial board and reviewer of a number of international scientific journals, member and co-chair of international grant programs on radiobiology.

Tom Cresswell is a Senior Research Scientist in Ecotoxicology at Australia's Nuclear Science and Technology Organisation (ANSTO). He uses nuclear techniques to understand the fate and impacts of a range of anthropogenic contaminants on aquatic organisms. Dr Cresswell has been invited to several key international meetings other than BRITE, including the Plastics Awareness Global Initiative (PAGI) held at UC San Diego in 2018 and several expert meetings organised by the International Atomic Energy Agency (IAEA) on using nuclear techniques in marine plastics research and using radioecological tracers to determine coastal and marine ecosystem health. Dr Cresswell is a board member of Science and Technology Australia (STA), a peak body representing over 70,000 STEM professionals across the country and holds executive board positions for the Society of Environmental Toxicology and Chemistry (SETAC) Australasia and the South Pacific Environmental Radioactivity Association (SPERA). He regularly participates in science outreach activities and STEM communication events to inspire the next generation of environmental scientists.

Gayle E. Woloschak is Professor of Radiation Oncology, Radiology, and Cell and Molecular Biology in the Feinberg School of Medicine, Northwestern University. Gayle received her B.S. in Biological Sciences, from Youngstown State University and a Ph.D. in Medical Sciences from the University of Toledo (Medical College of Ohio). She did her postdoctoral training at the Mayo Clinic, and then moved to Argonne National Laboratory until 2001. Her scientific interests are predominantly in the areas of Molecular Biology. Radiation Biology, and Nanotechnology studies, and she has authored over 200 papers. She is editor-in-chief for the International Journal of Radiation Biology, is past-President of the Radiation Research Society, is a member of the National Council of Radiation Protection and Measurements, is a member of Committee-1 for the International Commission on Radiation Protection and serves on a committee studying low dose radiation effects for UNSCEAR. She is also Associate Dean of Graduate Student and Postdoctoral Affairs in The Graduate School at Northwestern University.

Part I
Techniques for Biomarker Development

Chapter 1
Reducing Uncertainties in Live Monitoring of Radiation in Wildlife

Tom Cresswell

Abstract Inputs of radionuclides to the environment can result in minor increases in radiation levels for short periods of time, as with the release of radionuclides used in medical imaging, to major increases of radiation levels for long periods (>10s of years) as is potentially the case for the immediate area surrounding the Fukushima Daiichi NPP. Assessments of the uptake of radionuclides by wildlife have traditionally been conducted using lethal sampling techniques. Non-lethal and non-invasive techniques of assessing radiation levels in wildlife are becoming increasingly necessary for protecting animal welfare in research. This chapter explores the techniques used to determine levels of gamma radiation within live aquatic organisms in the laboratory and seeks to understand how such techniques may be applied to field monitoring. Important methodological principles of laboratory radiotracing are explored. The main factors discussed are animal rinsing, radioanalysis and geometry corrections. Examples of techniques in live aquatic animal radiotracer studies within the laboratory are provided to improve data quality control and demonstrate why each technique is crucial in interpreting the data from such studies. By using live radiation monitoring techniques of aquatic organisms in the field, an assessment can be made post accidental release of the potential for bioaccumulation and retention of radionuclides by native biota under site-specific conditions. This process will provide a clearer understanding of the acute and chronic effects of radionuclides to aquatic biota and associated risk management strategies to be determined.

1.1 Introduction

Elevated activities of radionuclides above background levels and their associated radioactivity can enter the natural environment through a number of ways; i) through extraction of mineralised resources and liberation of naturally-occurring radioactive

T. Cresswell (✉)
Australian Nuclear Science and Technology Organisation, ANSTO,
Lucas Heights, NSW, Australia
e-mail: Tom.Cresswell@ansto.gov.au

M. D. Wood et al. (eds.), *Biomarkers of Radiation in the Environment*, NATO Science for Peace and Security Series A: Chemistry and Biology,
https://doi.org/10.1007/978-94-024-2101-9_1

material (NORM) and technological-enhancement of that material (TENORM); ii) operational release of radionuclides from medical facilities iii) operation of nuclear facilities and subsequent generation of waste streams; iv) accidental releases of radionuclides from nuclear facilities such as occurred at the Chernobyl Nuclear Power Plant (NPP) in 1986 and at the Fukushima Daiichi NPP in 2011; v) testing of nuclear weapons; v) accidents involving the transportation of nuclear weapons via air (as occurred in 1966 over the Palomares village, Spain and in 1968 near the Thule Air Base, Greenland), land or sea. These inputs of radionuclides to the environment can result in minor increases in radiation levels for short periods of time, as is often the case with the release of radionuclides used in medical imaging, to major increases of radiation levels for long (>100 s of years) as was the case for the immediate area surrounding the Chernobyl NPP. The impact of radiation on wildlife is therefore dependent on the radionuclide released to the environment and the concentration of that radionuclide.

Once in the environment, radionuclides can impact wildlife in two main ways; indirectly through external gamma radiation providing a dose to wildlife and directly through accumulation of the radionuclide into an organism, providing potential gamma, beta and alpha radiation doses to the organism. Understanding the pathways of radionuclide bioaccumulation by organisms has been covered in many previous publications and is broadly considered to be via the respiratory pathway (i.e. through inhalation of a particle-associated radionuclide, via inhalation of an aerosol radionuclide for terrestrial animals and plants or via the respiration of a soluble radionuclide for aquatic organisms), the ingestion pathway (i.e. direct ingestion of particle-associated or soluble radionuclides in animals) or via passive absorption from the surrounding environment (i.e. accumulation into terrestrial flora from uptake via the soil or into aquatic flora from the external solution).

As discussed in Chaps. 11, 12, 13, 14 and 15, radiation can have varying effects on wildlife depending on the internal and external doses received so monitoring these exposures is important when assessing wildlife responses to radiation. This chapter primarily addresses the importance of monitoring bioaccumulated radiation within live organisms i.e. radionuclides that have crossed an external cellular membrane and have been transported to an internal organ where they are imparting a dose to the organism. Current field and laboratory measurement techniques will be identified and recommendations for using laboratory techniques in field monitoring will be provided. This chapter defines wildlife for the purposes of radiation protection as per the IAEA handbook on radionuclide transfer to wildlife (Howard et al., 2013); "all living organisms including all non-domesticated plants, animals and other organisms including feral species (i.e. non-native, self-sustaining populations)".

1.2 Monitoring Radiation Doses to Wildlife

Radiation dose exposure of wildlife has been conducted using personal dosimetry devices such as thermoluminescent dosimeters (TLDs). Aramrun et al., (2018) provide an evaluation of different dosimetry techniques available for specifically measuring the external exposure of wildlife. The authors of that paper suggest that LiF and Al_2O_3:C based TLDs, optically stimulated luminescent dosimeters (OSLD) and radiophotoluminescent dosimeters (RPLD) could be used to estimate doses to wildlife owing to specific criteria of dosimeter properties, detection limit, dose range and the suitability for a range of study organisms in field scenarios.

TLDs have been used to monitor the radiation exposure of mammals and birds, recently in conjunction with Global Positioning Systems (GPS) locator beacons or radio frequency tracking equipment by attaching the devices to the animals by way of a collar or bracelet. Hinton et al., (2015) developed a device that combined GPS technologies with electronic dosimeters and tested the tools in the laboratory exposed to uniform external radiation and in the field where they assessed their durability when worn on wildlife under natural conditions. The GPS-dosimeters were attached to wild pigs living on the U.S. Department of Energy's Savannah River Site in Aiken, South Carolina. This site is a large area (>750 km^2) former nuclear production facility. The GPS-dosimeters had a limit of reliable detection of approximately 3μSv and were found to be sufficiently robust and precise over a 45 day deployment on wild pigs to determine the location and short-term radiation dose received by free-ranging wildlife (Hinton et al., 2015).

Dosimeters made of LiF powder have been attached to live fish in the Columbia River, USA, downstream from effluent outfalls of the Hanford plutonium producing reactors to complement static dosimeters submerged in the river (Watson & Templeton, 1971). Dosimeters were attached just below the dorsal fin while others were implanted in the abdominal cavity through a slit in the abdominal wall, anterior to the pelvic fins. While these animals were not allowed to swim freely along the reaches of the river, they were kept in a fenced enclosure near the river shore where they had access to natural foods and were exposed to external sources of radiation in the river.

These techniques of dosimetry monitoring provide invaluable information about the heterogeneity of radiation sources in the environment as well as data to be able to predict long-term radiation exposures to animals. Monitoring the radiation exposure to flora is relatively easier than fauna as one is able to place TLDs or similar devices in situ (e.g. within the soil, at the soil surface, at increasing elevations from the soil surface, at the water surface or at increasing depths from the water surface).

There are some disadvantages of using dosimetry to monitor radiation in wildlife; these devices only provide information on external radiation doses to wildlife and are unable to provide information on any internalised/bioaccumulated radionuclides (unless gamma-emitting radionuclides of significant activities have been bioaccumulated and provide a dose through the cuticle to the attached dosimeter). To better understand the effects of radiation on wildlife, it is important to have an estimate of the internal (i.e. from bioaccumulated radionuclides) and external doses to the organism.

1.3 Determining Internal Radiation in Wildlife

By far, one of the most accurate measures of determining the internal radionuclide activity concentrations in an organism is through analysis of individual organs and tissues. This method allows for a detailed determination of the biodistribution of radionuclides and subsequent assessment of the internal radiation dose that organism would encounter. Monitoring programs of radionuclides in the environment often measure and report only the contaminant concentrations in edible parts of animal tissues (e.g. muscle), which will provide valuable information for human consumers but not necessarily for undertaking risk assessments to non-human biota. Yankovich et al. (2010) published a comprehensive series of look-up tables with whole-body to tissue-specific concentration ratios for a range of aquatic and terrestrial vertebrates and invertebrates. These values can then be used to determine tissue-specific and whole-body concentrations of contaminants for radiological risk assessments of biota using data generated from typical human-consumer tissues. Johansen et al. (2016) also provide a series of whole-body to tissue-specific ratios for $^{239,240}Pu$ concentration in mammalian wildlife following dispersal of plutonium by accidental-release tests in Maralinga, Australia.

Following the accident at the Fukushima Daiichi Nuclear Power Plant (FDNPP) in March 2011, the majority of aquatic biota assessments in the surrounding marine environment focused on using fish muscle tissue analysis. This was primarily in response to the requirement for a human-consumer risk assessment of seafood safety. However, Johansen et al. (2015) highlighted that when estimating doses to fish, the muscle data should not be used alone as radionuclides of strontium and plutonium were present in much greater concentrations in the bone and gastrointestinal (GI) tract of fish affected by the Fukushima accident and therefore muscle to tissue-specific ratios are important to define.

While the above tissue-specific assessments of radionuclide burdens provide the most thorough assessment of internalised radionuclide concentration within wildlife, these techniques require the euthanasia of the organism, along with time-consuming dissection, sample preparation and costly analysis. Initiatives such as the National Centre for Replacement, Refinement & Reduction of Animals in Research (NC3Rs; www.nc3rs.org.uk), are strongly promoting a move away from the destructive use of animals in research (Cook & Robinson, 2007) towards the development of new research techniques and approaches (Burden et al., 2015). This has been supplemented by international initiatives, such as the formation of the International Union of Radioecology (IUR) Task Group on Non-lethal Methods (http://goo.gl/GbgSnG).

It is therefore important to note that there are several types of non-lethal radiation detection options available, each with their own pros and cons. These options consist of the non-lethal, invasive determination of alpha, beta and gamma-emitting radionuclides via small mass tissue sampling (e.g. clippings of hair, fur, skin or bone) and blood samples and the non-lethal, non-invasive determination of gamma-emitting radionuclides using in-situ gamma spectrometry.

1.3.1 Non-lethal, Invasive Sampling for Radiation Determination in Wildlife

Tissue samples can be harvested from wildlife for radionuclide analysis using non-lethal techniques. Wood et al. (2011b) summarise techniques for estimating contaminant burdens in reptiles from the collection of osteoderms, scutes, skin, tail tissue, eggs and blood. Highly sensitive analysis methods (e.g. accelerator mass spectrometry) allows key radionuclides to be measured in very small, non-lethal tissue samples such as ear punches from mammals. Care must be taken when using non-lethally harvested tissue samples to infer internal organ/tissue concentrations as the correlations between these tissues may be species-specific (Burger et al., 2010) and each tissue will represent a different period of contaminant bioaccumulation. For example, radionuclide concentrations in the blood will represent current metabolic activity from very recent exposure; skin, eggs and hair may represent medium-term exposure, while osteoderms and tail tissue would likely represent long-term/historical exposure. The results from non-lethal samples may be correlated with other tissues, or with the whole-body burden. Activity concentration ratios among the main tissues of various wildlife organisms are summarised in knowledge of an animal's feeding, mating and migratory patterns is essential to interpreting results from contaminant burden of non-lethally harvested tissue. Also, skin punches and similar samples are prone to surface contamination which must be removed so that the analysis results represent absorbed (vs. adsorbed) contamination.

1.3.2 Non-lethal, Non-invasive Gamma Analysis of Radiation in Wildlife

Direct live monitoring of radiation in wildlife in situ has been undertaken in a small number of studies. Beresford et al. (1996) used a hand-held, lead-shielded 1.75" NaI detector linked to a single channel analyser to determine ^{137}Cs emissions from the hind-leg muscle of ewes and lambs in Cumbria, UK. The study used methods developed by Meredith et al. (1988) who determined that reporting the mean of triplicate 10 second counts on the most fleshy part of the buttock of the sheep was a sufficient to ensure good detection whilst being able to keep the sheep still during counting. Brisbin (1989) determined whole body burdens of radiocesium for hatchling alligators in a reservoir located on the United States Department of Energy's

Savannah River Plant (SRP), a pond which provided a closed-loop system of cooling water for one or more of the SRP's nuclear production reactors. Hatchlings were restrained in plastic tubes within a lead-shielded, whole-body counting chamber (10.2 × 15.2 cm h × w) and radioanalysed with a NaI detector where the centre of the alligator's body was centred immediately above the detector. Gamma emissions were counted during a 2000 second period and compared with same-day net counts from an aqueous phantom of similar body mass and geometry to the alligator (Brisbin, 1989).

In order to reduce the influence of background radiation when monitoring low levels of radiation in subjects, it is important to maintain the subject away from sources of contamination (e.g. soil or water) and ideally take long measurements of background radiation for subtraction at regular intervals. If monitoring for specific radionuclides e.g. ^{137}Cs, the gamma detection equipment can be set up with a reduced gamma energy window (400–800 keV) to ensure the target peak is incorporated (in this case 662 keV of ^{137}Cs), while reducing the influence of background radiation (Field, 2011). Repeat analyses of the same individual should be conducted in at least triplicate. To convert net count rates of ^{137}Cs to a radionuclide concentration, as was the case for sheep monitored in Wales and England by Beresford et al. (1996) and Field (2011), calibration curves can be created from live-animal net count rates compared directly to a subset of the same slaughtered animals whose muscle tissue has been analysed for Bq ^{137}Cs/kg of fresh tissue (Meredith et al., 1988).

While there have been several studies of live-animal gamma monitoring of terrestrial wildlife conducted as reported above, there are very few, if any studies reporting the live-monitoring of radiation in aquatic organisms. Following the same desire to reduce the number of animals sacrificed for research or monitoring purposes as described above, it is important to examine the potential for conducting live-animal radioanalysis of aquatic organisms for field monitoring purposes. This has become more necessary following the accident at the FDNPP in Japan in 2011, where radionuclides such as ^{134,137}Cs were released to the coastal environment. Furthermore, the majority of nuclear power plants around the world are situated adjacent to a large water body due to the requirement for great volumes of cooling water.

Monitoring of radiation within live aquatic organisms has been conducted in ecotoxicology laboratory studies since the 1970s (e.g. Fowler et al., 1971), providing opportunities to understand the kinetics of contaminant uptake and regulation by a range of organisms using non-destructive techniques (Cresswell et al., 2015, 2017a; Lanctôt et al., 2017a, b; Lee et al., 2015; Metian et al., 2011; Wang, 2001; Warnau & Bustamante, 2007).

1.4 Examples of Live-Animal Radioisotope Tracing Studies in the Laboratory

Live-animal radioisotope tracing in aquatic organisms usually involves exposing the animals to aqueous radiotracers (i.e. dissolved) that represent the contaminant(s) of interest. The animal is immersed in a media appropriate to its natural environment (e.g. freshwater, estuarine water or marine water), either as reconstituted water (i.e. deionised water with the addition of salts to achieve the representative physico-chemical parameters to be studied), or as filtered (often <0.45μm) field-collected water. The exposure media is then spiked with a radionuclide (often at 1–50 Bq/mL) such that there is no change to physico-chemical parameters and the spike represents field concentrations or lower of the contaminant (often at 0.01–1μg contaminant/L).

At pre-determined times during the exposure the animal is carefully removed from the exposure media, rinsed to remove weakly-sorbed radiotracer from the cuticle then placed into an enclosed vial and radioanalysed (counted) using a coaxial detector (usually HPGe, LaBr or NaI) underneath the organism, all housed within a lead shield. The detectors used are connected to a multi-channel analyser (MCA), which will determine the counts received from the organism as either counts per second (CPS) or counts per minute (CPM). After a short count period (often 10–600 seconds depending on the radioactivity concentration within the organism) the animal is returned to fresh exposure media and is counted again at a later stage, often with multiple repeat measurements taken during the exposure. After a predetermined exposure period, the animals can be transferred to 'clean' media without the inclusion of any active or non-active forms of the contaminant being studies. This allows for the determination of the retention of the contaminant within the animal and for the quantification of depuration/elimination rates via multiple counts over time in clean water. These aqueous exposures can be conducted with multiple contaminants (e.g. mixture exposures), providing the detector used has sufficient energy resolution to distinguish between the gamma energies of each radioisotope. Assimilation of contaminants from dietary sources can also be conducted using radioisotope tracers by conducting pulse-chase exposures. Conducting separate aqueous and dietary contaminant exposures with the same organisms using radioisotope tracers can be used to determine the predominant source of contaminant to an organism under controlled conditions (Casado-Martinez et al., 2009; Creighton & Twining, 2010; Cresswell et al., 2014a; Lee et al., 2015).

The aqueous radioisotope tracer exposure techniques described above have been used to determine the aqueous bioaccumulation kinetics of ^{54}Mn, ^{60}Co, ^{65}Zn, ^{75}Se, ^{109}Cd, ^{110m}Ag, ^{134}Cs and ^{241}Am in adult zebrafish (Reinardy et al., 2011), ^{109}Cd, ^{75}Se and ^{65}Zn bioaccumulation by tadpoles (Lanctôt et al., 2017a), ^{137}Cs, ^{65}Zn and ^{51}Cr bioaccumulation by green mussels (Qureshi et al., 2007), ^{226}Ra by tropical freshwater mussels (Doering and Bollhöfer, 2017) and ^{134}Cs and ^{241}Am by king scallops (Metian et al., 2011).

1.5 Reducing Uncertainties in Live Monitoring of Radiation in Aquatic Organisms

All published studies in the area of laboratory radioisotope tracing rely on three main methodological factors associated with live monitoring of aquatic organisms; animal rinsing, some method for immobilising animals during counting and corrections for irregular geometries during radioanalysis. The importance of these methods and errors associated with such procedures have been discussed by Cresswell et al., (2017b) and will be explored below for the application of live monitoring of radiation in aquatic wildlife.

1.5.1 Animal Rinsing Pre Radioanalysis

As described above, rinsing aquatic organisms that have been exposed to dissolved radionuclides in solution requires that the animals are rinsed to remove weakly-sorbed radiotracer from the cuticle. This is to ensure that the counts received by the detector reflect bioaccumulated (i.e. internalised) radiotracers only. This is also important when considering field organisms and the presence of contaminated dietary items within the gut; organisms should be depurated prior to radioanalysis to ensure only the radionuclides within animal tissues are being quantified. Cresswell et al. (2017b) propose that the rinse technique should be validated for each study by conducting successive radioanalysis – rinse – radioanalysis cycles until a stable count rate is reached. Rinse media should be either the standard medium the organism is exposed to (i.e. freshwater, estuarine water or marine water without the addition of the radionuclide and/or stable contaminant) and/or a pH-matched chelating solution (e.g. ethylenediaminetetraacetic acid; EDTA), which acts to increase the removal of weakly-sorbed radioelements from the cuticle (Cresswell et al., 2017b). This rinse technique and validation could be conducted relatively easily in field monitoring applications and any validation should be recorded and reported with the results.

It is important to note that the physico-chemical parameters of the exposure media can influence the sorption of radionuclides to the external surfaces of an aquatic animal. Cresswell et al. (2017b) reported the effect of rinsing empty shells of the freshwater snail *Potamopyrgus antipodarum* after exposure to ^{141}Ce in either synthetic freshwater or filtered river water for 2 hours. Shells were radioanalysed then subjected to successive rinse cycles of either Milli-Q water (>18 MΩ resistivity) or 50/100μM EDTA and re-analysed after each set of rinses. Shells that had been exposed to ^{141}Ce in synthetic freshwater and rinsed with either media had a reduction of radioactivity of 90% after three rinse cycles (i.e. only 10% of initial activity remained). However, shells that were exposed to ^{141}Ce in filtered river water only had a reduction of 10–20% (i.e. 80–90% of the initial radioactivity remained adsorbed to the shell), even after six rinse cycles. The data suggested that there was a stronger binding of ^{141}Ce to the empty shells in the presence of organic ligands in the natural river water such that physical removal (i.e. continuous rinsing) had a negligible impact on radioisotope removal (Cresswell et al., 2017b). When considering live monitoring of wildlife in the field, it is therefore important to have some understanding as to the radionuclide being analysed and its chemical behaviour in the waterbody to gain an insight into the likelihood for increased cuticle adsorption. This understanding will assist in the interpretation of whole-body radioanalysis counts (i.e. an estimate of the proportion of counts originating from the cuticle vs. from tissue bioaccumulation).

1.5.2 Accounting for Live Animal Movement

As discussed by Cresswell et al. (2017b), radioanalysis of any active object is subject to errors when the subject is mobile. As a radioactive source moves across the scintillation crystal, the solid angle subtended by the detector changes (Knoll, 2010) and the signal amplitude received by the multi-channel analyser will change, therefore affecting the resultant radioactivity reported during the count period. It is crucial that the organism remains as immobile as possible during radioanalysis and, as such, it may be appropriate to anaesthetise the animal. However, providing the correct dose of anaesthetic to each animal to immobilise for a short period of time (ideally 5–10 min) can be problematic and potentially affect the physiology of the animal post counting, therefore affecting bioaccumulation kinetics or survivability. Furthermore, administration of anaesthesia to aquatic organisms in the field would be logistically difficult. The following sections describe organism-specific techniques for restricting the movement of live aquatic animals during radioanalysis.

1.5.2.1 Amphibians

Larval and juvenile striped marsh frogs, *Limnodynastes peronei*, were exposed to ^{109}Cd, ^{75}Se and ^{65}Zn in synthetic freshwater to determine the bioaccumulation kinetics under solo or mixture exposures (Lanctôt et al., 2017a) and during exposure to different oxidation states of Se (SeIV and SeVI) to understand bioaccumulation kinetics as a function of selenium speciation (Lanctôt et al., 2017b). Tadpoles were radioanalysed in 10 mL polystyrene centrifuge tubes containing 5 mL of clean water (Fig. 1.1a) and were constrained at the bottom of the tube using plastic inserts that were fabricated by cutting the top and bottom ends of a 3 mL plastic transfer pipette and inserting a small piece of sponge (8 × 8 × 4 mm) at the tip to inhibit the tadpoles from swimming up the tube (Lanctôt et al., 2017a). Tubes were held in place using a polystyrene sample rack with a hole in the bottom such that the tube sat upright in the centre of the detector head (Fig. 1.1b).

During another study, tadpoles underwent metamorphosis into juvenile frogs/froglets and outgrew the counting vial described above. Instead, rinsed tadpoles

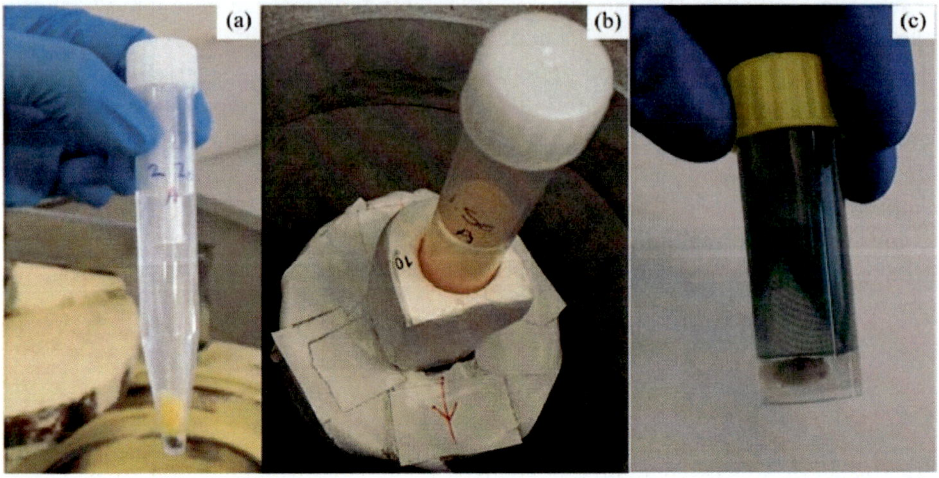

Fig. 1.1 Pictures of the set-up especially made for gamma-counting tadpoles (**a**), with the tube held in place by a polystyrene tube rack (**b**) and partially metamorphosed tadpoles/froglets (**c**) of the striped marsh frog, Limnodynastes peronei. (Photos courtesy of Chantal Lanctôt and Tom Cresswell)

were transferred to flat-bottomed 6 mL polystyrene tubes containing 4 mL of clean water (Fig.. 1.1c). Tadpoles were constrained to the bottom 4 mm of the tubes using plastic cylindrical inserts (H 40 × D 11 mm) closed at one end with 1 mm mesh to inhibit the tadpoles from swimming up the tube (Lanctôt et al., 2017b).

1.5.2.2 Ascidians

The tunicate ascidian *Styela plicata* was exposed to ^{109}Cd and ^{65}Zn in filtered seawater to determine bioaccumulation kinetics under solo or mixture exposures (Cresswell et al., unpublished, unpublished). Once the ascidians were removed from exposure solutions, they were carefully manually compressed to force unassimilated solution from the internal cavities prior to being rinsed. Individuals were placed in sealable polyethylene bags and radioanalysed in the same orientation and position on the detector to ensure reproducibility between counts.

1.5.2.3 Bivalves

Generally it is advisable that the subject/sample being radioanalysed fits entirely over the head of the gamma detector to ensure good counting efficiency. However, previous studies have been conducted using oysters with a greater diameter than the detector head. Lee et al., (2015) radioanalysed live Sydney rock oysters, *Saccostrea glomerata*, exposed to aqueous ^{65}Zn in seawater. The oysters were contained in sealable polyethylene bags and situated on the detector such that the digestive organs (main site of Zn bioaccumulation; Lee et al., 2015) were located over the centre of the detector. Four repeat radioanalyses of five oysters orientated in the same position on the detector resulted in a mean relative standard deviation of counts of 3% (Cresswell et al., 2017b), suggesting that reproducible results can be achieved when taking orientation and location of the animal into account.

1.5.2.4 Decapod Crustaceans

Spotted shore crabs, *Paragrapsus laevis*, have been radioanalysed live in the laboratory following exposure to ^{134}Cs and ^{85}Sr (Cresswell et al., unpublished). The crabs were placed in cylindrical 250 mL polycarbonate vials (slightly larger diameter to the width of the crab but the same diameter as the LaBr detector head) and were kept immobile using a moist sponge cut to a slightly greater diameter than the vial, pressed down onto the top shell of the crab as shown in Fig. 1.2. Care was taken with each radioanalysis to ensure that the crab was always in the same position within the vial (by aligning the distal end of the telson with the centre point of the vial) and that the crab remained in the same orientation on the detector each time (Cresswell et al., 2017b).

Prawns are less amenable to being out of water for sustained periods and have therefore been analysed in the presence of clean media. Cresswell et al. (2015) radioanalysed ^{109}Cd and ^{65}Zn in freshwater prawns, *Macrobrachium australiense*, by maintaining the animals in square 60 mL bottles (Nalgene) placed on their side with 30 mL synthetic river water and a lint-free tissue (Kimwipe) in the neck of the bottle. Where smaller prawns were found to have rotated inside the bottle during counting, the data was discarded and the animal was re-analysed.

Fig. 1.2 Pictures of the set-up especially made for gamma-counting spotted smooth shore crabs, *Paragrapsus laevis*. (Photos courtesy of Tom Cresswell from Cresswell et al., (2017b))

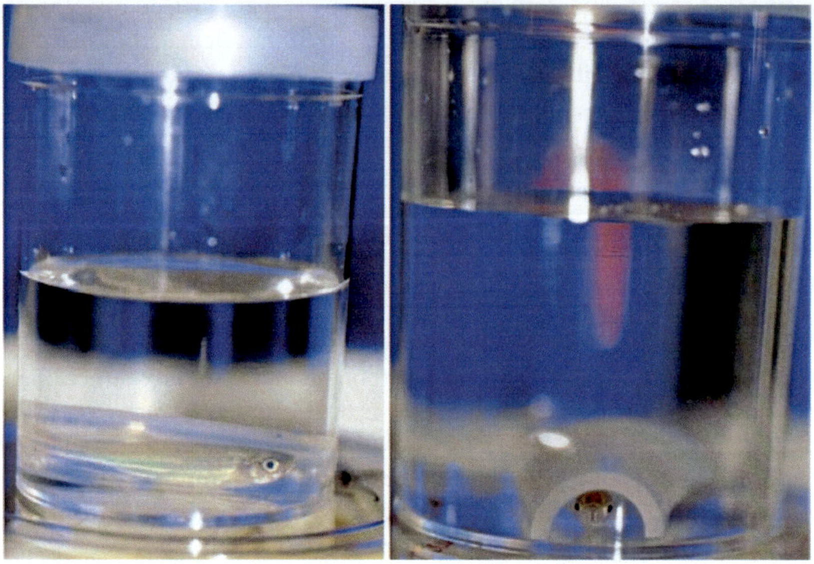

Fig. 1.3 Pictures of the set-up especially made for gamma-counting zebrafish, *Danio rerio* as described by (Reinardy et al., 2011). (Photos courtesy of Jean-Louis Teyssié from Cresswell et al. (2017b))

1.5.2.5 Fish

While it is relatively easy to constrain the movement of invertebrates or very small invertebrates as described above, it is more challenging when radioanalysing constantly moving organisms such as fish. Reinardy et al. (2011) provide a good example of live radioanalysis of zebrafish *Danio rerio* exposed to a range of aqueous gamma-emitting radiotracers. Small pieces of tubing are cut to the length of the internal diameter (ID) of a suitable vial (e.g. 250 mL volume, 60 mL ID), then halved along their length, creating a tunnel-like structure that sufficiently limits the movement of the fish while providing sufficient access to oxygenated water (Fig. 1.3). When working with demersal flat fish such as turbots (*Scophthalmus maximus*), individuals can often be radioanalysed without using a 'tunnel' due to their calm behaviour and given a suitable counting container such as a large petri dish (Cresswell et al., 2017b).

1.5.2.6 Gastropods

New Zealand mud snails (*Potamopyrgus antipodarum*; approx. 6–10 mm total length) exposed to nano ^{141}Ce were initially radioanalysed while inside a 6 mL polyethylene vial containing 3 mL of non-active media (Cresswell et al., 2017b). Snails were found to climb up the interior of the vial and away from the detector crystal, therefore reducing counts. Because this species of snail has an operculum (small "trapdoor" attached to the upper surface of the foot which serves to close the aperture of the shell and retain moisture within the shell), it was able to survive out of water and was therefore radioanalysed in the same vial without any media, keeping the snail close to the detector.

1.5.2.7 Other Factors to Reduce Uncertainty When Radioanalysing Live Aquatic Organisms

The reporting of the reproducibility of consecutive radioanalyses of the same individual organism is crucial for interpreting live animal radioanalysis data (Cresswell et al., 2017b). This validation data should be reported with the main quality control data in any publication as it is often the largest source of error for live monitoring of radiation in organisms. Consideration should also be given to the length of time the animal will be radioanalysed such that the dissolved oxygen (DO) concentrations do not reduce to a level that would cause stress to the organism. Pouil et al. (2015) conducted a study where the trophic transfer of ^{110m}Ag from prey items to turbot, *Scophthalmus maximus*, was investigated. Individual fish were placed in the counting vials and left in the dark for the duration of radioanalysis (25–60 mins). DO measurements of the water in the counting vials were 100% for the duration of the analysis. It is important to note that more energetic species would consume DO more rapidly and therefore pre- and post-radioanalysis DO measurements should be recorded.

1.5.3 Efficiency Calibration for Live Animal Radioanalysis

The process of converting a CPS reading from an MCA to meaningful units (e.g. Becquerel) requires the efficiency of the detector to be known (i.e. how many counts the detector reads for a given Bq or disintegrations per second) for a given geometry (shape and density) of source. This efficiency is calculated by placing a source with a known activity on or near the detector and comparing the known activity with the CPS received by the detector. However, the efficiency of detection will change for sources with equal Bq activities but different geometries. Consider a cylinder (e.g. 20 mL liquid scintillation vial) with 100 Bq activity homogeneously distributed throughout the shape. If that cylinder is placed on one end on top of a coaxial detector facing up and the counts registered by the MCA is 1 CPS, the absolute efficiency of detection is 1%. If an identical activity of 100 Bq was in the shape of a disc or petri dish (e.g. 90 mm diameter) and this source was placed on top of the detector, the counts registered may be 3 CPS, equivalent to an absolute efficiency of 3% for the same activity. This is due to having more of the activity closer to the detector crystal in the second instance. For the vast majority of gamma spectrometry analyses in the laboratory, samples are prepared in defined geometries, ensuring the activity within the sample is as homogeneous as possible, then compared against independently-certified standards in the same geometry. When radioanalysing live organisms, it is not possible to purchase a standard in the same geometry with the radionuclides of interest from commercial sources. Bespoke standards must therefore be created for the gamma radioanalysis of radiation in living organisms. The following discussion introduces three techniques to correct for irregular geometries including the fabrication of phantoms, the use of standard liquid geometries (Cresswell et al., 2017b) and the use of modelling software. Published laboratory radiotracer studies rarely describe how geometry-matched standards are prepared, which consequently leads to an uncertainty in the results so it is important when reporting data from live monitoring studies that these standards are well described.

1.5.3.1 Phantoms

A phantom corresponds to a "standard organism" bearing a known activity of the radioelements of interest and is used to simulate the geometry of the studied organism. Cresswell et al. (2017b) provide a comprehensive explanation for the preparation of aquatic organism phantoms. Briefly, a control or non-exposed individual from the population to be studied that is representative of size and weight of the individuals to be radioanalysed is euthanized and internal tissues (mainly viscera) are removed whilst maintaining the structure and exterior appearance. The organism is then soaked in a bath of diluted formalin solution (approximately 4%) for >12 h to preserve the phantom for the duration of the study. Once removed from formalin and air dried, an absorbent support (paper tissue) imbibed in dilute acid (vis. 2 M HCl) is spiked with the radioelements expected to be present in the samples then the support is placed strategically within the organism (in the cavity vacated by the removed viscera). The animal is subsequently sealed using tape and placed in the same vial/container used for the radioanalysis of live subjects without the addition of clean water so that leaching of the spiked radioelements from the organism to the surrounding media does not occur. The phantom is maintained in the same location and orientation on the detector each time of analysis for reproducibility and live subjects should be radioanalysed identically.

While the production of phantoms as above provides accurate and consistent efficiency calibrations within a laboratory setting, this may provide challenging for field applications, especially maintaining a phantom in good condition over extended periods of time. Furthermore, the above method assumes that the source of radionuclides within an organism is in a two dimensional plane and homogeneously distributed. In reality, different radionuclides will be sequestered to different internal organs depending on biochemical specificity so the phantom would not be representative of the live subjects. It is possible to position absorbent supports within a phantom to represent a specific organ where the radionuclide is accumulating, thus allowing for the efficiency calibration for a more defined geometry to be determined. Again, this method would be logistically challenging in the field.

Recent advances in 3D printing with a range of polymers has provided new opportunities for the production of custom phantoms of organisms being studied. If

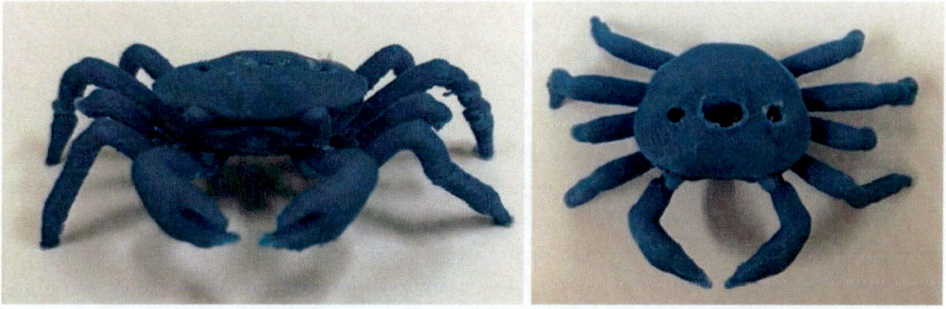

Fig. 1.4 3D printed phantoms of crabs used as phantoms for geometry correction during gamma spectrometry. Image on the right shows holes melted into the model that represent cavities for gills (left and right) and hepatopancreas (centre). Models carapace width is 22 mm at widest point. (Photos courtesy of Tom Cresswell)

a 3D computer model of the organism being studied is available for download (e.g. from https://www.thingiverse.com), the model can be re-sized to replicate the organism of interest. Alternatively, if a computer-tomography (CT) instrument is available, the study organism can be euthanized or anaesthetised then scanned to determine external dimensions and total volume for loading into 3D printer software. A 3D phantom can then be printed in a polymer (e.g. Polylactic acid; PLA) with a comparable density to the organism of study (e.g. 1 g/mL for most aquatic invertebrates) and cavities drilled in areas representing the location of main organs of radionuclide accumulation (Fig. 1.4). Solutions of radionuclides of interest with known activities can then be injected into the cavities, the cavities sealed with tape and the phantoms radioanalysed in the same containers and orientation as their live counterparts via gamma spectrometry. If the field study organism was well characterised ahead of time using the above techniques and the radionuclides of study were known, 3D printed phantoms produced in this manner could be useful in the field. However, knowledge of the site of internal accumulation (i.e. organ) a priori is required.

1.5.3.2 Creation of Standard Liquid Geometries

A second approach to dealing with irregular geometries is to prepare standard liquid geometries of exposed organisms using study organisms pre-exposed to the contaminants of interest. Three organisms are radioanalysed live individually (following the above methods), until significant counts are achieved. Individuals are euthanized, measured and weighed then digested in concentrated acid (e.g. aqua regia) in a microwave or on a hot plate. Once complete digestion has occurred, the digestant is transferred to a 20 mL LSC vial and made up to 10 mL with pure water (or a similar vessel depending on the standard being used). The digested solution is then radioanalysed and the CPS is compared to the CPS of the live organism to calculate a live to digested ratio. Once the live animal CPS have been converted to digested CPS, an absolute efficiency can be applied based on certified standards with known activities of the isotopes of study in exactly the same geometry (i.e. 10 mL homogenous solution in a 20 mL liquid scintillation vial). This then allows the conversion of digested animal CPS to Bq, which can be applied to the animals in the study environment (Cresswell et al., 2017b). The technique allows an accurate efficiency calibration of live organisms but does require the euthanasia of several organisms to produce the standards. Furthermore, the complete digestion of organisms with high lipid or calcium carbonate content requires specialised high-pressure laboratory microwave systems. While this method can be implemented well in the laboratory, it would be challenging to implement in the field due to the necessity of large power requirements for the heating/digestion step.

1.5.3.3 Modelling Software

A third technique to account for irregular geometries (efficiency calibrations) of live aquatic organisms during radioanalysis is the use of modelling software. Programs such as Canberra LABSOCS/ISOCS use NIST-traceable sources and the Monte Carlo N-Particle Transport Code (MCNP). The software allows for multiple geometries to be realised in a virtual graphical user interface where the source position relative to the detector, the surrounding material density (responsible for attenuation of gamma emissions prior to interaction with the detector's crystal), and any shielding from the container being used to house the organism during counting. Grządziel et al. (2017) used the Canberra ISOCS system with ISOCS shield systems collimators model ISOXSHLD from Canberra to calibrate an HPGE semiconductor detector. The study found that mean percentage difference of measurements predicted from ISOCS were 6% of all certified activities of the radioactive standard CBSS 2 (Czech Metrology Institute; covering gamma energies from 56 keV for ^{241}Am to 1836 keV ^{88}Y). This option for the efficiency calibration of a range of gamma spectrometers tasked with analysing a variety of irregular geometries is suited for field applications of the radioanalysis of organisms. However, the program primarily uses a 'box within a box' approach to estimate source location within a series of pre-defined geometric shapes (cylinders and prisms) that do not necessarily represent aquatic biota geometries. Regardless of the restrictions of this method, this is most likely the best approach to an efficiency calibration of a detector for field calibration of live aquatic biota radiation monitoring, so long as such model approaches are validated upon return to the laboratory.

1.5.4 Limits of Detection and Calibration for Field Monitoring Purposes

As discussed above, it is important to conduct radioanalysis of live aquatic wildlife in an area low background radiation to ensure sufficient counts. Similarly, it is important to determine the limit of detection (LOD) of a sample for a given detector in a given area of background radiation. Gilmore (2008) provides a comprehensive explanation for determining counting decision limits including detection limit ("what is the minimum number of counts I can be confident of detecting?") and the minimum detectable activity (MDA; "what is the least amount of activity I can be confident of detecting?"). It is important that both the LOD and the MDA are reported from radioanalytical results so that the final dataset can be utilised by others in a meaningful way where LOD values are present (Wood et al., 2011a).

When calibrating gamma spectrometers for field use, it is essential to note that living organisms will often have a heterogeneous distribution of the source within the sample compared those standards used for laboratory calibrations where the source is homogeneous. Mauring et al. (2017) developed a software tool called InSiCal (n Situ gamma spectrometry Calculator), which aims to make in situ calibration and activity calculation easier for soil analysis applications through a simple and intuitive graphical user interface. Methods for gamma spectrometry efficiency curves have been improved by Mrdja et al. (2017) and Persson et al. (2018) for in-situ measurements of radionuclide concentration in soil using Monte-Carlo simulations. While most new software and calibration methods have been designed for the field analysis of gamma radionuclides in soil, it is feasible that the same tools and techniques could be used for in-situ biota to improve calibration.

1.6 Detectors Suitable for Field Applications

The ideal gamma spectrometer for use in live monitoring of wildlife in the field should have several key features. It should be lightweight, robust, temperature stable (at a range of temperatures experienced at the study sight), has a degree of weather proofing, should have an integrated MCA with the ability to store multiple spectrums covering a broad energy range and should be able to be easily connected to a portable computer for more in-depth spectrum analysis and calibration. The following devices match these criteria:

(a) InSpector 1000 digital hand-held multi-channel analyser from Canberra (Mirion Technologies, USA). The device can accept several detectors including a stabilized NaI probe, a LaBr probe with sourceless stabilisation and a neutron probe.
(b) GAMMA-RAD5 integrated gamma ray spectrometer from Amptek Inc. (Bedford, Ma, USA). It is a single, integrated, portable module which includes a scintillator and PMT, a charge sensitive preamplifier, a digital pulse processor

and MCA. The integrated probe connects to a portable computer via either an Ethernet cable and/or USB cable, which provides both power and data transfer.

(c) AT6104DM and AT6104DM1 NaI Spectrometers from ATOMTEX (Minsk, Republic of Belarus). These spectrometers can monitor radiation levels in fresh and marine waters as well as bottom sediments and can be used at depths of up to 500 m. The probe connects to a portable computer where spectra can be visualised and analysed using proprietary software.

1.7 Summary and Conclusion

If the main goal in response to an accidental release of radionuclides is to determine the tissue radionuclide concentration in native biota for ecological risk assessments, why not use conventional lethal sampling and tissue analysis? By conducting live monitoring of aquatic wildlife in the field, there is the potential of undertaking kinetic experiments in situ with caged animals which provide a much better assessment of the bioaccumulation and retention potential for native aquatic wildlife under site-specific physico-chemical conditions. For example, native individuals of several taxa (e.g. bivalves, crustaceans, fish) could be live radioanalysed to determine starting/background radionuclide concentrations then placed within cages at the site of contamination. Regular live radioanalysis of the same individuals during exposure would provide an understanding of the bioaccumulation rates of organisms under the site-specific water conditions (e.g. temperature, pH, DO, dissolved and total organic carbon, total anion and cation concentration etc.), which would be a significant factor driving the bioaccumulation of each radionuclide. The organisms could then be transplanted to non-contaminated waters of similar physico-chemical conditions and radioanalysed regularly to determine the retention and/or depuration rate of each radionuclide. The knowledge gained from undertaking this process is crucial to aid in risk management to protect vital ecosystem services post accidental radionuclide release.

While laboratory studies of bioaccumulation kinetics provide some very useful data, they are conducted under very static and controlled conditions and often use surrogate species of those found in the field. A study of trace metal bioaccumulation from exposure to dissolved metals and to mine tailings under near identical laboratory conditions determined significant differences in bioaccumulated tissue metal concentrations among three species of the same genus of freshwater prawn (Cresswell et al., 2014b). The study demonstrated the issues with the use of surrogate species in the lab when attempting to determine metal bioaccumulation in field organisms, further illustrating the importance of conducting in-situ bioaccumulation studies with native wildlife in response to accidental releases.

Future live monitoring of gamma-emitting radiation in aquatic wildlife may well use gamma-ray imaging tools. Ziock (2017) provides a review of different approaches used for the indirect imaging of gamma-rays as designed for use in nuclear security applications. Coded-aperture cameras and Compton cameras can be used to overlay two-dimensional gamma-ray detection over an optical image such that the source(s) of radiation within a sample can be localised. Recent development of novel gamma ray imaging detectors at the Australian Nuclear Science and Technology Organisation (ANSTO) are implementing a single detector, rotating collimating mask design that allows for spectroscopic imaging over wide energy range (unpublished), which would allow for the rapid qualification and semi-quantification of multiple radionuclides in a two-dimensional space within a sample. While current technologies do not have the spatial resolution to resolve the internal biodistribution of radionuclides within small (<20 cm total length) organisms, the development is rapid and it is likely that a system will be available in the future to visualise and quantify internalised radionuclides within a range of organisms. The steps outlined above to minimise uncertainty when imaging gamma-rays from aquatic organisms should still be followed.

The techniques descried above allow for the rapid assessment of gamma-emitting radionuclides within aquatic organisms over time. However, it is important to recognise that where wildlife dose and subsequent effect originates solely from alpha or low energy beta emissions, these are impossible to detect with external live monitoring, particularly for radionuclides that accumulate in the bones of fish and mammals such as ^{226}Ra (Johansen & Twining, 2010) and plutonium isotopes (Johansen et al., 2016) or in tissues accumulating organically-bound tritium (OBT).

References

Aramrun, P., Beresford, N. A., & Wood, M. D. (2018). Selecting passive dosimetry technologies for measuring the external dose of terrestrial wildlife. *Journal of Environmental Radioactivity, 182*, 128–137. https://doi.org/10.1016/j.jenvrad.2017.12.001

Beresford, N. A., Barnett, C. L., Crout, N. M. J., & Morris, C. C. (1996). Radiocaesium variability within sheep flocks: Relationships between the 137Cs activity concentrations of individual ewes within a flock and between ewes and their progeny. *Science of Total Environment, 177*, 85–96. https://doi.org/10.1016/0048-9697(95)04863-4

Brisbin, I. L. (1989). Radiocesium levels in a population of American alligators: a model for the study of environmental contaminants in free-living crocodilians. In *In Proceedings of the 8th working meeting of the Crocodile Specialist Group of the Species Survival Committee of the International Union for Conservation of Nature and Natural Resources*. IUCN.

Burden, N., Mahony, C., Muller, B. P., Terry, C., Westmoreland, C., & Kimber, I. (2015). Aligning the 3Rs with new paradigms in the safety assessment of chemicals. *Toxicology, 330*, 62–66. https://doi.org/10.1016/j.tox.2015.01.014

Burger, J., Jeitner, C., Schneider, L., Vogt, R., & Gochfeld, M. (2010). Arsenic, cadmium, chromium, lead, mercury, and selenium levels in blood of four species of turtles from the Amazon in Brazil. *Journal of toxicology and environmental health Part A, 73*, 33–40. https://doi.org/10.1080/15287390903248877

Casado-Martinez, M. C., Smith, B. D., DelValls, T. A., & Rainbow, P. S. (2009). Pathways of trace metal uptake in the lugworm Arenicola marina. *Aquatic Toxicology, 92*, 9–17. https://doi.org/10.1016/j.aquatox.2008.12.010

Cook, A., & Robinson, V. (2007). Overview of the National Centre for the Replacement, Refinement and Reduction of Animals in Research (NC3Rs). *Comparative Biochemistry and Physiology, 146*, S73. https://doi.org/10.1016/j.cbpa.2007.01.081

Creighton, N., & Twining, J. (2010). Bioaccumulation from food and water of cadmium, selenium and zinc in an estuarine fish, Ambassis jacksoniensis. *Marine Pollution Bulletin, 60*, 1815–1821. https://doi.org/10.1016/j.marpolbul.2010.05.025

Cresswell, T., Simpson, S. L., Smith, R. E. W., Nugegoda, D., Mazumder, D., & Twining, J. (2014a). Bioaccumulation and retention kinetics of cadmium in the freshwater decapod Macrobrachium australiense. *Aquatic Toxicology, 148*, 174–183. https://doi.org/10.1016/j.aquatox.2014.01.006

Cresswell, T., Smith, R. E. W., Nugegoda, D., & Simpson, S. L. (2014b). Comparing trace metal bioaccumulation characteristics of three freshwater decapods of the genus. *Macrobrachium Aquatic Toxicology, 152*, 256–263. https://doi.org/10.1016/j.aquatox.2014.04.015

Cresswell, T., Simpson, S. L., Mazumder, D., Callaghan, P. D., & Nguyen, A. P. (2015). Bioaccumulation kinetics and organ distribution of cadmium and zinc in the freshwater decapod crustacean Macrobrachium australiense. *Environmental Science & Technology, 49*, 1182–1189. https://doi.org/10.1021/es505254w

Cresswell, T., Mazumder, D., Callaghan, P. D., Nguyen, A., Corry, M., & Simpson, S. L. (2017a). Metal transfer among organs following short- and long-term exposures using autoradiography: cadmium bioaccumulation by the freshwater prawn Macrobrachium australiense. *Environmental Science & Technology, 51*, 4054–4060. https://doi.org/10.1021/acs.est.6b06471

Cresswell, T., Metian, M., Golding, L. A., & Wood, M. D. (2017b). Aquatic live animal radiotracing studies for ecotoxicological applications: Addressing fundamental methodological deficiencies. *Journal of Environmental Radioactivity, 178–179*, 435–460. https://doi.org/10.1016/j.jenvrad.2017.05.017

Cresswell, T., Prentice, E., Howell, N., Callaghan, P., Metian, M., & Johansen, M. P. (unpublished). Bioaccumulation kinetics and internal distribution of the fission products radiocaesium and radiostrontium in an estuarine crab. *Journal of Hazardous Materials, 408*, 124453. https://doi.org/10.1016/j.jhazmat.2020.124453

Doering, C., & Bollhöfer, A. (2017). Water hardness determines 226Ra uptake in the tropical freshwater mussel. *Journal of Environmental Radioactivity, 172*, 96–105. https://doi.org/10.1016/j.jenvrad.2017.03.019

Field, A. (2011). An assessment of radiocaesium activity concentrations in sheep in restricted areas of England and Wales and potential consumer doses. Food Standards Agency. Aviation House, 125 Kingsway, London. WC2B 6NH. UK.

Fowler, S. W., Small, L. F., & Dean, J. M. (1971). Experimental studies on elimination of zinc-65, cesium-137 and cerium-144 by euphausiids. *Marine Biology, 8*, 224–231. https://doi.org/10.1007/BF00355220

Gilmore, G. R. (2008). Statistics of counting. In *Practical gamma-ray spectrometry* (pp. 101–129). John Wiley & Sons Ltd. https://doi.org/10.1002/9780470861981.ch5

Grządziel, D., Kozak, K., Mazur, J., & Mroczek, M. (2017). Application of ISOCS system in the laboratory efficiency calibration. *Journal of Environmental Radioactivity*. https://doi.org/10.1016/j.jenvrad.2017.09.017

Hinton, T. G., Byrne, M. E., Webster, S., & Beasley, J. C. (2015). Quantifying the spatial and temporal variation in dose from external exposure to radiation: A new tool for use on free-ranging wildlife. *Journal of Environmental Radioactivity, 145*, 58–65. https://doi.org/10.1016/j.jenvrad.2015.03.027

Howard, B. J., Beresford, N. A., Copplestone, D., Telleria, D., Proehl, G., Fesenko, S., Jeffree, R., Yankovich, T., Brown, J., Higley, K., Johansen, M., Mulye, H., Vandenhove, H., Gashchak, S., Wood, M. D., Takata, H., Andersson, P., Dale, P., Ryan, J., et al. (2013). The IAEA handbook on radionuclide transfer to wildlife. *Journal of Environmental Radioactivity, 121*, 55–74. https://doi.org/10.1016/j.jenvrad.2012.01.027

Johansen, M. P., & Twining, J. R. (2010). Radionuclide concentration ratios in Australian terrestrial wildlife and livestock: data compilation and analysis. *Radiation and Environmental Biophysics, 49*, 603–611. https://doi.org/10.1007/s00411-010-0318-9

Johansen, M. P., Ruedig, E., Tagami, K., Uchida, S., Higley, K., & Beresford, N. A. (2015). Radiological dose rates to marine fish from the Fukushima Daiichi accident: The first three years ccross the North Pacific. *Environmental Science & Technology, 49*, 1277–1285. https://doi.org/10.1021/es505064d

Johansen, M. P., Child, D. P., Caffrey, E. A., Davis, E., Harrison, J. J., Hotchkis, M. A. C., Payne, T. E., Ikeda-Ohno, A., Thiruvoth, S., Twining, J. R., Beresford, N. A., et al. (2016). Accumulation of plutonium in mammalian wildlife tissues following dispersal by accidental-release tests. *Journal of Environmental Radioactivity, 151*, 387–394. https://doi.org/10.1016/j.jenvrad.2015.03.031

Knoll, G. F. (2010). *Radiation detection and measurement* (4th ed.). John Wiley & Sons.

Lanctôt, C., Cresswell, T., & Melvin, S. D. (2017a). Uptake and tissue distributions of cadmium, selenium and zinc in striped marsh frog tadpoles exposed during early post-embryonic development. *Ecotoxicology and Environmental Safety, 144*, 291–299. https://doi.org/10.1016/j.ecoenv.2017.06.047

Lanctôt, C. M., Melvin, S. D., & Cresswell, T. (2017b). Selenium speciation influences bioaccumulation in Limnodynastes peronii tadpoles. *Aquatic Toxicology, 187*, 1–8. https://doi.org/10.1016/j.aquatox.2017.03.009

Lee, J. H., Birch, G. F., Cresswell, T., Johansen, M. P., Adams, M. S., & Simpson, S. L. (2015). Dietary ingestion of fine sediments and microalgae represent the dominant route of exposure and metal accumulation for Sydney rock oyster (Saccostrea glomerata): A biokinetic model for zinc. *Aquatic Toxicology, 167*, 46–54. https://doi.org/10.1016/j.aquatox.2015.07.020

Mauring, A., Vidmar, T., Gäfvert, T., Drefvelin, J., & Fazio, A. (2017). InSiCal – A tool for calculating calibration factors and activity concentrations in in situ gamma spectrometry. *Journal of Environmental Radioactivity*. https://doi.org/10.1016/j.jenvrad.2017.10.011

Meredith, R. C. K., Mondon, K. J., & Sherlock, J. C. (1988). A rapid method for the in vivo monitoring of radiocaesium activity in sheep. *Journal of Environmental Radioactivity, 7*, 209–214. https://doi.org/10.1016/0265-931X(88)90028-8

Metian, M., Warnau, M., Teyssié, J.-L., & Bustamante, P. (2011). Characterization of 241Am and 134Cs bioaccumulation in the king scallop Pecten maximus: Investigation via three exposure pathways. *Journal of Environmental Radioactivity, 102*, 543–550. https://doi.org/10.1016/j.jenvrad.2011.02.008

Mrdja, D., Bikit, K., Forkapic, S., Bikit, I., Slivka, J., & Hansman, J. (2017). Improvement of in-situ gamma spectrometry methods by Monte-Carlo simulations. *Journal of Environmental Radioactivity*. https://doi.org/10.1016/j.jenvrad.2017.11.005

Persson, L., Boson, J., Nylén, T., & Ramebäck, H. (2018). Application of a Monte Carlo method to the uncertainty assessment in in situ gamma-ray spectrometry. *Journal of Environmental Radioactivity, 187*, 1–7. https://doi.org/10.1016/j.jenvrad.2018.02.003

Pouil, S., Warnau, M., Oberhänsli, F., Teyssié, J.-L., & Metian, M. (2015). Trophic transfer of 110m Ag in the turbot Scophthalmus maximus through natural prey and compounded feed. *Journal of Environmental Radioactivity, 150*, 189–194. https://doi.org/10.1016/j.jenvrad.2015.08.016

Qureshi, R. M., Mashiatullah, A., Yaqoob, N., Akhtar, P., Chaghtai, F., Jabbar, A., & Warnau, M. (2007). Bioaccumulation of 137Cs, Zn and Cr[VI] in the green mussel Perna viridis: Influence of salinity and temperature. *Environmental Bioindicators, 2*, 245–252. https://doi.org/10.1080/15555270701693596

Reinardy, H. C., Teyssie, J.-L., Jeffree, R. A., Copplestone, D., Henry, T. B., & Jha, A. N. (2011). Uptake, depuration, and radiation dose estimation in zebrafish exposed to radionuclides via aqueous or dietary routes. *Science of Total Environment, 409*, 3771–3779. https://doi.org/10.1016/j.scitotenv.2011.06.057

Wang, W. X. (2001). Comparison of metal uptake rate and absorption efficiency in marine bivalves. *Environmental Toxicology and Chemistry, 20*, 1367–1373. https://doi.org/10.1002/etc.5620200628

Warnau, M., & Bustamante, P. (2007). Radiotracer techniques: A unique tool in marine ecotoxicological studies. *Environmental Bioindicators, 2*, 217–218. https://doi.org/10.1080/15555270701714822

Watson, D.G., & Templeton, W.L. (1971). Thermoluminescent dosimetry of aquatic organisms, report. January 1, 1971; Richland, Washington. (digital.library.unt.edu/ark:/67531/metadc866677/: Accessed October 4, 2017), University of North Texas Libraries, Digital Library, digital.library.unt.edu; crediting UNT Libraries Government Documents Department.

Wood, M. D., Beresford, N. A., & Copplestone, D. (2011a). Limit of detection values in data analysis: Do they matter? *Radioprotection, 46*, S85–S90. https://doi.org/10.1051/radiopro/20116728s

Wood, M. D., Beresford, N. A., Yankovich, T. L., Semenov, D. V., & Copplestone, D. (2011b). Addressing current knowledge gaps on radionuclide transfer to reptiles. *Radioprotection, 46*, S521–S527. https://doi.org/10.1051/radiopro/20116792s

Yankovich, T., Beresford, N. A., Wood, M. D., Aono, T., Andersson, P., Barnett, C. L., Bennett, P., Brown, J. E., Fesenko, S., Fesenko, J., Hosseini, A., Howard, B. J., Johansen, M. P., Phaneuf, M. M., Tagami, K., Takata, H., Twining, J. R., Uchida, S., et al. (2010). Whole-body to tissue concentration ratios for use in biota dose assessments for animals. *Radiation and Environmental Biophysics, 49*, 549–565. https://doi.org/10.1007/s00411-010-0323-z

Ziock, K. P. (2017). Principles and applications of gamma-ray imaging for arms control. *Nuclear Instruments and Methods in Physics Research Section A: Accelerators, Spectrometers, Detectors and Associated Equipment.* https://doi.org/10.1016/j.nima.2017.04.001

Chapter 2
Synchrotron Light Facilities and Applications in Life Sciences

V. M. Tsakanov

Abstract The synchrotron radiation emitted by high-energy ultrarelativistic electrons is one of the most powerful tools for investigating matter. The radiation has a broad continuum spectrum covering radiation wavelengths from infrared to hard X-ray ranges. The radiation is emitted in a narrow cone providing a high brilliance collimated photon beam with a small divergence and size from a high intensity electron source. The continuous spectrum of synchrotron radiation provides more than five orders of magnitude increase in intensity and more than ten orders of magnitude increase in brilliance than more conventional sources, such as VUV lamps and X-ray tubes. In advanced synchrotron light facilities the radiation is produced in bending magnets, undulators and wigglers, enabling dedicated tunable polarized photon beams. The radiation has a pulsed time structure allowing the execution of time-resolved experiments along with scattering, spectroscopy and imaging experiments. These features of synchrotron light have provided a continuous growth of synchrotron radiation usage in diverse fields of life, materials and environmental sciences during the last decades. In this chapter, a brief introduction is provided to synchrotron light sources, the main features of their associated radiation, experimental techniques and applications in life sciences.

Keywords Synchrotron radiation · Insertion devices · Radiation spectrum · Experimental technique · Beamline instrumentation · Life and environmental sciences

---

V. M. Tsakanov (✉)
CANDLE Synchrotron Research Institute, Yerevan, Armenia
e-mail: tsakanov@asls.candle.am

M. D. Wood et al. (eds.), *Biomarkers of Radiation in the Environment*, NATO Science for Peace and Security Series A: Chemistry and Biology,
https://doi.org/10.1007/978-94-024-2101-9_2

2.1 Introduction

During most of the twentieth century, intensive studies were made into the substructure of the atom and its core, called the nucleus. This led to discoveries of many new subatomic particles such as neutrons, mesons, leptons, quarks, and neutrinos and to the nature of their properties and interactions, which in those instances were more dominant than the electromagnetic interactions. These discoveries were made possible through advancements in the technology of particle accelerators, which propel electrons and protons to energies many orders of magnitude greater than their values within stable atoms. One of these devices was the electron synchrotron, which was especially important in providing the capability of determining the structure of the atomic nucleus. These devices had one rather distressing feature which was that the circulating electrons would radiate more and more of their energy into electromagnetic radiation as their energies increased, subsequently limiting the energy to which electrons could be raised.

This radiation emitted by electrons confined to circular orbits by magnetic fields is named "synchrotron radiation" following its first experimental observation at 70 MeV energy electron synchrotron of General Electric (Elder et al., 1947), although the feature that accelerated charged particles' radiate energy is a general consequence of electromagnetic theory (Jackson, 1998). The spectrum of the radiation, its intensity, and direction relativc to the direction of the electron are all predictable (Wiedemann, 2003; Hoffmann, 2004). For a physical device, such as a synchrotron of several GeV electron energy using 1 Tesla field strength magnets, the spectrum is predominantly in the ultraviolet and soft X-ray range of the electromagnetic spectrum. The power of the emitted radiation increases as the fourth power of the electron energy and the direction of the radiation concentrated in the direction of the radiating electron.

During the period of intense development of high energy circular electron accelerators, in the later part of the twentieth century, it became apparent that the intensity of this synchrotron radiation is much higher than the radiation intensity from more conventional devices, such as X-ray tubes. This feature attracted materials scientists, biologists and others to consider the usage of high energy physics facilities like DESY (Lemke & Labs, 1967) and SPEAR (Doniach et al., 1997) for conducting their studies on a "parasitic" basis whenever these facilities were operative (Winick, 1994). Later this family of electron accelerators was classified as the first generation synchrotron light facilities. A 6 GeV ARUS electron synchrotron in Armenia was among them, where three X-ray beamlines were constructed in early 70's to support the solid state, materials science and radiation biophysics research fields (Williams & Winick, 2015).

By the 1980s, the demand for usage of synchrotron radiation had mushroomed to the point where second generation synchrotron light sources were built as "dedicated" facilities. Thus, a whole new industry was spawned from radiation deemed to be a "waste product" by researchers engaged in the studies into sub-atomic physics.

The advancement from the early "parasitic" mode of operation to today's "state of the art" dedicated facilities has been a rapid evolution and has culminated in a revolution in many diverse areas of research. They include biology, chemistry, pharmacology, geology, materials and environmental sciences, medicine and many other related fields.

In response to user demands for a higher photon flux and brightness, the new third generation facilities have been developed which incorporate a number of new features, including higher circulating currents of electrons, small electron beam phase space, special devices, such as wigglers or undulators inserted into the magnetic structure of the storage ring, and well instrumented photon beam lines with monochromatic, highly collimated and precise properties (Hoffmann, 2004; Wiedemann, 2003; Willmott, 2011).

The current tendency in synchrotron light source development is aimed to even high brilliance photon beams with electron beam phase space at the level of X-ray diffraction limit. This is achieved by using the sophisticated design of the magnetic lattice with multi-band achromats, longitudinal magnetic field gradient, coupling of the horizontal-vertical particles motion and accelerator technology advancement (Borland et al., 2016; Sargsyan et al., 2016; Wanzengerg et al., 2017). With the exponential growth in the usage of synchrotron radiation, the number built in the past twenty years has increased to over 50 facilities throughout the world.

Along with the synchrotron light facilities, the experimental techniques has greatly advanced, enabling researchers to deeper exploit the matter at the cell, molecular and atomic levels. Finally, a strong user community has been established worldwide to propel the frontier research in life, materials and environmental sciences.

2.2 Synchrotron Light and Sources

In synchrotron light facilities the electrons are accelerated to very high energies of several GeV and stored in a ring for many hours by a magnetic guide field. The electrons in the storage ring experience transverse acceleration due to magnetic field in bending magnets and insertion devices like undulators and wigglers with periodic magnetic fields. Due to acceleration, the electrons emit electromagnetic radiation and for highly relativistic particles, most of the radiation is emitted in a forward direction, concentrated in a very small cone with an opening angle of $1/\gamma$, where γ is the electron energy in units of its rest energy. For electrons $\gamma \approx 1957E$, where E is the electron energy in GeV. Thus for 3 GeV electrons, the radiation opening angle is approximately 0.16 mrad.

The typical schematic design of the third generation light sources is similar to the one of CANDLE synchrotron light facility project in Armenia (Tsakanov et al., 2002). The facility consists of 3 GeV energy electron storage ring, full energy booster synchrotron and 100 MeV linear accelerator (Fig. 2.1). To replenish the energy lost to synchrotron radiation, accelerating cavities are located in the ring at

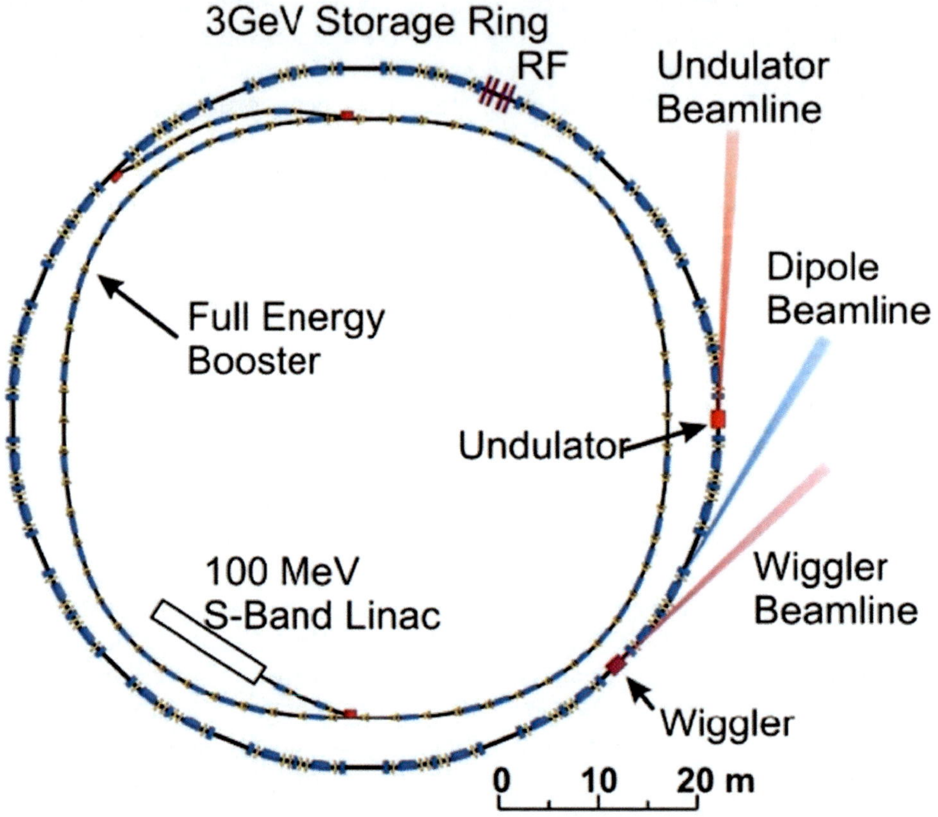

Fig. 2.1 The general layout of CANDLE facility

frequency of about 500 MHz. The electron beam pulse structure consists of 20 psec rms duration individual bunches spaced with 2 nsec time gap.

Important figures of merits for synchrotron radiation are the spectral flux and brightness. The flux is the number of photons emitted into an angular fan per unit time and is the appropriate merit for applications where little beam collimation is required and the sample transverse size is sufficiently large so as to intercept the entire photon beam.

The brightness is the flux per unit phase space and the high brightness is required for experiments that involve samples or optics with very small phase space acceptance or techniques that exploit beam coherence. High brightness of the photon beam is achieved by the electron beam with small transverse size and divergence.

In the CANDLE design, in total 12 straight sections of 4.8 m in length are planned for insertion devices – undulator and wiggler magnets. The photon beams from the dipoles and the conventional insertion devices cover the energy range of 0.01–50 keV with high spectral flux and brightness. Figure 2.2 presents the CANDLE spectral flux and brightness for dipole (1.35 T), undulator (0.3 T) and wiggler (1.3 T, 2 T) sources.

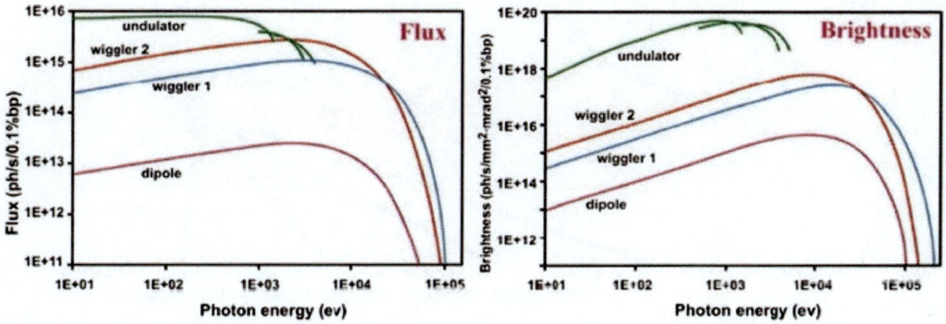

Fig. 2.2 CANDLE photon beam spectral flux (left) and brightness (right) from dipoles, undulator and two types of wigglers

The spectrums of radiation in the bending and wiggler magnets are continuous. Due to the number of magnetic poles and high magnetic field the radiation from wiggler magnets is more intensive and shifted to hard X-ray region.

In undulators, the electrons perform purely sinusoidal oscillations in weak periodic magnetic fields. The radiation cones emitted during each oscillation period overlap and interfere, thus, the intensity of undulator radiation is much higher than for wigglers and bending magnets. The radiation wavelength is given by the undulator period length reduced by a factor due to Lorentz contraction and the Doppler effect. Thus, for the cm's undulator period length the radiation wavelength is in VUV and soft X-Ray regions. The synchrotron radiation is linear polarized in the plane of acceleration. The radiation observed above or below the midplane has an elliptical polarization. The circular polarized radiation can be obtained by using helical undulators, where the electrons move on spiral trajectory. More details about the electron beam physics and synchrotron radiation characteristics can be found elsewhere (Duke, 2000; Hoffmann, 2004; Wiedemann, 1999; Wiedemann, 2003).

2.3 Experimental Techniques

The unique properties of the synchrotron radiation: broad spectrum, high spectral flux and brightness, tunability, polarization and the pulsed structure, make this type of radiation a powerful tool for experimentalists to study the micro-world. The radiation spectrum covering the range of dimensions from atomic level to biological cells makes synchrotron radiation very effective for research in physics, biology, medicine, chemistry, material and environmental sciences by selecting the wavelength required for the particular experiment. Figure 2.3 presents the general schematic layout of a typical synchrotron radiation beamline, that consists of source (dipole, wiggler or undulator magnets), optics that guides the radiation to the sample and detector (Abashian, 2002).

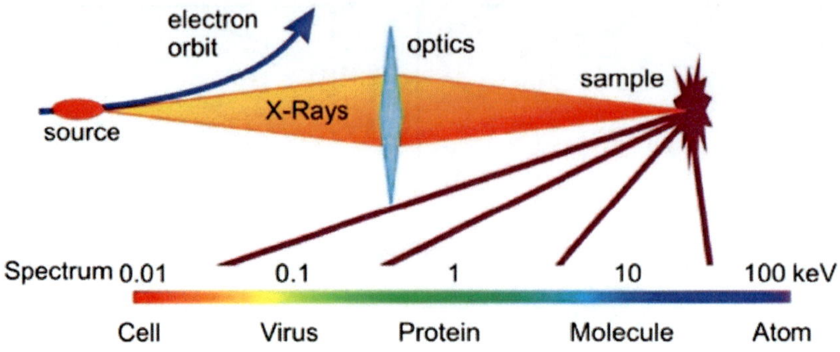

Fig. 2.3 Schematic layout of the of synchrotron radiation beamline

The complexity of the fields under consideration (DNA and ligands, proteins and nucleic acids, chemical dynamics, crystal structure etc) requires a complete description of the sample properties and the state that includes four primary characteristics: namely the energy, momentum, position and dynamics. The corresponding techniques that realize these features using synchrotron radiation are the spectroscopy, scattering, imaging and time-resolved experiments (Mobilio et al., 2014; Willmott, 2011; Winick & Doniach, 1980. The time resolved experiments exploit the pulsed structure of the synchrotron radiation for the sample study in time domain.

Spectroscopy is the technique to determine the emitted or absorbed by sample particle energy under the synchrotron light expose. This technique is used to study the characteristics of chemical bonding and electron motion. The X-ray spectroscopy technique involves: X-ray absorption spectroscopy (XAS), Extended X-ray absorption fine structure (EXAFS), Near-edge X-ray absorption fine structure (NEXAFS), X-ray absorption near-edge structure (XANES), X-ray emission spectroscopy (XES), Resonant inelastic X-ray scattering (RIXS), X-ray magnetic circular dichroism (XMCD), X-ray photoemission spectroscopy (XPS), X-ray fluorescence spectroscopy (XFS).

Scattering observes the diffracted light pattern of a sample as a function of incident and scattered angle, polarization, and wavelength. This technique provides information about the material structure, chemical composition and physical properties. The X-ray scattering technique involves: X-ray diffraction (XRD), X-ray powder diffraction, X-ray standing wave (XSW), multi-wavelength anomalous diffraction (MAD), Small-angle X-ray scattering (SAXS), X-ray Raman scattering, inelastic X-ray scattering (IXS), X-ray emission scattering (XES).

Imaging is a technique to determine sample image with the fine special resolution and is used for visualizing cellular structures in a wide range of biological and medical studies. This technique involves: X-ray diffraction imaging (XDI), diffraction enhanced imaging (DEI), scanning transmission X-ray microscopy (STXM), X-ray tomography and topography, phase contrast imaging, photoelectron emission microscopy (PEEM), computer-aided tomography (CAT), X-Ray lithography.

The beamlines and corresponding instrumentation are an integrated part of the synchrotron light facility design, that define the quality of the machine and efficiency of photon source usage. As an example, the first group of the CANDLE project beamlines implies the General Diffraction and Scattering Beamline from dipole source, X-ray Absorption Spectroscopy Beamline from dipole, Soft X-ray Spectroscopy Beamline from undulator, Imaging and Small Angle X-ray Scattering Beamlines from the wigglers.

General Diffraction and Scattering Beamline. The beamline is based on the dipole source and aimed to produce focused or unfocused tunable hard X-rays (5-30 keV) sequentially serving two experimental stations: 1- for structural study of low or high temperature polycrystalline materials, thin films and multi-layers; 2 - for single crystal structure determination, charge density studies and anomalous dispersion experiments. Figure 2.4 presents the schematic layout of the beamline optical elements (mirrors and double crystal monochromator-DCM).

A 10 keV photon beam profile simulation along the beamline is given in Fig. 2.5. The initial beam (a), the beam reflected by the mirror M1 (b), the focused beam after the DCM (c) and the focused beam at the end station (d) are shown. The simulations are performed by ray tracing code SHADOW (Sanchez del Rio et al., 2011).

XAS Beamline. The XAS beamline will cover a photon energy range up to 35 keV with sufficient intensity in soft X-ray region. The spectrum covers the K edges of elements such as Si, S, P and Cl, which are of high technical interest. Using double crystal monochromator and gold coated reflection mirror, the beamline will be able to operate in hard X-ray region allowing users to measure EXAFS of all elements either at K or L3 edges.

Imaging Beamline. Using the radiation from 3 T permanent wiggler this beamline will provide a high flux "white" or tuneable monochromatic coherent radiation in 6–120 keV photon energy range at about 150 m from the source. The experimental program will include: phase contrast and diffraction-enhanced imaging; hard X-ray microscopy; holographic imaging and tomography; micro-focusing; X-ray topography, diffractometry; micro-fluorescence and high resolution inelastic scattering.

SAXS Beamline. The beamline utilizes a non-destructive method to study the nanoscale structure of any type of material ranging from new composite nanosystems to biological macromolecules. The primary elements of the beamline include

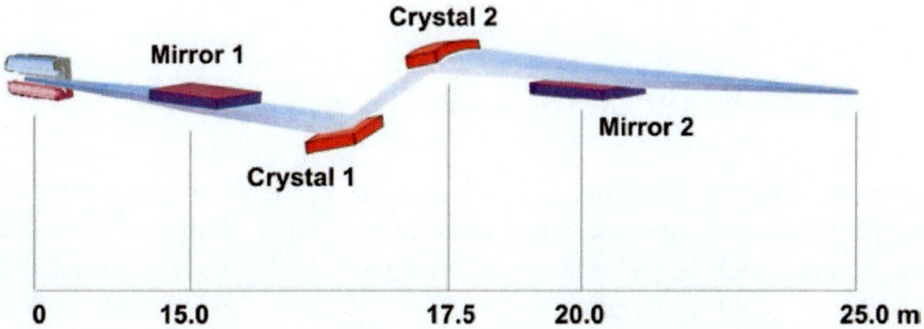

Fig. 2.4 Schematic layout of the diffraction and scattering beamline

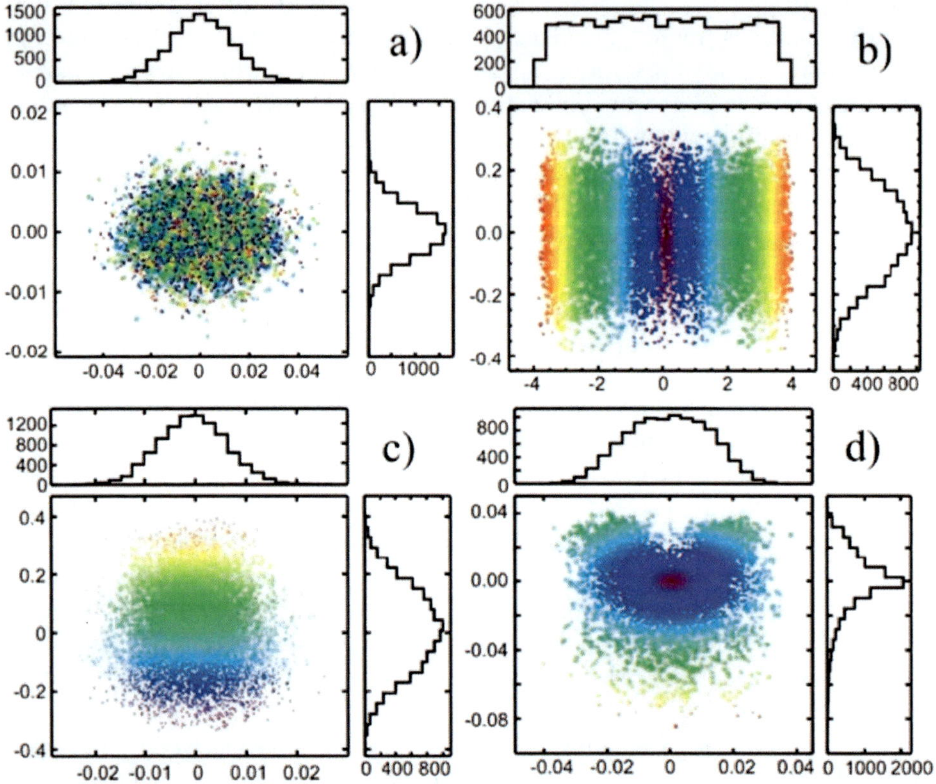

Fig. 2.5 The spatial distribution of photon beam along the general scattering and diffraction beamline

a pin-hole geometry X-ray scattering camera, high resolution and high heat load monochromators.

Soft X-ray Spectroscopy Beamline. The high brightness photon beam from the undulator will support this beamline addressing complex problems in materials, environmental and biological sciences. The beamline implies two types of microscopes: a zone plate based scanning transmission X-ray microscope and a photoelectron emission microscope.

2.4 Applications in Life Sciences

Synchrotron radiation usage in life sciences is one of the most propounded applications in synchrotron light facilities. Application fields include a wide range of diverse branches in biology, medicine, chemistry, ecology, food, pharmacology etc.

Medical imaging and radiation therapy. The use of synchrotron radiation in medical research has become an important application field at synchrotron radiation facilities (Ando & Uyama, 1998; Bravin, 2007; Suortti & Thomlinson, 2003). The high brightness, tunability and coherence distinguish these sources from standard clinical and research instruments. The highlights for medical therapy and diagnosis by synchrotron radiation usage include:

- Angiography
- Bronchography
- Mammography
- Computed Tomography
- Microbeam Radiation Therapy
- Photon Activation Therapy

The high brilliance and tunability of synchrotron X-ray beams can dramatically improve the speed, clarity and safety of diagnostic tools, such as coronary angiography. The coronary angiography is an X-ray procedure in which coronary vessels are made visible through the injection of iodine as a contrast medium. Two monochromatic beams with an energy above and below the K-absorption edge of iodine at 33.17 KeV are used to record two images simultaneously. The difference image of arteries is greatly enhanced allowing usage of much lower iodine concentrations and lower X-ray doses, compared to conventional angiography. With such low iodine level, the contrast agent can safely be introduced through an arm vein. Monochromatic X-rays of sufficient intensity to visualize coronary arteries with an extremely low iodine mass density are only provided by synchrotron radiation.

Macromolecular Crystallograph. A macromolecular crystallography is dedicated to the determination of the 3-dimensional structure of large biological molecules using X-ray diffraction (Drenth, 2006; Helliwell, 2005). The scientific and application research of macromolecular crystallography using synchrotron radiation include a broad fields of life sciences like:

- Enzyme Mechanism
- Supra-molecular structure
- Molecular Recognition
- Nucleic Acids
- Structural Genomics
- Drug design

The synchrotron radiation is an effective instrument to study the structures of DNA, RNA, or large molecular assemblies, such as ribosomes and viruses. After sequencing the human genome, the next scientific challenge is to elucidate the structure and function of the proteins encoded by the genes. Following November 21, 2017 Protein Data Bank (http://www.rcsb.org/pdb), the number of biological macromolecular structures (proteins, DNA, RNA and protein nucleic acid complexes) deposited are 135,600 (Fig. 2.6), from which more than 90% have been deposited using synchrotron radiation.

The knowledge about these structures has profound implications for overall understanding of life processes in general and for the understanding and treatment of disease. The results of high-resolution structure analyses extracted from X-ray diffraction studies of macromolecular crystals provide invaluable information for modeling drug-receptor binding.

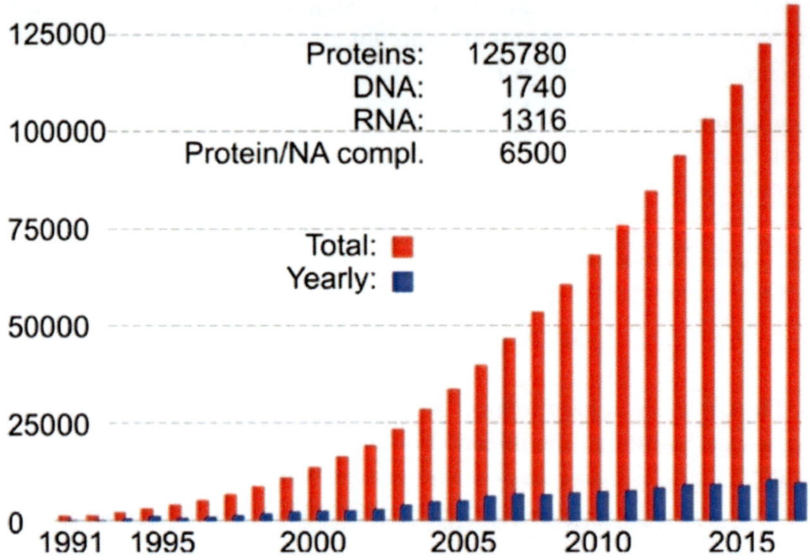

Fig. 2.6 Yearly growth of deposited macromolecular structures. Protein Data Bank, 21 November 2017

Because proteins are large and flexible, protein crystals tend to be small, imperfect, and weakly diffracting compared to crystals of small molecules. The intensity of the synchrotron beam allows data to be collected from the weakly diffracting protein crystals. The well-collimated X-ray beams can be focused to a size comparable to that of the samples used, typically on the order of 10–100 mm. Collimation can also be particularly important for crystals with large cell dimensions where separation of reflections on the detector can be difficult.

Another significant advantage of synchrotron radiation is that it provides the opportunity to select the energy of the X-ray beam. Tunability allows to perform experiments that utilize the small variations in the intensity of diffracted reflections at different energies due to the resonant scattering of a heavy atom. Two popular techniques: SAD (single-wavelength anomalous dispersion) phasing and MAD (multi-wavelength anomalous dispersion) phasing rely on this effect.

Environmental sciences. Many spheres of human activity are result in negative impact on the environment. Among them, it is necessary to mention the chemical industry, which can be a source of highly toxic xenobiotics; civil and weapon nuclear technologies - source of radionuclides; metallurgy - source of heavy metals; transport - products of fuel combustion; agriculture - pesticides and fertilizers. All these compounds enter the human organism through water, air or food. Many of them are already dangerous at very low concentrations. The topics of environmental sciences are numerous:

- Determination of heavy metals, radionuclides, toxic organic compounds;
- Investigation of environmental transport and accumulation places of these contaminants;
- Investigation of their utilization, identification and dead-end products;

- Investigation of contaminants' influence on ecosystems;
- Development of tools for environment monitoring;

These problems are closely interconnected and have a strong connection with biological sciences: biochemistry, biotechnology, microbiology, etc. How and which contaminants are distributed in the environment, how these contaminants interact with soil, plants, way of their migration with ground waters, changes related to this effects are not an exhaustive list of the problems related to environmental science (De Giudici et al., 2015; Hettiarachchi et al., 2017).

The solution of these problems requires an investigation of physical, chemical, biological processes at the level of molecular scale in addition to the macroscopic one. The study should be performed both in native conditions and under the effect of contaminants. Only such comprehensive approach allows to have a complete understanding and description of the fundamental mechanisms, which underlie the processes in the environment. The experimental techniques used in synchrotron radiation application in environmental sciences involve XAS, EXAFS, XANES, XFS, SXTM and X-ray tomography.

2.5 Summary

This chapter presents a brief description of the basics of synchrotron radiation and sources, techniques and instrumentation, as well as a number of applications in the field of life sciences. Synchrotron radiation facilities, experimental methods and applications are rapidly developing areas enabling frontier research across the entire range of basic and applied sciences, and their discussion goes far beyond the scope of this chapter.

This work was supported by the RA MES State Committee of Science, in the frames of the research project N:16AR-1c002.

References

Abashian, A., Aghasyan, M., Amatuni, G., Arakelyan, V., Avagyan, V., Avakian, R., Ayvazyan, V., Gagiyan, H., Grigoryan, A., Grigoryan, B., Ivanian, M., Jalalyan, V., Harutiunyan, S., Harutiunian, V., Kirakosyan, A., Khachatryan, V., Laziev, E., Martirosyan, Y., et al. (2002). Design study of CANDLE 3 GeV synchrotron light source, ASLS-CANDLE R-001-02. http://candle.am/design_report

Ando, M., & Uyama C. (Eds) (1998). Medical applications of synchrotron radiation, Springer, 200 p.

Borland M., Hettel, R. O., Leemann, S. C., & Robin, D. (2016). Accelerator physics challenges in the design of multi-bend-achromat-based storage ring, Proc. of NAPAC 2016, Chicago, USA, 278–283.

Bravin, A. (2007). The biomedical programs at the ID17 beamline of the ESRF. In V. Tsakanov & H. Wiedemann (Eds.), *"Brilliant light in life and material sciences", NATO security through science series B; physics and biophysics* (pp. 225–239). Springer.

De Giudici, G., Lattanzi, P., & Medas, D. (2015). Synchrotron radiation and environmental sciences. In S. Mobilio, F. Boscherini, & C. Meneghini (Eds.), *Synchrotron radiation* (pp. 661–676). Springer.
Doniach, S., Hodgson, K., Lindau, I., Pianetta, P., & Winick, H. (1997). Early work with synchrotron radiation at Stanford. *Journal of Synchrotron Radiation, 4*, 380–395.
Drenth, J. (2006). *Principles of protein X-ray crystallography* (332 p). Springer.
Duke, P. J. (2000). *Synchrotron radiation* (251 p). Oxford University Press.
Elder, F. R., Gurewitsch, A. M., Langmuir, R. V., & Pollock, H. C. (1947). Radiation from electrons in a synchrotron. *Physical Review, 71*(11), 829–830.
Helliwell, J. R. (2005). *Macromolecular crystallography with synchrotron radiation* (595 p). Cambridge University Press.
Hettiarachchi, G. M., Donner, E., & Doelsch, E. (2017). Application of synchrotron radiation-based methods for environmental biogeochemistry: Introduction to the special section. *Journal of Environmental Quality, 46*(6), 1139–1145.
Hoffmann, A. (2004). *The physics of synchrotron radiation* (323 p). Cambridge University Press.
Jackson, J. D. (1998). *Classical electrodynamics* (3rd ed., 832 p). Wiley.
Lemke, D., & Labs, D. (1967). The synchrotron radiation of the 6-GeV DESY machine as a fundamental radiometric standard. *Applied Optics, 6*(6), 1043–1048.
Mobilio, S., Boscherini, F., & Meneghini, C. (Eds.). (2014). *Synchrotron radiation: Basics, methods and applications* (799 p). Springer.
Sanchez del Rio, M., Canestrari, N., Jiang, F., & Cerrina, F. (2011). SHADOW3: A new version of the synchrotron X-ray optics modelling package. *Journal of Synchrotron Radiation, 18*, 708–716.
Sargsyan, A., Zanyan, G., Sahakyan, V., & Tsakanov, V. (2016). Sub-nm emittance lattice design for CANDLE storage ring. *Nuclear Instruments and Methods A, 832*, 249–253.
Suortti, P., & Thomlinson, W. (2003). Medical applications of synchrotron radiation. *Physics in Medicine and Biology, 48*(13), R1–R35.
Tsakanov, V., et al. (2002). CANDLE: A new project for 3 GeV intermediate energy light source in the Republic of Armenia. *The Review of Scientific Instruments, 73*(3), 1411–1413.
Wanzenberg, R., et al. (2017). Research activities towards a conversion of PETRA III into a diffraction limited synchrotron light source, Proc. of IPAC 2017, Copenhagen, Denmark, 3077–3080.
Wiedemann, H. (1999). *Particle accelerator physics* (Vol. 1,2). Springer.
Wiedemann, H. (2003). *Synchrotron radiation* (274 p). Springer.
Williams, G. P., & Winick, H. (2015). Introduction to the special issue on pioneers in synchrotron radiation. *Synchrotron Radiation News, 28*(4), 2–3.
Willmott, P. (2011). *An introduction to synchrotron radiation. Techniques and applications* (352 p). Wiley.
Winick, H. (Ed.). (1994). *Synchrotron radiation sources – A primer* (507 p). World Scientific.
Winick, H., & Doniach, S. (Eds.). (1980). *Synchrotron radiation research* (754 p). Plenum Press.

Chapter 3
Elemental Imaging in Biology Using Synchrotron X-Ray Fluorescence Microscopy

Koshonna Brown, Tatjana Paunesku, and Gayle E. Woloschak

Abstract X-ray Fluorescence Microscopy (XFM), also known as Synchrotron Radiation based X-Ray Fluorescence (SRXRF) or Microprobe Synchrotron X-ray fluorescence (mSXRF), is a powerful and versatile technique for the investigation of elemental content in biological samples. Whole cells can be imaged with resolutions better than 100 nm and their elemental content 3D reconstructed despite a cell thickness of 10 microns or more; with some compromises in the spatial resolution even samples as thick as 100 s of microns can also be imaged in 3D. The resultant elemental map is quantitative – concentrations of the different elements are detected simultaneously pixel by pixel, as the fluorescence radiation emitted is proportional to the atom's concentration within the sample. Detection limits as low as 0.1–5 ppm can be achieved for metals. With new technical developments such as "fourth generation" synchrotrons, faster detectors and even "X-ray focusing" optics, it is likely that XFM will continue to develop toward ever higher resolution and speed of data acquisition. While XFM can be used for detection of radionuclides in biological material, due to extremely low background for such elements in samples collected in non-contaminated areas, radionuclide quantities are generally low and imaging them is difficult. Moreover, radioactive decay and resultant elemental transitions further decrease numbers of atoms of interest that are available for detection. With the increase in brightness, new generations of synchrotrons and their further updates can be expected to improve sensitivity of radionuclide detection.

K. Brown · T. Paunesku · G. E. Woloschak (✉)
Departments of Radiation Oncology, Radiology, and Cell and Molecular Biology, Robert E. Lurie Comprehensive Cancer Center, Feinberg School of Medicine, Northwestern University, Chicago, IL, USA
e-mail: g-woloschak@northwestern.edu

M. D. Wood et al. (eds.), *Biomarkers of Radiation in the Environment*, NATO Science for Peace and Security Series A: Chemistry and Biology,
https://doi.org/10.1007/978-94-024-2101-9_3

3.1 Introduction

X-ray Fluorescence Microscopy (XFM) also known as Synchrotron Radiation based X-Ray Fluorescence (SRXRF) or Microprobe Synchrotron X-ray fluorescence (mSXRF) is a technique that detects X-rays emitted by atoms irradiated with hard X-ray photons of an energy higher or equal to the binding energy of the inner shell electrons of the sample's atoms (Paunesku et al., 2012; Petibois, 2010). For biological samples, incoming X-ray energy is most often set at 10 keV. The characteristic X-rays from each element in the sample are emitted and elemental K alpha and K beta X-rays produced by elements between Si and Zn are especially suited to create an elemental map of a biological sample. Such elemental map is also quantitative - concentrations of all different elements are detected simultaneously pixel by pixel, as the fluorescence radiation emitted is proportional to the atom's concentration within the sample. XFM imaging allows detection of metals with detection limits as low as 0.1–5 ppm while most other micro-analytical techniques have 10 ppm as a detection limit. Moreover, XFM recognizes chemical elements regardless of their chemical environment. While histological staining for some chemical elements is possible and can provide detailed information, it is relatively crude and destructive; it is also often specific to the chemical speciation of an element and only semi-quantitative (Berg & Shi, 1996). The sensitivity of existing histochemical staining methods is often low – existing methods frequently detect only relatively highly concentrated metal deposits, making a detailed relative distribution of the cellular contents impossible (Jin et al., 2017).

Customarily, X-rays sources are considered soft or hard, based on their penetrance (Petibois 2010). The preferred hard X-ray energies for work with biological samples are 10 kilo electron Volts (keV) or higher. Imaging at this energy is compatible with large focal depth and penetration of 10 microns or more. At the same time, this X-ray energy excites the K line fluorescence for the majority of biologically relevant trace elements. However, all elements from the periodic table can be detected using X-ray fluorescence. Presence of chemical elements not naturally present in biological material is detected particularly well, due to low background for such chemical elements in biological samples. Many different X-ray synchrotron facilities were used for studies of biological samples, such as Advanced Photon Source (APS) and National Synchrotron Light Source (NSLS) in USA, Diamond Light Source in UK, European Synchrotron Radiation Facility (ESRF) in France, Berliner Elektronenspeicherring-Gesellschaft für Synchrotronstrahlung (BESSY II) and Deutsches Elektronen-Synchrotron (DESY) in Germany, Super Photon Ring-8 GeV (SPring8) and Photon Factory in Japan, Shanghai Synchrotron Radiation Facilities (SSRF) in China and Australian Synchrotron (AS).

In the context of radiation biology and exploration of radionuclide content in biological samples, XFM can be a very powerful screening technique. On one hand, most radioactive elements are not native to biological samples, their background is negligible and for that reason any amount of a radionuclide found in biological samples is likely to be accurate. On the other hand, XFM is equally suitable to elemental detection of radionuclides as well as all of their decay products. Nevertheless, examples of XFM use for investigation of radioactive samples are relatively few (Paunesku et al. 2012).

3.2 XFM Exploration of Elemental Concentration, Distribution and Translocation in Differentiation and Normal Homeostasis

Metals and non-metal trace elements such as iron (Fe), copper (Cu), zinc (Zn), selenium (Se) etc. are essential to most forms of life. They serve as components of proteins and nucleic acids that are critical to many biological processes necessary for survival (Berg & Shi, 1996; Chandra et al., 1989; Jin et al., 2017; Whitaker, 2006; Yruela, 2013). These trace elements have important roles in cell proliferation, differentiation, and development. During cellular differentiation and biological development trace elements localize in specific areas of tissue or subcellular locations and are then redistributed; often, such redistribution signals new stage of development and new biological processes. Previous studies examining the roles of metals in cellular development made use of chelating techniques, or metal removal, to assess functional differences in growth and differentiation as a result of changes in metal concentration or localization. For many elements however, chelation would not be a suitable approach for exploration because of the side-effects of many chelating agents and the fact that many chelating agents bind multiple elements at the same time. In such cases information about the roles of elements in cellular processes may be gleaned from information about their distribution, concentration, and translocation under different circumstances. However, this information too remains largely unknown due to the lack of analytical techniques that can image elements with sufficient sensitivity and resolution. XFM provides a solution for this problem. XFM can act as a definitive *in situ* elemental imaging technique and provide data that need not be supplemented by optical microscopy with special dyes. Few examples below describe how XFM can be used to monitor subcellular elemental redistribution as cells undergo differentiation as well as monitor elemental distribution in homeostasis in whole animals.

XFM was used to support immunocytochemistry-based investigations of human the embryonic stem cells line H9. The loss of pluripotency is a key step in cellular development and differentiation of stem cells; this process is marked not only by changes in protein expression but also by alterations in elemental content. As H9 cells began to differentiate under the influence of retinoic acid, the Oct4 protein, characteristic for non-differentiated cells decreased while the Zn concentration increased. This process was monitored qualitatively by fluorescent microscopy, using the dye FluoZin-3,AM and quantitatively by XFM work conducted at the APS (Wolford et al., 2010). Interestingly, high resolution XFM images in this study showed that Zn is excluded from the chromosomes of mitotic cells.

In a separate study investigating elemental changes in the course of the cell cycle affirmed the latter finding using the mouse NIH 3 T3 fibroblast cell line (McRae et al., 2013). Moreover, a complete elemental overview of cells undergoing mitosis has shown that Cu and Fe are also excluded from the cell region with condensed mitotic chromosomes. At the same time, Cu and Zn concentrations immediately adjacent to chromosomes were shown to be higher than in the remainder of the cytoplasm.

Angiogenesis is the process of forming of blood vessels through endothelial cell division and subsequent differentiation. Growth of blood vessels is stimulated by Cu (Finney et al., 2009); one of the earliest indications of this mechanism came from XFM studies on microvascular endothelial cells conducted at the APS. These cells were induced to mimic early angiogenesis and a shift of Cu from the nuclear periphery to the cellular periphery extended to filopodia extensions, process extensions, and cell to cell contacts during tubulogenesis. Importantly, XFM scans demonstrated that copper re-localization in these cells was not matched by distribution changes of any of the other elements (Finney et al., 2007).

In addition to their intrinsic interest, these XFM studies also opened a new field of study, elementalomics: the study of the dynamic spatial regulation of transition metals such as Cu and Zn. Prior to this work only fluxes of Calcium (Ca), Potassium (K), and Sodium (Na) were considered as appropriate subjects for the study of dynamic elemental regulation in cells. Historically, Fe, Cu and Zn were considered as comparatively static. It is possible that this view was based on the inability to sensitively measure and spatially localize these metals. Traditionally, metal sensing fluorophores have been used for imaging of elemental distribution in live cells by optical microscopy; however, the data produced in such studies are subject to different types of artifacts and difficult to interpret quantitatively. Often, procedures for optical imaging require a significant degree of sophistication to ensure accuracy such as, for example Ca++ imaging by FURA (O'Connor & Silver, 2007; Roe et al., 1990). In other cases, the problem with metal ion imaging lies in their inaccessibility; the affinity of endogenous metalloproteins for metal atoms studied is often higher than the affinity of the metal sensing fluorophore. Thus, these compounds interact only with the cellular pool of 'free' or 'available' metal. The selectivity of optical imaging fluorophores for free copper was used in an interesting way in combination with XFM to show the differences between dynamic Cu redistribution against the backdrop of "stable" cellular Cu associated with Cu binding proteins (Dodani et al., 2011). Coppersensor-3 is a fluorescent probe that images only labile copper pools in cells; this dye was used in combination with Ca chelator BAPTA in optical microscopy to show the interconnectedness between Ca and labile Cu. At the same time – a complete elemental overview of samples prepared in parallel for XFM provided information about distribution of all Cu in these cells, including Cu stably bound to proteins and inaccessible to Coppersensor-3 dye.

XFM was also used to examine the cellular and subcellular localization and quantity of Se during development of spermatids (male germ cells) in the process of spermatogenesis. Se is concentrated in late spermatids due to elevated levels of selenoproteins (Kehr et al., 2009). Furthermore, high-resolution scans revealed that

Se is specifically enriched near the lumen side of elongating spermatids, where the tail is developed. During spermatogenesis, phosphorus is inversely correlated with Se and it was concentrated at the place where the head of the sperm forms. In mature sperm extracted from mouse testis however, Se was detectable in the head and co-localized with P and Zn. Se was also accumulated in the midpiece of the sperm tail where it co-localized with Cu and Fe. It is of interest to note that in these samples Se was well above levels considered to be toxic indicating the ability of sperm cells to handle excessive Se amounts (Kehr et al., 2009).

Differences in Se content have been found to be associated with healthy ovarian follicles as well. XFM at the AS was used to measure Cu, Fe, Zn, Se and Br distribution in bovine ovaries focusing on follicles at different stages of development (from pre-ovulatory to those undergoing atresia). Significant elemental differences between follicle groups were found and Se appeared to be the element with greatest concentration difference between large pre-ovulation follicles and their smaller counterparts (Ceko et al., 2015). The authors postulate that selenoprotein glutathione peroxidase 1, an innate repressor of oxidative stress, may be responsible for these differences.

Some of the medium and lower resolution XFM setups allowed imaging of whole animals and study of elemental distribution in normal organisms. For example, XFM was used at the AS to examine the distribution of K, Ca, Mn, Fe, Zn in anesthetized and immobile, whole, hydrated adult *Caenorhabditis elegans* without any additional processing. Genetically identical and developmentally synchronous animals were used to show similarity in elemental content and distribution across specimens. For example, the anterior part of the body was enriched for Ca and Fe, while Mn enrichment was found towards the animal's posterior (James et al., 2013).

Similarly, XFM was used to visualize the elemental distribution in whole zebrafish embryos at the APS (Bourassa et al., 2014; Bourassa et al., 2016). In these studies samples were embedded in Lowicryl K4M resin as solid substrate. This approach allowed both good preservation of the elemental distribution due to low temperatures during embedding process and enabled 3D imaging at room temperature. A special setup for sample rotation coupled with fast raster scanning permitted imaging of a few millimeters large sample – 48 h old zebrafish embryo, with X-ray beam spot of 3.5 micron over an angular space of 180 degrees at 3 degree intervals; this data was reconstructed tomographically using a combination of several approaches. Ultimately, qualitatively new data were obtained, offering unique insight into the trace metal distribution at key stages of embryonic development. For example, the total Zn content of the entire 48 h old embryo was 8.2 ng and this information was in good agreement with prior "bulk" elemental studies. However, this study also documented a clear segregation between Zn and Fe into different parts of the animal with the highest Fe concentration in the yolk syncytial layer, for example.

Special attention should be given to studies of human tissue samples because they form a baseline against which patient samples from different diseases can be compared. A particularly interesting work with 37 samples of brain tissue from the region of substantia nigra (Surowka et al., 2015) was done at BESSY II. In this work samples from healthy individuals were grouped based on age into three groups (61–70, 71–80 and 81–90 years old) and screened for P, S, Cl, K, Ca, Fe, Cu and Zn content with the idea that some of the aging related changes may be associated with gradual development of neurodegenerative diseases (Surowka et al., 2015). Interestingly, neuronal bodies and extraneuronal matter did not match each other with regard to elemental concentrations; moreover, while extraneuronal elemental concentrations did not vary much with age, age and concentrations of redox-relevant elements Fe and Cu in neuronal bodies followed inverse trends.

An excellent example of human tissue study is a recent XFM exploration of human placental tissue conducted at the APS (Punshon et al., 2015). In addition to information about this important tissue, this work provides a comparison of different sample preparation techniques: (i) fixation in formalin; (ii) fixation in a mixture of 3% glutaraldehyde and 1% paraformaldehyde in 0.1 M phosphate buffer; (iii) and fixation in a mixture of 3% glutaraldehyde and 1% paraformaldehyde in 0.1 M HEPES buffer, with or without prior freezing at −80 °C and thawing; finally, these samples were also embedded in LR resin. Two instruments were used for this work, a "standard" microprobe with a beam spot size of 500 nm and the Bionanoprobe instrument with a spot size of 30 nm. X-ray fluorescence data were analyzed (as always at the APS) using MAPS software developed by Stefan Vogt (Vogt, 2003). Greatest variations in elemental abundance in this study, dependent on sample preparation approach, were noted for Ca and Cu. As recognized by this team – use of different preparation approaches is often necessitated by clinical work conditions; variations in sample "quality" with regard to particular elements should be recognized if change of sample preparation approach is not possible.

3.3 Use of XFM for Examination of Pathological Elemental Misbalance in Disease

Information on elemental distribution in samples from different disease conditions has the potential to provide insights in normal and toxic variations in concentration, trafficking, distribution, and storage of different elements. For instance, it is known that during the degeneration of dopaminergic neurons in Parkinson's Disease (PD) proteins accumulate and form inclusions in substantia nigra region of the brain. Up to 24 h postmortem samples of substantia nigra from PD patient were sectioned at 20 micron thickness, dried at −30 °C and imaged with 15 × 15 micron steps at BESSY II (Szczerbowska-Boruchowska et al., 2012). Elemental maps of dopaminergic nerve cells and extraneuronal spaces for P, S, Cl, K, Ca, Fe, Cu, Zn, Bromine (Br) and Rubidium (Rb) were generated for Parkinson's disease sample and compared to control. PD neurons had significantly higher content of S, Cl, Ca, Fe, and Zn in substantia nigra compared to control groups. Additionally, P, S, Cl, Zn and Rb were increased in areas outside of nerve cell bodies in PD samples.

Another interesting example of work done with human samples utilized large sample instrument at the AS where specimens of tens of centimeters can be accommodated. Chronic anemia sufferers who depend on blood transfusion often experience iron overload that can lead to cardiac toxicity. Recently, investigators at AS were able to image a 5 mm tick cross section of a postmortem heart of a transfusion dependent patient (House et al. 2014). XFM iron maps with 500 × 500 micron resolution displayed the iron distribution of the heart in great detail. A gradient in Fe distribution was found, with concentrations decreasing from the outer to the inner side of the heart muscle; however, it should be noted that this sample was stored in formalin prior to sectioning and in that process, some of the Fe could have been redistributed.

A different Fe investigation study was done with livers from patients with chronic liver diseases. The histochemistry stain Berlin blue is used most often to determine Fe in such samples, however, work by Kinoshita and others who used XFM at SPring-8 has documented that the latter approach shows superior sensitivity. Fe accumulation was noted especially in periportal area of the liver samples from patients with chronic hepatitis C (CHC) or cirrhosis. Interestingly, even in healthy livers hepatocytes in the periportal area showed higher Fe concentration, suggesting that these cells may be exposed to some degree of iron-induced free radical damage (Kinoshita et al., 2010).

Menkes disease (MD) is caused by mutations in the Cu-transporting ATPase gene (ATP7A). This protein transports Cu from the cytosol to the trans-Golgi network and enables synthesis of cuproenzymes. In addition, under elevated Cu conditions this protein translocates to plasma membrane and begins with efflux of Cu across the plasma membrane in order to maintain cellular Cu homeostasis. XFM was used to study many aspects of these processes both in patient samples and in experimental model systems. For example, cell culture studies focused on ATP7A were done with patient fibroblasts and mouse 3 T3-L1 cells in which this protein was knocked out using CRISPR/Cas9 (Bhattacharjee et al., 2016). In both cases XFM mapping (conducted at the APS) confirmed increased Cu concentrations throughout the cell volume – in cytosol, nuclei and mitochondria. Interestingly, this study found that ATP7A removal had the most profound effect on mitochondria as reflected by glutathione oxidation and increased concentration of H_2O_2. An exploratory study of Menkes disease focused primarily on patient samples was done at SPring-8 and the Photon Factory in Japan. A portion of the work was done on cultured keratinocyte cells lines from two Menkes disease patients and four matched healthy individuals; this work documented that patient cells could not maintain physiologic Cu concentrations under a mild Cu overload. Patient samples form intestine, kidney and spinal cord were also included in this work. Excess Cu was found in the mucosal epithelial cells of the intestine, in tubular cells of the kidney, endothelial cells of blood and lymph vessels as well as some parts of the spinal cord (Kinebuchi et al., 2016).

In cytosol, Cu incorporation into secretory enzymes is mediated by Cu chaperone protein Antioxidant-1 (Atox1); in the nucleus, this protein acts as a Cu-dependent transcription factor. This protein is necessary for development of neovasculature and wound healing through regulation of vascular endothelial growth factor 1. Effects of Cu deprivation and absence of Atox 1 protein in $Atox1^{-/-}$ mice on wound healing were investigated recently through a series of *in vitro*, *ex vivo* and *in vivo* experiments (Das et al., 2016). XFM imaging at the APS was done to corroborate Cu concentration differences and confirm that presence of Cu in the absence of Atox1 protein does not suffice for successful wound healing process.

Another interesting example of XFM use with human samples comes from an NSLS XFM study of cadaveric human eyes (Flinn et al., 2014). In the human eye, age-related macular degeneration was associated with metal-rich deposits, but the concentration and distribution of metals in deposits was not known. Interestingly, this study established that Ca and Zn were present in all deposits, Fe only occasionally, while Cu concentration in deposits could not be differentiated from the background.

Use of XFM for work with animal models of elemental misbalance is particularly popular, not the least because animal models allow controlled genetic manipulations. For example, James and others used XFM at AS to investigate deposition of Fe, Cu and Zn in extracellular plaques forming in brains of animals with cerebral amyloidosis caused by expression of mutant human amyloid precursor protein (APP) and presenilin 1 (James et al., 2016). This work follows studies at the APS and NSLS where whole brain sections from animals of the same genotype but of different ages were scanned by XFM and Zn accumulation found to be pronounced only in animals with fully developed plaques, with a period of more diffuse increased Fe accumulation prior to plaque formation (Leskovjan et al., 2011). Interestingly, while plaques could be detected by infrared spectroscopy (Leskovjan et al., 2011; Miller et al., 2006), they could be identified by Compton inelastic scattering as well (James et al., 2016).

3.4 XFM Analyses of Elemental Content in Cancer

Despite the fact that the majority of XFM studies with cells in culture are done with cancer cells, relatively few studies compared trace element status in different cancers, possibly because imaging of patient samples requires high throughput and large area scans for comprehensive, statistically significant comparisons. Nevertheless, it is probable that such work will increase in volume with the development of faster data acquisition approaches.

Work done at the APS by Chandler and others has shown subcellular Zn accumulation in breast cancer depends on cancer subtype (Chandler et al., 2016). Zn accumulation in luminal breast tumors was found at the tissue periphery, while it was more evenly distributed in the basal breast tumors. The authors continued onward to investigate Zn transporters by microarray analysis and extend the study to breast cancer cells *in vitro*. Ultimately, Zn comparison between luminal type breast cancer cell line T47D, basal-like cell line MDA-MB-231 and non-malignant breast cell line MCF10A has shown different staining patterns and Zn accumulation according to FluoZin-3 dynamic labeling. It should be noted that Zn concentration in non-malignant cell line was the lowest. This was in keeping with previous work done at the APS where benign breast cancer screening by XFM was done with 252 matched pairs of benign breast cancer samples (Cui et al., 2007); in that work increased Zn concentrations were associated with an increased cancer risk. It is interesting to note that a study of esophageal samples, also conducted at the APS associated increased Zn concentrations with decreased risk of subsequent cancer development (Abnet et al., 2005).

Recently, XFM was used to examine and compare trace metals concentration at the micro- and nanometer scale in patient derived human glioblastoma cells (11ST, 36ST, 86ST) and commercially available cell culture cell line (U87MG). The work was done at the APS using flash-frozen cells at the high resolution cryogenic instrument Bionanoprobe and at the microprobe at sector 2ID-D equipped with a cryo-jet. Quantities of P and Zn were found to be increased in the cell nuclei while Fe, Mn and Cu accumulated in cytoplasm and perinuclear region. Fe and Mn accumulated in different regions of U87MG cells. In the patient derived cell lines 11ST, 36ST and 86ST Fe accumulated mostly in perinuclear space with Zn and K. Interestingly, the patient derived 86ST cell line presented the highest content for most metals in comparison to other cell lines. In these cells Cl, Fe, Ni, Cu and Zn were increased at least twice compared to other samples (Ducic et al., 2017).

Cancer related animal model studies have used XFM as well. For example, in order to identify the relationship between selenium intake and breast cancer progression Seleno-L-Methionine supplement was tested in mice with developed 4 T1 murine mammary carcinoma. Oral treatment was done for 28 days and the effect of SeMet on tumor growth was assessed. XFM was done at the SSRF and Se accumulation detected in tumor tissues, surprisingly, overlapping with the highest Fe distribution (and possibly tumor vasculature). Immunohistochemistry for Se binding proteins: glutathione peroxidase 1 and Selenium-binding protein 1 showed inverse pattern in treated and untreated tumors, the latter protein more increased in animals exposed to Seleno-L-Methionine supplement. Tumor growth in these mice was decreased (Song et al., 2015).

3.5 XFM Investigation of (Experimental) Elemental Overload

Successful use of XFM to investigate elemental homeostasis served as a basis to inspire studies of elemental "overload" either to explore processes involved in elemental storage or to attempt to exploit accumulated elements in order to achieve better diagnostic imaging using only "native" elemental content of cells.

For example, the idea to use Mn to increase native tissue contrast for magnetic resonance imaging (MRI) was entertained by researchers interested in functional imaging of pancreatic β-cells during activation. In an *in vitro* study conducted at the

APS XFM was used to visualize and quantify Mn in pancreatic β-cells and their subcellular regions. As a model system MIN-6 insulinoma cells were used. When these cells were grown in standard tissue culture conditions with only a trace amount of Mn present in the media, this element was present in low 22μM concentration and equally distributed across the cell. Exposure of cells to a higher Mn concentration (50μM $MnCl_2$) in medium with 2 mM glucose did not activate cells but resulted in non-glucose-dependent Mn uptake (250μM cellular Mn concentration). However, when cells were activated by 16 mM glucose in the presence of exogenous $MnCl_2$ cellular Mn concentration increased to 590μM. Interestingly, Mn distribution in these cells was uneven – with Mn both in the cytoplasm and accumulated in a perinuclear region, possibly corresponding to the Golgi apparatus and involving the secretory pathway (Leoni et al., 2011).

A different group of researchers focused on iron accumulation in cells exposed to inorganic iron (Fe(III) citrate, Fe(III) chloride and Fe(II) sulfate) and iron associated with its natural carrier protein - transferrin (Mihucz et al., 2016). Two cell lines (colorectal cancer cell lines HT-29 and HCA-7) with different expression of transferrin receptors were studied using several X-ray techniques – XFM at the Diamond Light Source and X-ray absorption near edge structure (XANES) spectroscopy at HASYLAB beamline at DESY synchrotron. While XFM mapping demonstrated Fe accumulation in cytoplasmic regions, XANES could be used to complement this study and determine relative ratios of Fe(III) sulfate vs. ferritin in HT-29 cells treated with different sources of iron. Comparisons with known XANES profiles for pure chemical/biological compounds was done using ATHENA software package analysis (Ravel & Newville, 2005).

In a recent study by our group, a liver sample coming from a rat exposed to yttrium ^{90}Y containing glass spheres, 20–40μm in diameter was sectioned and imaged by XFM. Primary purpose of the study was to look for signs of sphere sequestration. An example is shown in Fig. 3.1. This tissue is showing spheres with high silica content (due to the presence of glass) and a significant yttrium signal. Interestingly, some areas of this tissue also show weak non-spherical Si and Y shapes as well as a highly increased sulfur, zinc and iron content compared to surrounding parenchyma (white circles). These features most likely correspond to Kupfer cells.

3.6 XFM Exploration Beyond Mammals: Elemental Concentration, Distribution and Speciation in Bacteria, Eukaryotic Single Cell Organisms and Plants

In addition to work with mammalian cells and other samples discussed so far, XFM is also often used to quantify elements in bacteria and other single cell organisms, all of them of greatest importance for maintenance of biosphere. For instance, the marine cyanobacterium *Trichodesmium erythraeum* is composed of groups of single

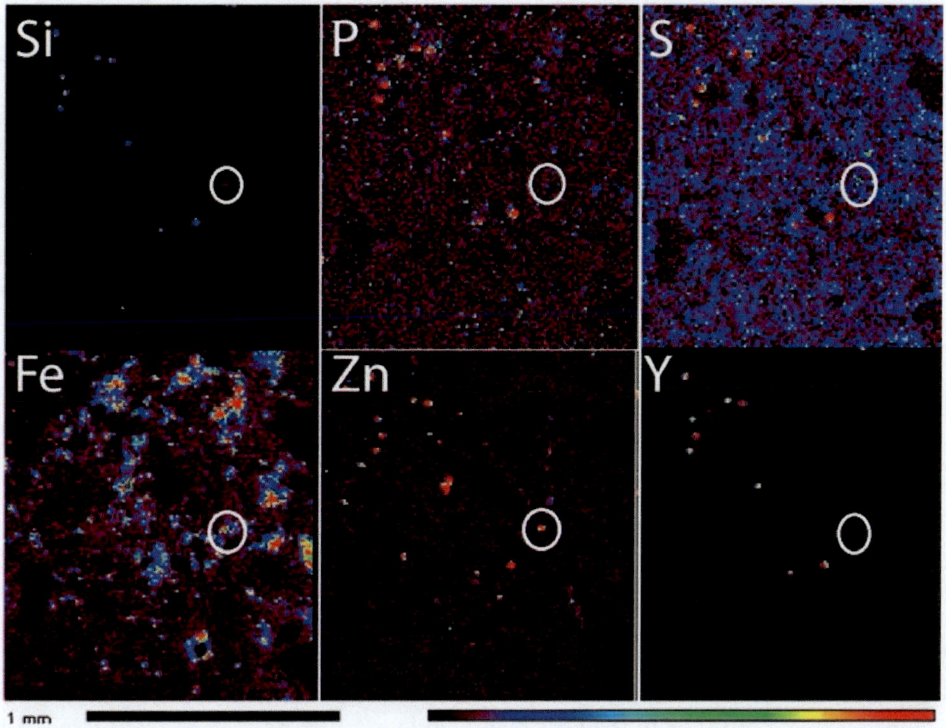

Fig. 3.1 A detail of rat liver tissue elemental map obtained with 300 nanometer spot size (imaged at sector 2 microprobe at APS), with incoming scanning X-ray energy of 21 keV. Scale bar –100 microns, color bar indicates elemental quantity as represented by false colors, from black (no signal) to red (highest signal)

cells – trichomes, and has a special adaptation that allows it to sequester CO_2 by photosynthesis and do nitrogen fixation at the same time. If both processes were in the same cell, oxygen produced by photosynthesis would interact with iron from the enzyme nitrogenase and interfere with the nitrogen-fixing process, therefore, two functions are segregated into separate compartments – trichomes. The spatial distribution of Fe, S, and P in trichomes was mapped by XFM during exponential and growth phases of this organism. About 16% of trichomes sections (up to 25 cells) have two-fold more Fe and S and two-fold less P compared to neighboring trichome sections. This ability to re-allocate elements indicates that *Trichodesmium* is a community of cells with specialized regions of multiple cells with unique chemical composition and specialized "tasks" (Nuester et al., 2014).

Sulfate-reducing bacteria can produce sulfide that reacts with metals to form insoluble products such as ZnS. Spherical aggregates of ZnS (sphalerite) 2–5 nm diameter in size are formed within natural biofilms dominated by sulfate-reducing bacteria of the family Desulfobacteriaceae. Zinc concentration in these biofilms can be as much as 100-fold higher than the Zn concentrations of groundwater in which the biofilm is growing. XFM analysis done at the APS showed that ZnS aggregates in addition to these core atoms, also contain arsenic (As), Se, and Fe but exclude lead despite of its presence in the groundwater. This finding suggests that microbes and microbial biofilms control metal concentrations in groundwater (Labrenz et al., 2000).

Traditionally, the metal composition of bacterial cells has been measured by bulk elemental analysis and averaged, with little interest in possible alterations of elemental composition in response to substrate sufficiency vs. starvation vs. toxic stress. XFM at the APS was employed to determine changes in the elemental composition of bacterium *Nitrosomonas europaea* ATCC 19,718 caused by growth and/or exposure to Cu(II). Exposure to Cu(II) caused statistically significant increase in cellular Cu concentration, while different growth status had little effect on Cu and Fe concentration. Ratio of Fe and Cu to other elements in this organism was higher than in *Pseudomonas fluorescens*. This suggests that that the former bacterium may have higher demand for Fe and Cu because of enzymes involved in nitrogen fixation need metal cofactors for electron transport (Yu et al., 2011).

Phytoplankton in the coastal seas of West Antarctica leads to substantial loss of iron from marine ecosystems because it incorporates it into biogenic silica (Si). XFM was used to analyze Fe and Si quantity and distribution in these organisms and confirm this hypothesis. In each specimen Fe and Si were co-localized, confirming structural incorporation of iron into cell walls made of silica (Ingall et al., 2013).

Arbuscular mycorrhiza is a symbiosis between plants and fungi. Cooperation between the species improves the supply of water and macro-elements such as phosphate and nitrogen to the host plant, while as much as 20% of plant-fixed biomass supplies the fungus as a source of energy. XFM at SPring-8 was used to find that exposure to cadmium (Cd) leads to accumulation of this toxic element in cells and cell walls of both symbiotic organisms (Nayuki et al., 2014). Further research may be able to establish how to reduce or increase Cd accumulation when plant production is used for the purposes of food production or heavy metal decontamination, respectively.

Fe is an essential element for plant growth and development; however, its potential interaction with oxygen produced by photosynthesis requires compartmentalization and tightly regulated Fe acquisition and allocation. To evaluate Fe in tomato plant roots and cucumber leaves, XFM was done at DESY. Inside root, Fe was found in the outer layer - the rhizodermis. The Fe concentration there was two to three orders of magnitude more than in the inner portion of this organ where it averaged 10μg/g. In the cucumber leaf the Fe concentration was highest in the midrib, with 600μg/g, up to 60 fold higher than in the cells of parenchyma (Terzano et al., 2013).

Similar to the compartmentalization of iron, Ca^{++} distribution in *Arabidopsis thaliana* is tightly regulated. Proteins called cation exchangers (CAXs) are involved in this process. XFM imaging of wild type and mutant *Arabidopsis* seeds was done at the NSLS. Elemental maps showed that altered Ca localization within cells and reduced partitioning into organelles in mutated seeds. Interestingly, changes in Ca distributions were different in CAX1 and CAX3 single and double mutants, suggesting that these two genes play distinct roles in different seed compartments. In addition, K, Mn and Fe were also redistributed in mutant seeds (Punshon et al., 2012).

XFM at the SSRF was done to investigate accumulation of Pb in plant *Elsholtzia splendens* — a copper accumulator plant exposed to Pb. This toxic element was mostly accumulated in the roots but also stems and leaves. In latter organs, Pb was restricted to the vascular bundles and epidermis. A significant positive correlation between Pb, Ca, K, and Zn distribution was found, and correlation between Pb and Ca was the greatest. This insight can be exploited either to decrease or increase accumulation of Pb in this species (Zhang et al., 2011).

Production of biofuels and chemicals from plants requires chemical pretreatments that could be made simpler if iron was present in plant material. To enrich plant matter with iron, transgenic *Arabidopsis* plants expressing heterologous ferritin gene were engineered. These plants accumulate iron under both control and iron-fertilized growth conditions. Growth in the presence of Fe fertilizer led to an increase in plant height and dry weight by more than 10%. Fe accumulation in these plants occurred in cell walls as determined by XFM studies conducted at the APS (Lin et al., 2016).

Finally, of special interest for this review, XFM was also used for investigation of bacteria mediated mobilization of actinides from contaminated soil (Francis & Dodge, 2015). This work found that microbial activity may change solubility of plutonium, for example, and change its availability in contaminated environment.

3.7 Conclusions

XFM is a powerful and versatile technique for the investigation of elemental content in biological samples. Whole cells can be imaged with resolutions better than 100 nm and their content 3D reconstructed even for a cell thickness of 10 microns or more; with some compromises in the spatial resolution samples as thick as 100 s of microns can also be imaged in 3D. With new technical developments such as "fourth generation" synchrotrons, faster detectors and even "X-ray focusing" optics it is likely that XFM will continue to develop toward ever higher resolution and speed of data acquisition. While XFM can be used for detection of radionuclides in biological material, due to extremely low background for such elements in samples collected in non-contaminated areas, radionuclide quantities are generally low. Moreover, radioactive decay and resultant elemental transitions further decrease numbers of atoms of interest that are available for detection. With the increase in brightness, new generations of synchrotrons and their further updates can be expected to improve sensitivity of detection.

References

Abnet, C. C., et al. (2005). Zinc concentration in esophageal biopsy specimens measured by x-ray fluorescence and esophageal cancer risk. *Journal of the National Cancer Institute, 97*(4), 301–306.

Berg, J. M., & Shi, Y. (1996). The galvanization of biology: A growing appreciation for the roles of zinc. *Science, 271*(5252), 1081–1085.

Bhattacharjee, A., et al. (2016). The activity of Menkes disease protein ATP7A is essential for Redox balance in mitochondria. *The Journal of Biological Chemistry, 291*(32), 16644–16658.

Bourassa, D., et al. (2014). 3D imaging of transition metals in the zebrafish embryo by X-ray fluorescence microtomography. *Metallomics, 6*(9), 1648–1655.

Bourassa, D., et al. (2016). MicroXRF tomographic visualization of zinc and iron in the zebrafish embryo at the onset of the hatching period. *Metallomics, 8*(10), 1122–1130.

Ceko, M. J., et al. (2015). Quantitative elemental analysis of bovine ovarian follicles using X-ray fluorescence imaging. *Metallomics, 7*(5), 828–836.

Chandler, P., et al. (2016). Subtype-specific accumulation of intracellular zinc pools is associated with the malignant phenotype in breast cancer. *Molecular Cancer, 15*, 2.

Chandra, S., et al. (1989). Quantitative imaging of free and total intracellular calcium in cultured cells. *Proceedings of the National Academy of Sciences of the United States of America, 86*(6), 1870–1874.

Cui, Y., et al. (2007). Levels of zinc, selenium, calcium, and iron in benign breast tissue and risk of subsequent breast cancer. *Cancer Epidemiology, Biomarkers & Prevention, 16*(8), 1682–1685.

Das, A., et al. (2016). Endothelial antioxidant-1: A key mediator of copper-dependent wound healing in vivo. *Scientific Reports, 6*, 33783.

Dodani, S. C., et al. (2011). Calcium-dependent copper redistributions in neuronal cells revealed by a fluorescent copper sensor and X-ray fluorescence microscopy. *Proceedings of the National Academy of Sciences of the United States of America, 108*(15), 5980–5985.

Ducic, T., et al. (2017). Structural and elemental changes in glioblastoma cells in situ: Complementary imaging with high resolution visible light- and X-ray microscopy. *Analyst, 142*(2), 356–365.

Finney, L., et al. (2007). X-ray fluorescence microscopy reveals large-scale relocalization and extracellular translocation of cellular copper during angiogenesis. *Proceedings of the National Academy of Sciences of the United States of America, 104*(7), 2247–2252.

Finney, L., et al. (2009). Copper and angiogenesis: Unravelling a relationship key to cancer progression. *Clinical and Experimental Pharmacology & Physiology, 36*(1), 88–94.

Flinn, J. M., et al. (2014). Correlations in distribution and concentration of calcium, copper and iron with zinc in isolated extracellular deposits associated with age-related macular degeneration. *Metallomics, 6*(7), 1223–1228.

Francis, A. J., & Dodge, C. J. (2015). Microbial mobilization of plutonium and other actinides from contaminated soil. *Journal of Environmental Radioactivity, 150*, 277–285.

House, M. J., et al. (2014). Mapping iron in human heart tissue with synchrotron x-ray fluorescence microscopy and cardiovascular magnetic resonance. *Journal of Cardiovascular Magnetic Resonance, 16*, 80.

Ingall, E. D., et al. (2013). Role of biogenic silica in the removal of iron from the Antarctic seas. *Nature Communications, 4*, 1981.

James, S. A., et al. (2013). Direct in vivo imaging of essential bioinorganics in Caenorhabditis elegans. *Metallomics, 5*(6), 627–635.

James, S.A., et al. (2016). Iron, copper, and zinc concentration in Abeta plaques in the APP/PS1 mouse model of Alzheimer's disease correlates with metal levels in the surrounding neuropil. *ACS Chemical Neuroscience*.

Jin, Q., et al. (2017). Preserving elemental content in adherent mammalian cells for analysis by synchrotron-based x-ray fluorescence microscopy. *Journal of Microscopy, 265*(1), 81–93.

Kehr, S., et al. (2009). X-ray fluorescence microscopy reveals the role of selenium in spermatogenesis. *Journal of Molecular Biology, 389*(5), 808–818.

Kinebuchi, M., et al. (2016). Diagnostic copper imaging of Menkes disease by synchrotron radiation-generated X-ray fluorescence analysis. *Scientific Reports, 6*, 33247.

Kinoshita, M., Uchida, T., Sato, A., Nakashima, M., Nakashima, H., Shono, S., Habu, Y., Miyazaki, H., Hiroi, S., & Seki, S. (2010). Characterization of two F4/80-positive Kupffer cell subsets by their function and phenotype in mice. *Journal of Hepatology, 53*(5), 903–910. https://doi.org/10.1016/j.jhep.2010.04.037

Labrenz, M., et al. (2000). Formation of sphalerite (ZnS) deposits in natural biofilms of sulfate-reducing bacteria. *Science, 290*(5497), 1744–1747.

Leoni, L., et al. (2011). Beta-cell subcellular localization of glucose-stimulated Mn uptake by X-ray fluorescence microscopy: Implications for pancreatic MRI. *Contrast Media & Molecular Imaging, 6*(6), 474–481.

Leskovjan, A. C., et al. (2011). Increased brain iron coincides with early plaque formation in a mouse model of Alzheimer's disease. *NeuroImage, 55*(1), 32–38.

Lin, C. Y., et al. (2016). Directed plant cell-wall accumulation of iron: Embedding co-catalyst for efficient biomass conversion. *Biotechnology for Biofuels, 9*, 225.

McRae, R., Lai, B., & Fahrni, C. J. (2013). Subcellular redistribution and mitotic inheritance of transition metals in proliferating mouse fibroblast cells. *Metallomics, 5*(1), 52–61.

Mihucz, V. G., et al. (2016). Iron overload of human colon adenocarcinoma cells studied by synchrotron-based X-ray techniques. *Journal of Biological Inorganic Chemistry, 21*(2), 241–249.

Miller, L. M., et al. (2006). Synchrotron-based infrared and X-ray imaging shows focalized accumulation of cu and Zn co-localized with beta-amyloid deposits in Alzheimer's disease. *Journal of Structural Biology, 155*(1), 30–37.

Nayuki, K., et al. (2014). Cellular imaging of cadmium in resin sections of arbuscular mycorrhizas using synchrotron micro X-ray fluorescence. *Microbes and Environments, 29*(1), 60–66.

Nuester, J., Newville, M., & Twining, B. S. (2014). Distributions of iron, phosphorus and sulfur along trichomes of the cyanobacteria Trichodesmium. *Metallomics, 6*(6), 1141–1149.

O'Connor, N., & R.B. (2007). Silver, ratio imaging: Practical considerations for measuring intracellular Ca2+ and pH in living cells. *Methods in Cell Biology, 81*, 415–433.

Paunesku, T., et al. (2012). X-ray fluorescence microscopy for investigation of archival tissues. *Health Physics, 103*(2), 181–186.

Petibois, C. (2010). Imaging methods for elemental, chemical, molecular, and morphological analyses of single cells. *Analytical and Bioanalytical Chemistry, 397*(6), 2051–2065.

Punshon, T., et al. (2012). The role of CAX1 and CAX3 in elemental distribution and abundance in Arabidopsis seed. *Plant Physiology, 158*(1), 352–362.

Punshon, T., et al. (2015). High-resolution elemental mapping of human placental chorionic villi using synchrotron X-ray fluorescence spectroscopy. *Analytical and Bioanalytical Chemistry, 407*(22), 6839–6850.

Ravel, B., & Newville, M. (2005). ATHENA, ARTEMIS, HEPHAESTUS: Data analysis for X-ray absorption spectroscopy using IFEFFIT. *Journal of Synchrotron Radiation, 12*(Pt 4), 537–541.

Roe, M. W., Lemasters, J. J., & Herman, B. (1990). Assessment of Fura-2 for measurements of cytosolic free calcium. *Cell Calcium, 11*(2–3), 63–73.

Song, H., Ren, X., & Liu, P. (2015). Distribution and inhibition effect of Seleno-L-methionine on 4T1 mouse mammary carcinoma. *International Journal of Physiology, Pathophysiology and Pharmacology, 7*(2), 76–86.

Surowka, A. D., et al. (2015). Synchrotron radiation based X-ray fluorescence shows changes in the elemental composition of the human substantia nigra in aged brains. *Metallomics, 7*(11), 1522–1531.

Szczerbowska-Boruchowska, M., Krygowska-Wajs, A., & Adamek, D. (2012). Elemental micro-imaging and quantification of human substantia nigra using synchrotron radiation based x-ray fluorescence--in relation to Parkinson's disease. *Journal of Physics. Condensed Matter, 24*(24), 244104.

Terzano, R., et al. (2013). Spatially resolved (semi)quantitative determination of iron (Fe) in plants by means of synchrotron micro X-ray fluorescence. *Analytical and Bioanalytical Chemistry, 405*(10), 3341–3350.

Vogt, S. (2003). MAPS: A set of software tools for analysis and visualization of 3D X-ray fluorescence data sets. *Journal de Physique IV, 104*, 635–638.

Whitaker, M. (2006). Calcium microdomains and cell cycle control. *Cell Calcium, 40*(5–6), 585–592.

Wolford, J. L., et al. (2010). Loss of pluripotency in human embryonic stem cells directly correlates with an increase in nuclear zinc. *PLoS One, 5*(8), e12308.

Yruela, I. (2013). Transition metals in plant photosynthesis. *Metallomics, 5*(9), 1090–1109.

Yu, R., et al. (2011). Elemental profiling of single bacterial cells as a function of copper exposure ands growth phase. *PLoS One, 6*(6), e21255.

Zhang, J., et al. (2011). Lead tolerance and cellular distribution in Elsholtzia splendens using synchrotron radiation micro-X-ray fluorescence. *Journal of Hazardous Materials, 197*, 264–271.

Chapter 4
Data and Biomaterial Archives in Radioecology and Radiobiology; the Importance of STOREing

Paul N. Schofield, Ulrike Kulka, Soile Tapio, Gayle Woloschak, Michael Gruenberger, Shin Saigusa, Mandy Birschwilks, and Bernd Grosche

Abstract In this commentary we consider the importance of ready access to ecological data, existing resources and approaches for radiological datasets and material, and wider public policy developments in regard to data access and reuse. We describe the development and operation of the STORE database for radiobiology, radioecology and epidemiology as a central data sharing resource, and finally we consider the issues of financial and scientific sustainability of material resource archives and the future of resources like STORE.

P. N. Schofield (✉) · M. Gruenberger
Department of Physiology Development and Neuroscience, University of Cambridge, Cambridge, UK
e-mail: pns12@cam.ac.uk

U. Kulka · M. Birschwilks · B. Grosche
Bundesamt fuer Strahlenschutz, Neuherberg, Germany

S. Tapio
Helmholtz Zentrum Muenchen, German Research Center for Environmental Health GmbH, Institute of Radiation Biology, Neuherberg, Germany

G. Woloschak
Departments of Radiation Oncology, Radiology, and Cell and Molecular Biology, Feinberg School of Medicine, Northwestern University, Chicago, IL, USA

S. Saigusa
Institute of Radiation Epidemiology, Radiation Effects Association, Tokyo, Japan

M. D. Wood et al. (eds.), *Biomarkers of Radiation in the Environment*, NATO Science for Peace and Security Series A: Chemistry and Biology,
https://doi.org/10.1007/978-94-024-2101-9_4

4.1 Introduction

The natural environment is under persistent and increasingly complex challenge from anthropogenic contaminants including radionuclides, heavy metals, organic pesticides, endocrine disruptors and mimetics. Increasingly these are found to interact to generate a combinatorial challenge; the multi-stressor problem (Vanhoudt et al., 2012). Consequently environmental protection, arguably inseparable from that of the human population, depends on accurate data and sample collection; the production of theoretical models on which safety assessments may be made, and improvements in our understanding of the contribution of these agents to compromising the integrity of the biotic ecosphere (Mothersill et al., 2018, 2019; Salbu, 2009).

Both longitudinal, geospatial and niche-specific collection of both data and material for analysis are critical in our attempts to achieve these aims. However, the internationally distributed nature of studies, their details, and how to access data currently lacks a coherent or common platform or clearing house to allow investigators and regulators to discover datasets and material that may often, especially in the case of environmental samples, be completely unique. In this way the domain of ecology, more so than many others, is dependent on the archiving, discovery and access to data and samples in order to carry out comparisons, data aggregation/integration, and novel modes of analysis. Data and sample reuse are therefore critical to the field.

The issues of data archiving and reuse have been under considerable scrutiny in recent years, resulting in the formulation of the FAIR guidelines for Open Data (Wilkinson et al., 2016). Resulting from extensive consultation between funding agencies, journals and scientists, these guidelines have been adopted by many major funding agencies, the European Commission and formally by the countries of the OECD and G20 group of nations (Arzberger et al., 2004; Mons et al., 2017). Findability, Accessibility, Interoperability, and Reusability represent the four principles of Open data and are essential for effective data governance and management (Sansone et al., 2018). The advantages of data sharing are overwhelming, amongst which are improved reproducibility, accountability, and the added value, both scientific and financial, of reusing data for purposes for which it was not originally intended; aggregating with other datasets, or conducting novel analysis in the light of new methods or paradigms (Cook-Degan, 2007). For the individual this also provides increased recognition and often collaborations or further developments of studies that they had not anticipated.

There has always been acceptance in the ecological sciences that release of primary data by investigators is an important norm and, while this is not always respected, it is fair to say that this community has an excellent track record in comparison to many (Michener, 2015). The development of very large datasets in recent years has increased the willingness to share, though there are still some issues, common with other disciplines (e.g. Blumenthal et al., 2006), which inhibit full and free sharing. These include protectionism, concerns that flaws in analysis might be revealed, lack of time, expertise or funding for preparation of data for upload, lack of appropriate sharing platforms, concerns over intellectual property protection, and loss of "ownership". These are common to many disciplines and, although radioecologists were a small category of respondents in a recent survey of data sharing in a large European radiobiology project, CONCERT, the responses received broadly reflected these common findings (Madas & Schofield 2019).

Responsibility for encouragement of data sharing rests significantly on funding agencies and journals. Current funder policies vary according to agency but although data sharing is encouraged, mandatory sharing is not stipulated by most agencies funding ecological research. A notable exception for example is the UK NERC, though the European Commission is currently undertaking a data sharing pilot study (Horizon 2020 guidelines, 2016). A summary of many funding agency and journal policies can be found on the FAIR sharing website; (https://fairsharing.org/) and (McQuilton et al., 2016).

Journals are increasingly developing policies to conform to the criteria laid out in the TOP (Transparency, Openness and Reproducibility) guidelines (Nosek et al., 2015) with the aspiration of requiring mandatory sharing of data and resources as a condition of publication. So far however, with the exception of PLoS journals (Bloom et al., 2014), there seems to have been little impact on the availability of primary publication data (Federer et al., 2018) and in the domain of radioecology data sharing is encouraged by many journals, often implementing a publisher's blanket policy, such as Elsevier, but we have been unable to identify a journal in the area of radioecology or radiation biology for which deposition into a public database is mandatory.

The problem of data availability has recently been raised by Beresford et al. (2020) and reflects a common issue about the provision of summary data alone, or in some cases no primary data at all. Withholding of primary data not only slows the progress of science, for example withholding unique contamination datasets, but also makes intercomparison and aggregation of datasets impossible. This adds to the uncertainty about reliability of conclusions where it is impossible to replicate the analysis. Where this kind of problem impacts on regulatory activities and safety assessments, with potentially huge implications for humans and environmental safety, as well as major economic impacts, there is an additional imperative for the community to ensure that the highest standards are met.

4.2 Environmental and Ecological Data

4.2.1 Environmental Information Data Centre

Several structured databases have been established specifically for the general domain of ecology. The UK Natural Environment Research Council's Environmental Information Data Centre is hosted by the Centre for Ecology & Hydrology (CEH) and provides access to data and tools related to integrated research in terrestrial and freshwater ecosystems and their interaction with the atmosphere. It is a well structured data resource with a high degree of FAIR compliance (NERC Data Centre; http://eidc.ceh.ac.uk/), This database currently contains 15 radioecology datasets (data accessed 5.11.18) from a wide range of studies.

4.2.2 The Radioecology Exchange

The European Radioecology Alliance (ALLIANCE) was created in in September 2012, and has developed a framework and strategic plan for radioecology research which is now continued under COMET (COordination and iMplementation of a pan-European instrumenT for radioecology), a Coordination and Support Action funded by the EC/ Euratom FP7. As part of the COMET infrastructure the Radioecology Exchange was created to act as a platform portal for radioecological data (Muikku et al., 2018). The Radioecology Exchange contains a wide range of datasets from six European countries and Japan from the STAR Network of Excellence, and is an important resource for radioecology (https://radioecology-exchange.org/content/radioecology-data).

4.2.3 Other Dedicated Databases

The FREDERICA database (Copplestone et al., 2008) contains data on the effects of radiation on non-human biota curated from around 1200 papers with approximately 30,000 data points. The data contains details of exposures, biological effects, environmental conditions, life cycle and pathway of exposure.

The Wildlife Transfer Database (Copplestone et al. 2013) (https://www.wildlifetransferdatabase.org/) provides parameter values for use in environmental radiological assessments to estimate the transfer of radioactivity to non-human biota.

The **PROBA UIAR database** contains radionuclide spatial distribution data from the Chernobyl exclusion zone (Kashparov et al., 2018) and can be found both in the NERC data centre (Kashparov et al., 2017) and the STORE database (see below: https://doi.org/10.20348/STOREDB/1087).

The US Earth Observation system data and information database, EOSDIS (https://earthdata.nasa.gov/), which supports "discovery and processing of earth science data from satellite, aircraft and field campaigns" is also a source of some radiation ecology associated data but much less used by the community.

Radnet is the United States environmental radiation monitoring service (Wolbarst et al., 2008) which is run by the US environmental protection agency (EPA). It monitors the radionuclide content of air, precipitation and drinking water in the environment; in some cases in real time, and has historical records of ambient environmental radiation going back to the 1940s. Further information may be found on (https://www.epa.gov/radnet/radnet-databases-and-reports).

4.3 Biological and Inorganic Sample Archives

4.3.1 Radioecology Exchange Samples Register

Biomaterials from non-human biota and inorganic matter including water and air are generally archived as part of specific data gathering, often over protracted periods of time with the aim of gathering longitudinal data from the same site. This means that samples are scattered across the community and discovery of relevant material depends on familiarity with published studies. In an attempt to produce a clearing house for such samples collections the Radioecology Alliance has collected lists of available samples on its website, mainly derived from European studies. These include samples derived from air (mainly filters), water, soil and building materials, as well as biological material. The data records for these archives may be found on https://radioecology-exchange.org/content/sample-archives along with the appropriate contact details. Work is underway to curate these collections for the STORE database in order to improve accessibility and discovery for other investigators.

4.3.2 Sample Bank of Fukushima Animals, Japan

Following the Fukushima Daiichi Nuclear Power Plant (FNPP) accident, a sample bank of animals affected was established. Organs of domestic livestock in the evacuation zone, within a 20-km radius from FNPP, were sampled between August 29, 2011 and March 21 2013. Organs (1270) and peripheral blood samples (200) from 302 exposed cows have been archived, and analysis on radionuclide content carried out (Fukuda et al., 2013). Organs were either stored as formalin fixed, paraffin embedded blocks or frozen at −80 °C (Takahashi et al., 2015). More recently the sample bank has been augmented by the collection of organs from more than 400 Japanese macaques (Urushihara et al., 2018 and M. Fukumoto. Pers. Comm.). Detailed environmental dosimetry, geographical distribution and other data are available on request (manabu.fukumoto.a8@tohoku.ac.jp).

4.4 STORE DB; a Database for Radiobiology, Radioecology and Epidemiology

While there already exist public databases dedicated to particular domains or data types, such as Array Express (RNA expression studies (Kolesnikov et al., 2015)), and PRIDE (proteomics (Jarnuczak & Vizcaino, 2017)), domain specific databases which carry a wide range of data types relevant to studies on one theme, e.g. Mouse genome informatics (MGI), (genomic, variant and phenotypic data on mice (Eppig, 2017)) are much more rare. There are huge advantages in domain-specific databases, notably that of expert curation, data structure and, specifically, domain metadata. As long as domain-specific metadata are consistent with recognised standards and therefore allow data discovery and integration with other datasets, such databases can be important resources for a community.

Following the development of the ERA database between 1999 and 2011 (Gerber et al., 1996; 2006; Gerber & Wick, 2004; Tapio et al., 2008; Birschwilks et al., 2011), with the aim of sustaining legacy data from very large scale animal exposure experiments, it became apparent that there was a need for a database that would be available for the deposition and sharing of contemporary as well as legacy data that could be accessed by anyone in the community.

In response to this need, the STORE database was initiated under European Commission funding in 2009 and has been sustained through successive grants until the present. STORE provides a platform for all types of data, organised on a project basis. The "Study" provides a root directory into which datasets and individual data items can be loaded in a hierarchical fashion, in principle allowing for the multiple outputs of a project, protocols, raw data, processed data, etc. to be filed in order to document a complete project if desired. This structured clustering of data has advantages over the approaches taken by commercial data-agnostic repositories that are centred only on the data entry itself and although it is not mandatory to structure data entries in this way it is very helpful for large integrated projects as repository for the research methods and outputs to be archived and shared. This is particularly helpful when referencing data and protocols in publications, as STORE generates stable accession identifiers and digital object identifiers (DOIs) which can be referenced in publications rather than depositing information as journal supplementary data. STORE is also used to archive links to datasets in other databases, and is completely integrated with the ERA database. Where bio- or inorganic matter material collections are entered, any web presence, database of material or other formal point of contact such as a curator may be recorded and these collections are described together with any publications.

Data and datasets are tagged with metadata terms taken from the Ontology for Biomedical Investigations (Bandrowski et al., 2016) and the Experimental factor ontology (Malone et al., 2010). There are ongoing efforts to add further terms to these ontologies for radiation biology, but this is work in progress. Use of established semantic standards for data tagging will become important in making STORE FAIR, and allowing programmatic access and data discovery in the future. STORE provides persistent digital object identifiers and accession IDs which use a persistent namespace formally registered with identifiers.org at the EBI. It is recognised by the FAIRSharing initiative (McQuilton et al., 2016) and *re3data* (Pampel et al., 2013).

Although curatorial help and training are available currently from the STORE team, users can generate a user account using their ORCID identifier, upload, tag and describe their data themselves in an intuitive GUI. Deposition and access to data are free to individual investigators and institutions. Data is stored live for a guaranteed period of 7 years after the most recent access, after which it will be stored successively for another 7 years and so on. If data are not accessed for longer than 7 years they will be taken offline and archived as a way of making sure that STORE has sufficient capacity to take new data while retaining everything that has ever been entered. STORE is available on http://www.storedb.org (Fig. 4.1) through an html interface, although programmatic access is also planned in the near future as recommended in the FAIR guidelines (Wilkinson et al., 2018).

The database is physically located within the secure BfS network platform and the BfS has undertaken to maintain the database indefinitely which means that data will be secure and accessible for the foreseeable future. Currently STORE contains more than 3000 data objects across a wide range of data types and has 95 registered data depositors.

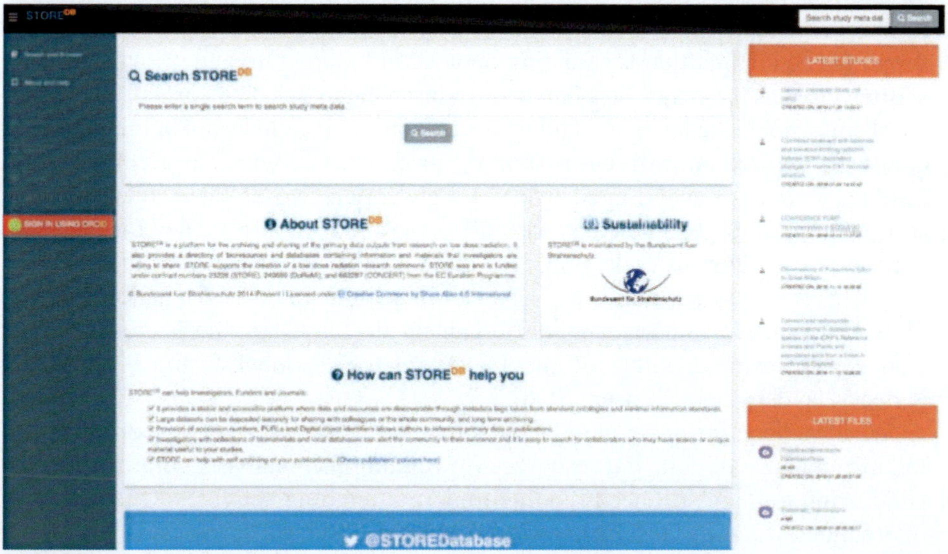

Fig. 4.1 A screenshot of the front page of the STORE database; http://www.storedb.org. 30.1.19

4.5 Database and Bioresource Sustainability

The adoption of open data policies by journals and funding agencies does not seem to have made significant impact on sharing, and there remains considerable resistance even in communities where sharing is regarded as a community norm (Piwowar, 2011; Savage & Vickers, 2009; Tenopir et al., 2011, 2015). Even when data is deposited, for example in journal supplementary information sites, it is often not complete or, due to missing information, effectively unusable. In a recent study only 56% of the data purportedly available from sampled ecological studies was found to be inadequate and effectively unusable (Roche et al., 2015; Roche, 2017).

The development of supplementary information sites for journals over the last 20 years is no longer regarded as an adequate repository for primary data, as many of these repositories are unstructured and lose data (Alsheikh-Ali et al., 2011; Anderson et al., 2006), undiscoverable or data not actually submitted, in contradiction to explicit journal policies (Federer et al., 2018). It is clear therefore that stable repositories, such as provided by STORE and other public databases constitute an essential part of biological data infrastructure.

Sustaining the infrastructure for data sharing is a major challenge for which there is as yet no satisfactory formula applicable to all communities (Chandras et al., 2009; Kaiser, 2016; Reiser et al., 2016; Sansone et al., 2018; Schofield et al., 2010). A recent study (Attwood et al., 2015) of the long-term sustainability of databases reported that 62.3% of 326 databases listed in a 1997 directory, DBCat, were no longer operating after 18 years. Database longevity was strongly associated with long-term sources of financial support from local institutions or central government funding, and databases which failed to be sustained were funded by short term grant support. The competition for funding between infrastructure and primary research is often cited as the major problem in sustaining databases, and particularly biomaterial collections and archives. This essential conflict is found even in funding agencies with avowed infrastructure funding programmes, which nevertheless limit funding periods to at the most five years and often two or three. It is relatively straightforward to get a new database off the ground or raise money for the establishment of a materials collection, it is quite another to sustain it in to the future, even with extensive use and value, as recently demonstrated by the major reduction of funding to five important model organism databases by the National Human Genome Research Institute of the United States National Institutes of Health (Check-Hayden, 2016; Kaiser, 2016).

We have been fortunate with STORE that the BfS have taken responsibility for the maintenance of data, but resources for running the database, curating data and conducting training still need funding to permit ongoing activities. One model for international funding, to our knowledge unique in the biosciences, is the United Kingdom BBSRC and the United States NSF collaborative funding scheme which coordinates funding strategies and procedures for international projects and infrastructures and supports researchers wishing to apply for UK-US collaborative research funding. In another Europe model, many international databases and resources are being supported by the ELIXIR intergovernmental organisation, which integrates and sustains infrastructure and resources for bioinformatics across Europe (Durinx et al., 2016). The ELIXIR model requires national funders to contribute to resources, providing additional funds to coordinate and integrate them nationally and offers a potential model for the funding of databases like STORE which are of international value. Alternative models where users pay for access, such as the commercial sharing resources explicitly discriminate against nationals of countries insufficiently wealthy to access the data. This model was discussed and to a degree implemented in a multi-tier model by the Arabidopsis database community (International Arabidopsis Informatics 2010, Reiser et al., 2016), but with serious concern about exclusion of some significant stakeholders for example from third world countries; there would be similar issues with radioecology data as many countries with serious radioecological challenges have poorly funded science.

4.6 Conclusions

The foresight shown by the European Commission and the BfS in supporting sharing infrastructures in the overall field of radiation biology has resulted in the discipline being somewhat ahead of the pack in having a dedicated sharing platform open to the whole community. Similarly the ALLIANCE consortium has been able to amply demonstrate the importance of coordinating access to ecological data and materials in the Radioecology Exchange. Nevertheless, the field faces the challenge of failure to share data and materials. In many cases this is an issue of training in data management and community norms, or financial constraints on distribution of materials. However, it is also clear that in some cases active decisions are being made to withhold primary data. Addressing these issues of training and culture would seem to be some of the most important that we face and require support from the community, funding agencies and stakeholders to improve the reproducibility of analyses and realise the added value gained from access to the results of publicly funded data. Sustainability of the platforms is another issue to be addressed urgently, and while different models for sustainability exist it is clear that without the coordinated input of governmental and other funding agencies this is not likely to be viable in the long run.

Investigators in the area of radioecology and radiation protection have additional responsibilities to those in many other areas of the biosciences. The maintenance of public trust in our research and our inputs into the activities of regulators and public policy bodies are a critical element in our work. The safety of humans and the ecosphere within which we live is directly dependent on maintaining this this trust and openness.

Acknowledgements Thanks go to the partners of the initial STORE project (HMGU, Germany; SUBI, Russian Federation; The University of Edinburgh-Edinburgh Cancer Research Centre, UK; Imperial College London, UK; RIVM, The Netherlands; IBBL, Luxemburg). Further thanks go to Simon Bouffler, Janet Tawn, Nick Beresford, and Jonathan Bard for their advice, skill and wisdom. PNS thanks Carmel Mothersill and Francois Brechignac for their enthusiastic encouragement to enter the world of radiation ecology. The authors are deeply indebted to Bundesamt fuer Strahlenschutz for guaranteeing the sustainability of STORE.

Disclosure Statement The authors report no conflicts of interest.

FundingThe ERA database was developed within the Euratom FP**6** contract **28725** (ERA-PRO). STORE was funded by Euratom FP**7** contract **232628** (STORE), and partly by **249689** (DoReMi). Current funding is from CONCERT; this project has received funding from the Euratom research and training programme **2014–2018** under grant agreement No **662287**.

References

Alsheikh-Ali, A. A., Qureshi, W., Al-Mallah, M. H., & Ioannidis, J. P. (2011). Public availability of published research data in high-impact journals. *PLoS One, 6*, e24357.

Anderson, N., Tarczy-Hornoch, P., & Bumgarner, R. (2006). On the persistence of supplementary resources in biomedical publications. *BMC Bioinformatics, 7*, 260.

Arzberger, P., Schroeder, P., Beaulieu, A., Bowker, G., Casey, K., Laaksonen, L., Moorman, D., Uhlir, P., & Wouters, P. (2004). An international framework to promote access to data. *Science, 303*, 1777.

Attwood, T. K., Agit, B., & Ellis, L. B. M. (2015). Longevity of biological databases. *EMBnet Journal, 21*. https://doi.org/10.14806/ej.21.0.803

Bandrowski, A., Brinkman, R., Brochhausen, M., Brush, M. H., Bug, B., Chibucos, M. C., Clancy, K., Courtot, M., Derom, D., Dumontier, M., et al. (2016). The ontology for biomedical investigations. *PLoS One, 11*, e0154556.

Beresford, N. A., Scott, E. M., & Copplestone, D. (2020). Field effects studies in the Chernobyl Exclusion Zone: Lessons to be learnt. *Journal of Environmental Radioactivity, 211*, 105893. https://doi.org/10.1016/j.jenvrad.2019.01.005

Birschwilks, M., Gruenberger, M., Adelmann, C., Tapio, S., Gerber, G., Schofield, P. N., & Grosche, B. (2011). The European radiobiological archives: Online access to data from radiobiological experiments. *Radiation Research, 175*, 526–531. Epub 2011/01/27.

Bloom, T., Ganley, E., & Winker, M. (2014). Data access for the open access literature: PLOS's data policy. *PLoS Biology, 12*, e1001797.

Blumenthal, D., Campbell, E. G., Gokhale, M., Yucel, R., Clarridge, B., Hilgartner, S., & Holtzman, N. A. (2006). Data withholding in genetics and the other life sciences: Prevalences and predictors. *Academic Medicine, 81*, 137–145.

Chandras, C., Weaver, T., Zouberakis, M., Smedley, D., Schughart, K., Rosenthal, N., Hancock, J. M., Kollias, G., Schofield, P. N., & Aidinis, V. (2009). Models for financial sustainability of biological databases and resources. *Database: The Journal of Biological Databases and Curation, 2009*, bap017.

Check-Hayden, E. (2016). Funding for model-organism databases in trouble. *Nature*. https://doi.org/10.1038/nature.2016.20134

Cook-Deegan, R. (2007). The science commons in health research: Structure, function, and value. *The Journal of Technology Transfer, 32*, 133–156. Epub 2006/12/07.

Copplestone, D., Hingston, J., & Real, A. (2008). The development and purpose of the FREDERICA radiation effects database. *Journal of Environmental Radioactivity, 99*, 1456–1463.

Copplestone, D., Beresford, N. A., Brown, J. E., & Yankovich, T. (2013). An international database of radionuclide concentration ratios for wildlife: Development and uses. *Journal of Environmental Radioactivity, 126*, 288–298. Epub 2013/07/03.

Durinx, C., McEntyre, J., Appel, R., Apweiler, R., Barlow, M., Blomberg, N., Cook, C., Gasteiger, E., Kim, J. H., Lopez, R., et al. (2016). Identifying ELIXIR core data resources. *F1000Research, 5*. Epub 2017/03/30.

Eppig, J. T. (2017). Mouse Genome Informatics (MGI) resource: Genetic, genomic, and biological knowledgebase for the laboratory mouse. *ILAR journal/National Research Council, Institute of Laboratory Animal Resources, 58*, 17–41.

Federer, L. M., Belter, C. W., Joubert, D. J., Livinski, A., Lu, Y.-L., Snyders, L. N., & Thompson, H. (2018). Data sharing in PLOS ONE: An analysis of data availability statements. *PLoS One, 13*, e0194768.

Fukuda, T., Kino, Y., Abe, Y., Yamashiro, H., Kuwahara, Y., Nihei, H., Sano, Y., Irisawa, A., Shimura, T., Fukumoto, M., et al. (2013). Distribution of artificial radionuclides in abandoned cattle in the evacuation zone of the Fukushima Daiichi nuclear power plant. *PLoS One, 8*, e54312. Epub 2013/02/02.

Gerber, G. B., & Wick, R. R. (2004). The European Radiobiology Archives (ERA), present state and future developments. *Radiation Protection Dosimetry, 112*, 529–530. Epub 2004/12/30.

Gerber, G., Watson, C., Sugahara, T., & Okada, S. (1996). International radiobiology archives of long-term animal studies I. *Descriptions of Participating Institutions and Studies*. https://doi.org/10.20348/STOREDB/21119/21185

Gerber, G. B., Wick, R. R., Kellerer, A. M., Hopewell, J. W., Di Majo, V., Dudoignon, N., Gossner, W., & Stather, J. (2006). The European Radiobiology Archives (ERA)--content, structure and use illustrated by an example. *Radiation Protection Dosimetry, 118*, 70–77. Epub 2005/10/26.

H2020 Guidelines on FAIR Data Management. (2016). http://ec.europa.eu/research/participants/data/ref/h2020/grants_manual/hi/oa_pilot/h2020-hi-oa-data-mgt_en.pdf

International Arabidopsis Informatics Consortium. (2010). An international bioinformatics infrastructure to underpin the Arabidopsis community. *Plant Cell, 22*, 2530–2536. Epub 2010/09/03.

Jarnuczak, A. F., & Vizcaino, J. A. (2017). Using the PRIDE database and ProteomeXchange for submitting and accessing public proteomics datasets. *Current Protocols in Bioinformatics, 59*, 13 31 11–13 31 12. Epub 2017/09/14.

Kaiser, J. (2016). Funding for key data resources in jeopardy. *Science, 351*, 14.

Kashparov, V., Levchuk, S., Zhurba, M., Protsak, V., Khomutinin, Y., Beresford, N.A., Chaplow, J.S. (2017). Spatial datasets of radionuclide contamination in the Ukrainian Chernobyl Exclusion Zone. In: *NERC Environmental Information Data Centre*. https://doi.org/10.5285/782ec845-2135-4698-8881-b38823e533bf (accessed 5.11.18).

Kashparov, V., Levchuk, S., Zhurba, M., Protsak, V., Khomutinin, Y., Beresford, N. A., & Chaplow, J. S. (2018). Spatial datasets of radionuclide contamination in the Ukrainian Chernobyl Exclusion Zone. *Earth System Science Data, 10*, 339–353.

Kolesnikov, N., Hastings, E., Keays, M., Melnichuk, O., Tang, Y. A., Williams, E., Dylag, M., Kurbatova, N., Brandizi, M., Burdett, T., et al. (2015). ArrayExpress update—Simplifying data submissions. *Nucleic Acids Research, 43*, D1113–D1116.

Madas, B. G., & Schofield, P. N. (2019). Survey on data management in radiation protection research. *Radiation Protection Dosimetry, 183*(1), 233–236. https://doi.org/10.1093/rpd/ncy250

Malone, J., Holloway, E., Adamusiak, T., Kapushesky, M., Zheng, J., Kolesnikov, N., Zhukova, A., Brazma, A., & Parkinson, H. (2010). Modeling sample variables with an Experimental Factor Ontology. *Bioinformatics (Oxford, England), 26*, 1112–1118. Epub 2010/03/03.

McQuilton, P., Gonzalez-Beltran, A., Rocca-Serra, P., Thurston, M., Lister, A., Maguire, E., & Sansone, S.-A. (2016). BioSharing: Curated and crowd-sourced metadata standards, databases and data policies in the life sciences. *Database, 2016*, baw075–baw075.

Michener, W. K. (2015). Ecological data sharing. *Ecological Informatics, 29*, 33–44.

Mons, B., Neylon, C., & Velterop, J. (2017). Cloudy, increasingly FAIR; revisiting the FAIR data guiding principles for the European Open Science Cloud. *Inform Services Use, 37*, 49–56. Epub 49.

Mothersill, C., Abend, M., Brechignac, F., Iliakis, G., Impens, N., Kadhim, M., Moller, A. P., Oughton, D., Powathil, G., Saenen, E., et al. (2018). When a duck is not a duck; a new interdisciplinary synthesis for environmental radiation protection. *Environmental Research, 162*, 318–324.

Mothersill, C., Abend, M., Brechignac, F., Copplestone, D., Geras'kin, S., Goodman, J., Horemans, N., Jeggo, P., McBride, W., Mousseau, T. A., et al. (2019). The tubercular badger and the uncertain curve: The need for a multiple stressor approach in environmental radiation protection. *Environmental Research, 168*, 130–140. Epub 2018/10/09.

Muikku, M., Beresford, N. A., Garnier-Laplace, J., Real, A., Sirkka, L., Thorne, M., Vandenhove, H., & Willrodt, C. (2018). Sustainability and integration of radioecology-position paper. *Journal of Radiological Protection, 38*, 152–163. Epub 2017/11/22.

Nosek, B. A., Alter, G., Banks, G. C., Borsboom, D., Bowman, S. D., Breckler, S. J., Buck, S., Chambers, C. D., Chin, G., Christensen, G., et al. (2015). Scientific standards. Promoting an open research culture. *Science, 348*, 1422–1425.

Pampel, H., Vierkant, P., Scholze, F., Bertelmann, R., Kindling, M., Klump, J., Goebelbecker, H.-J., Gundlach, J., Schirmbacher, P., & Dierolf, U. (2013). Making research data repositories visible: The re3data.org registry. *PloS one, 8*, e78080.

Piwowar, H. A. (2011). Who shares? Who doesn't? Factors associated with openly archiving raw research data. *PLoS One, 6*, e18657.

Reiser, L., Berardini, T. Z., Li, D., Muller, R., Strait, E. M., Li, Q., Mezheritsky, Y., Vetushko, A., & Huala, E. (2016). Sustainable funding for biocuration: The Arabidopsis Information Resource (TAIR) as a case study of a subscription-based funding model. *Database: the Journal of Biological Databases and Curation, 2016*, baw018.

Roche, D. G. (2017). Evaluating science; open-data policy. *Science, 357*, 654.

Roche, D. G., Kruuk, L. E., Lanfear, R., & Binning, S. A. (2015). Public data archiving in ecology and evolution: How well are we doing? *PLoS Biology, 13*, e1002295. Epub 2015/11/12.

Salbu, B. (2009). Challenges in radioecology. *Journal of Environmental Radioactivity, 100*, 1086–1091.

Sansone, S. A., Cruse, P., & Thorley, M. (2018). High-quality science requires high-quality open data infrastructure. *Scientific data, 5*, 180027.

Savage, C. J., & Vickers, A. J. (2009). Empirical study of data sharing by authors publishing in PLoS journals. *PLoS One, 4*, e7078.

Schofield, P. N., Eppig, J., Huala, E., de Angelis, M. H., Harvey, M., Davidson, D., Weaver, T., Brown, S., Smedley, D., Rosenthal, N., et al. (2010). Research funding. Sustaining the data and bioresource commons. *Science, 330*, 592–593.

Takahashi, S., Inoue, K., Suzuki, M., Urushihara, Y., Kuwahara, Y., Hayashi, G., Shiga, S., Fukumoto, M., Kino, Y., Sekine, T., et al. (2015). A comprehensive dose evaluation project concerning animals affected by the Fukushima Daiichi nuclear power plant accident: Its set-up and progress. *Journal of Radiation Research, 56*(Suppl 1), i36–i41. Epub 2015/12/18.

Tapio, S., Schofield, P. N., Adelmann, C., Atkinson, M. J., Bard, J. L., Bijwaard, H., Birschwilks, M., Dubus, P., Fiette, L., Gerber, G., et al. (2008). Progress in updating the European radiobiology archives. *International Journal of Radiation Biology, 84*, 930–936.

Tenopir, C., Allard, S., Douglass, K., Aydinoglu, A. U., Wu, L., Read, E., Manoff, M., & Frame, M. (2011). Data sharing by scientists: Practices and perceptions. *PLoS One, 6*, e21101.

Tenopir, C., Dalton, E. D., Allard, S., Frame, M., Pjesivac, I., Birch, B., Pollock, D., & Dorsett, K. (2015). Changes in data sharing and data reuse practices and perceptions among scientists worldwide. *PLoS One, 10*, e0134826.

Urushihara, Y., Suzuki, T., Shimizu, Y., Ohtaki, M., Kuwahara, Y., Suzuki, M., Uno, T., Fujita, S., Saito, A., Yamashiro, H., et al. (2018). Haematological analysis of Japanese macaques (Macaca fuscata) in the area affected by the Fukushima Daiichi nuclear power plant accident. *Scientific Reports, 8*, 16748.

Vanhoudt, N., Vandenhove, H., Real, A., Bradshaw, C., & Stark, K. (2012). A review of multiple stressor studies that include ionising radiation. *Environmental Pollution, 168*, 177–192. Epub 2012/05/29.

Wilkinson, M. D., Dumontier, M., Aalbersberg, I. J., Appleton, G., Axton, M., Baak, A., Blomberg, N., Boiten, J. W., da Silva Santos, L. B., Bourne, P. E., et al. (2016). The FAIR guiding principles for scientific data management and stewardship. *Scientific data, 3*, 160018.

Wilkinson, M. D., Sansone, S. A., Schultes, E., Doorn, P., Bonino da Silva Santos, L. O., & Dumontier, M. (2018). A design framework and exemplar metrics for FAIRness. *Scientific data, 5*, 180118. Epub 2018/06/27.

Wolbarst, A. B., Griggs, J., Lee, H. N., Ren, T., Hudson, T., White, J. D., & Zhu, C. (2008). Comparison of environmental radiation monitoring programs in China and the United States. *Health Physics, 94*, 501–511. Epub 2008/05/13.

Part II
Low-Dose Effect Mechanisms

Chapter 5
Modelling Direct and Indirect Effects of Radiation: Experimental, Clinical and Environmental Implications

Sarah C. Brüningk and Gibin G. Powathil

Abstract Radiotherapy is a commonly used treatment for cancer and is usually given in varying doses. Mathematical modelling of radiation effects traditionally means the modelling or estimation of cell-kill due to its direct exposure to irradiation and sometimes ignoring other multiple direct/indirect effects. However, advances in molecular biology have expanded this classical view and it is now realized that in addition to cell-death, signals produced by irradiated cells can further influence the behavior of non-irradiated cells or organisms in several ways. Consequently, it has now wider implications in multiple areas making it relevant for further exploration, both experimentally and mathematically. Here, we provide a brief overview of a hybrid multiscale mathematical model to study the direct and indirect effects of radiation and its implications in clinical radiotherapy, experimental settings and radiation protection.

5.1 Introduction

Radiotherapy (RT) is currently part of the treatments of 40% of cancer patients in the UK. This comprises patients receiving external beam RT, particle irradiation, molecular radiotherapy and brachytherapy. Independent of the treatment modality used, a treatment plan or dose simulation is performed for each patient based on computed tomography or standard x-ray images to conform the radiation dose delivered to the tumour and to spare the surrounding normal tissues and organs at risk. Although through this process the physical dose delivered may be calculated

S. C. Brüningk (✉)
Joint Department of Physics, The Institute of Cancer Research and The Royal Marsden NHS Foundation Trust, London, UK
e-mail: brueningk@bsse.ethz.ch

G. G. Powathil
Department of Mathematics, Swansea University, Swansea, UK

M. D. Wood et al. (eds.), *Biomarkers of Radiation in the Environment*, NATO Science for Peace and Security Series A: Chemistry and Biology,
https://doi.org/10.1007/978-94-024-2101-9_5

with high precision using state of the art dose calculations, the physical dose will always remain a surrogate for the biological effects induced. Understanding and quantifying these biological effects is, however, critical for prediction and evaluation of treatment response. Quantification may range spatially from sub-cellular level (DNA damage), to the cellular scale in terms of radiation induced cell kill, to tissue and organ level (organ functionality), the overall patient response (survival), and finally the impact of radiation on a treated population as a whole. This implies that also the temporal scale on which these effects are observed and impact may vary from minutes to years. Computational modelling may help to better understand, quantify and predict the biological response on any of these spatio-temporal scales and may therefore bridge the gap between biological effects observed in vitro and in vivo. In contrast to analytical dose-response models, systems biology simulations model the biological response in a time and space dependent manner allowing for a detailed description of the dynamic behaviors. As such, systems biology represents a holistic approach that describes complex biological systems using a number of interdependent components that result in the function and behavior of the system. Depending on the spatio-temporal scales of interest, the interaction of genes, proteins, cells or larger sub-structures within a tissue or organism may be described. For example, in the case of radiation response modelling for cancer therapy applications models may start from describing individual cells with intra- and intercellular processes being governed by a set of predefined rules or a system of ODEs (ordinary differential equations). By simulating a large number of these interacting cells, insight into emergent tissue level phenomena can be achieved.

5.2 Multiscale Modelling to Study Radiation Effects

5.2.1 Mathematical Model: Multiscale Approach

A number of modelling approaches exists within the framework of systems biology simulations, but in general two groups can be defined: discrete and continuum-based approaches. Continuum-based models describe the system entities by a set of (partial) differential equations. Cells are not modelled directly but are described in terms of cell densities that fluctuate with time. Multiple cell types may be described by interconnected sets of equations. Distributions of nutrients, oxygen, or signaling molecules are modelled by iteratively solving the respective reaction-diffusion equation.

The most basic approach of discrete models is a cellular automaton (CA). CA models represent cells as discrete voxels/pixels on a fixed grid and cellular processes are driven by a set of pre-defined rules and cell properties. As such, CA requires defining a regular lattice of a fixed size, on which initially some pixel/voxel are occupied by 'cells' (initialization step). Each cell can be in one of several predefined states, these could be different cell cycle stages, or simply 'alive' vs. 'dead'. The stages of all cells are then updated in each time step according to the rules defined. For example, cell cycle progression can be controlled by assigning an individual clock to each cell that is incremented in each simulation time step until a threshold time is reached upon which the cell divides into two daughter cells. In general, all rules are pre-defined in CA, meaning that they remain unchanged throughout the simulation. Given their simple implementation, CA models have widely been used to simulate a number of different scenarios, such as tumor progression (Anderson & Chaplain, 1998; Kansal et al., 2000; Patel et al., 2001; Turner & Sherratt, 2002), or cellular treatment response modelling in vitro (Ribba et al., 2006; Richard et al., 2007). A comprehensive review on the use of CA for cancer modelling applications has been provided by Moreira and Deutsch (2002).

An alternative individual based approach is using a multiscale cellular Potts model (CPM) or the Glazier-Graner-Hogeweg (GGH) approach (Glazier et al., 2007). The GGH model is very well-suited to model individual objects (such as cancer cells) and interactions with the cellular properties that evolve with respect to time and space (Andasari et al., 2012). Each cell is considered as a collection of lattice pixels with a common index marker and is represented as a spatially extended domain on a fixed lattice, either a 3D Cartesian lattice or a 3D hexagonal lattice. The cell's behaviors and interactions are defined through the local minimization of effective energies depending on cell and pixel configurations. By minimizing energy via the Modified Metropolis Algorithm, CPM recovers a linear relation between force and a cell's velocity and hence allows for the translation of lab-measurable cellular characteristics into model parameters. Regardless of choice of computational approach, each method should give qualitatively similar results for the same biologically-determined classes of objects, behaviors and interactions. Any observed discrepancies between methods can used to veto and/or improve modelling methods (Powathil et al., 2015).

The advantage of CA and CPM approaches is their detailed description and their capability to describe the progression of individual cells and the population as a whole. Depending on the detail of cellular properties accounted for, the number of cells and/or the simulated time span may, however, be limited. To limit computational cost there has to be a trade-off between the level of detail and the overall number of cells per simulation. Hybrid models between CA or CPM and continuum models have been developed to overcome some of the limitations of these models. In hybrid models the entire multiscale is simulated over a finite period of time and the simulation of cells is controlled in a CA fashion as individual grid points. In addition to this, micro environmental dynamics such as oxygen or nutrient diffusion are modelled using a continuum-based approach. At every time step, a finite difference scheme is evaluated and the corresponding nutrient concentration levels are updated and assigned at each grid position. Timescale increments may, however, be quite different for these two model combinations (hours for cell division vs. subsecond increments for finite differences) requiring careful synchronization upon implementation. Recently, Powathil et al. developed a hybrid multiscale cellular automaton approach to model cancer progression and used the model to study the effects of cell-cycle dependent chemotherapeutic drugs alone and in combination with radiation therapy (Powathil et al., 2012a, 2013). Examples of hybrid CA and CPM are shown in Fig. 5.1.

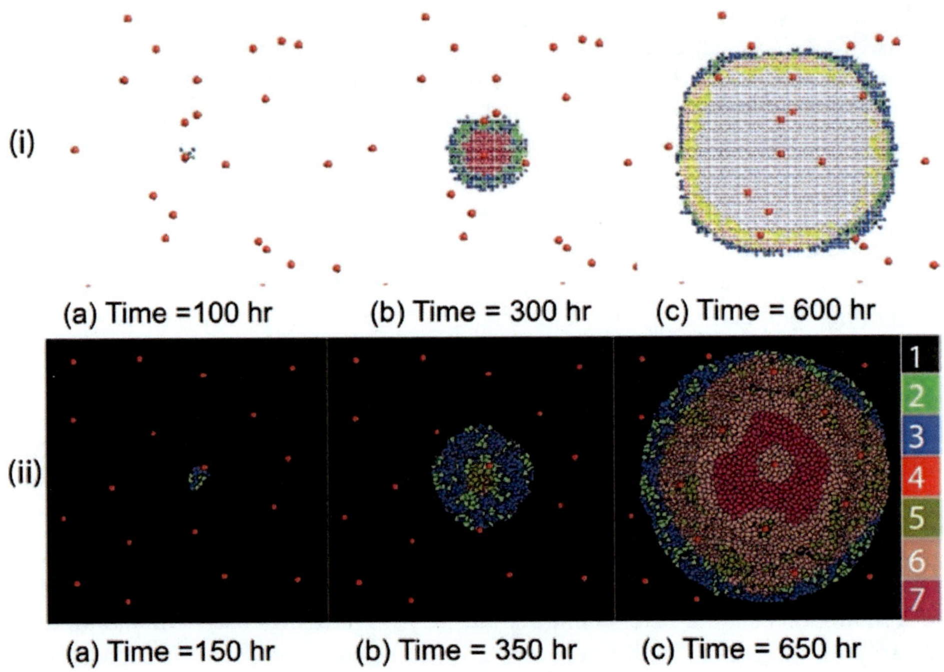

Fig. 5.1 Figure adapted from (Powathil et al., 2015). Spatial cell and vessel distributions of simulations using a hybrid multiscale CA (i) or a CPM (ii) are shown. In both model, colours represent various stages of cell-cycle and the simulation is started from a single cell at 0 h. These are for the (i): G1 (blue), S-G2 -M (green), resting (magenta), hypoxic cells in G1 (rose), hypoxic cells in S-G2 -M (yellow) and hypoxic cells in resting (silver). (ii) Plots from hybrid CPM(using Compucell3D) and the colour legend shows the types of the tumour cells; For (ii) the color legend indicates the cell type: 1- surrounding medium, 2- G2 phase, 3- G1 phase, 4- vessel cross sections, 5- hypoxic G2 phase, 6- hypoxic G1 phase and 7- resting cells

5.2.2 Multiscale Model Implementation

As mentioned previously, there is always a tradeoff to be made between the number of cells and detail of the inter- and intracellular processes modelled. Depending on the problem to be simulated, the choice of model is the essential first step towards an efficient implementation. A direct implementation of a CA model is ideally optimized for computational performance to maximize its quality and significance. The easiest way to implement a CA is by defining a fixed sized array of designated cell positions. The respective cellular or micro environmental properties (occupancy, cell cycle stage, oxygen level, etc.) are directly assigned to each element of this array. Each time frame loops over all array elements and potentially updates relevant properties. Although this describes a very simple implementation, it means that a fixed number of positions have to be processed regardless of their occupancy by cells. Obviously, this represents a very poor implementation in terms of computational efficacy with the limiting factor of the simulation being the overall size of the array. One option to speed up this calculation has been proposed by Poleszczuk and

Enderling (2014) who continuously increase the computational domain upon colony growth. However, this may come at the cost of having to reorganize data structures in memory throughout the simulation. A more scalable method which aims for large scale problems on high performance computing hardware was proposed by Brüningk et al. (2018). In this work, the spatial cell grid and the physical properties of the cells are stored separately in memory in a list-like structure which grows along the cells at constant costs. The logic separation of the data allows for a quicker pass-through of the cell arrangement and facilitates a parallel or distributed implementation of the CA.

The computational simulations of the multiscale CPM are implemented in the CompuCell3D framework developed by Glazier et al. (see http://www.compucell3d.org for full details) (Powathil et al., 2015, 2016). Here, cells are represented by a collection of pixels in a 2D lattice and as in hybrid CA model, the cellular proliferation is governed by the internal cell-cycle dynamics modelled using the kinetic equations as given in Fig. 5.2. These set of equations are solved in each Monte Carlo time step using SBML solvers such as Bionetsolver (Powathil et al., 2015). The evolution of concentration fields such as oxygen concentration is incorporated into the CompuCell3D as a diffusive chemical field that follows the respective partial differential equation as given in Fig. 5.2. Further details of both the simulation approaches that are used to study the multiscale mathematical model and the relevant parameter values can be found in recent papers by Powathil et al. (2012a, 2013, 2014).

5.2.3 *Applications of Systems Biology Simulations*

In the previous sections a number of modelling approaches have been introduced. Each of these models has its specific advantages and shortcomings, and which modelling approach is suited for the purpose is highly dependent on the question to be answered. Multiscale models can provide a better understanding of the underlying biological reaction chain and its influence on the emerging system response. As such, these simulations allow studying and weighting the importance of different contributing mechanisms and pathways to the overall response using a sensitivity analysis. A (variance-based) sensitivity analysis (Satelli et al., 2010) may point out which model parameters are most influential and thus crucial for defining the uncertainty of the simulation result. This is important since the analysis may flag up parameters that are essential to be validated experimentally, whereas rough estimates may be sufficient in other cases saving time and resources.

Moreover, systems approaches, allow analyzing the influence of a heterogeneous subset of different cell types and to study their interaction. Although it may be possible to experimentally validate single cell type based simulations, experiments involving multiple cell types are difficult and simulations may help to better predict more realistic scenarios of heterogeneous populations. Within such more realistic scenarios, it is then possible to identify and optimize radiation schedules for cancer

Cell-cycle dynamics

$$\frac{d[\mathrm{CycB}]}{dt} = k_1 - (k_2' + k_2''[\mathrm{Cdh1}] + [\mathrm{p27/p21}][HIF])[\mathrm{CycB}],$$

$$\frac{d[\mathrm{Cdh1}]}{dt} = \frac{(k_3' + k_3''[\mathrm{p55cdc_A}])(1 - [\mathrm{Cdh1}])}{J_3 + 1 - [\mathrm{Cdh1}]} - \frac{k_4[\mathrm{mass}][\mathrm{CycB}][\mathrm{Cdh1}]}{J_4 + [\mathrm{Cdh1}]},$$

$$\frac{d[\mathrm{p55cdc_T}]}{dt} = k_5' + k_5''\frac{([\mathrm{CycB}][\mathrm{mass}])^n}{J_5^n + ([\mathrm{CycB}][\mathrm{mass}])^n} - k_6[\mathrm{p55cdc_T}],$$

$$\frac{d[\mathrm{p55cdc_A}]}{dt} = \frac{k_7[\mathrm{Plk1}]([\mathrm{p55cdc_T}] - [\mathrm{p55cdc_A}])}{J_7 + [\mathrm{p55cdc_T}] - [\mathrm{p55cdc_A}]} - \frac{k_8[Mad][\mathrm{p55cdc_A}]}{J_8 + [\mathrm{p55cdc_A}]} - k_6[\mathrm{p55cdc_A}],$$

$$\frac{d[\mathrm{Plk1}]}{dt} = k_9[\mathrm{mass}][\mathrm{CycB}](1-[\mathrm{Plk1}]) - k_{10}[\mathrm{Plk1}],$$

$$\frac{d[\mathrm{mass}]}{dt} = \mu[\mathrm{mass}]\left(1 - \frac{[\mathrm{mass}]}{m_*}\right).$$

Oxygen dynamics

$$\frac{\partial K(x,t)}{\partial t} = \nabla.(D_K(x)\nabla K(x,t)) + r(x)m(x) - \phi K(x,t)\mathrm{cell}(x,t)$$

Radiation effects

$$S(d) = \exp\left[\gamma\left(-\alpha.\mathrm{OMF}.d - \beta(\mathrm{OMF}.d)^2\right)\right]$$

where

$$\mathrm{OMF} = \frac{\mathrm{OER}(pO_2)}{\mathrm{OER}_m} = \frac{1}{\mathrm{OER}_m}\frac{\mathrm{OER}_m\, pO_2(x) + K_m}{pO_2(x) + K_m}$$

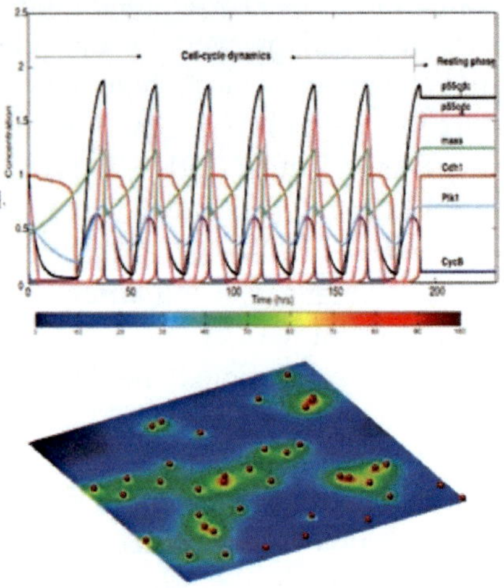

Fig. 5.2 Figure showing various processes involved in the simulation. Top: Plot of the concentration profiles of the various intracellular proteins and the cell-mass over a period of 200 h for one automaton cell in the model. This is obtained by solving the system of equations shown under 'Cell-cycle dynamics', with the relevant parameter values from Powathil et al. (2012a). Middle: Representative realization of the spatial distribution of oxygen (K(x, t)) obtained by solving the reaction-diffusion equation. This equation describes the dynamic change of oxygenation due to oxygen diffusion (first term), oxygen production at vessel locations (second term), and oxygen consumption by the cells (third term). Bottom: Radiation effects are modeled by calculating the surviving fraction S as a function of dose delivered, cell cycle stage (γ), and oxygenation level (OMF). Here the oxygen modification factor (OMF) is calculated as the ratio of the oxygen enhancement rations (OER) at the position of the cell (OER(pO2)), and the maximum OER, OERM. Here, pO2 (x) is the oxygen concentration at position x, and Km is the oxygen level at which the OER is at half its maximum level

treatments, potentially in combination with other treatment modalities such as targeted drugs or chemotherapy. In particular, it may be possible to include patient specific information allowing for a personalized treatment optimization. In the following we will give two examples of multiscale models and their application.

5.3 Modelling Cellular Response to Radiotherapy: Simulation and Validation

In this section, we describe the components of a simulation framework that models radiation response at a cellular level in the range of therapeutic doses as used clinically for radiotherapy treatments (1–5 Gy, 1–4 Gy/min). Such a framework consists of four main parts: A simulation of the cellular microenvironment, an implementation of normal cell growth within this environment, treatment delivery, and response modelling. Different modelling approaches are chosen for each of these components to minimize computation time (hybrid CA model).

5.3.1 Setting the Scene: Modelling Cellular Growth

Normal cell growth is simulated using the CA approach. Cells are modelled individually as elements of a two or three dimensional lattice allowing for modelling of simple in vitro experiments or a small part of a tumor in vivo. Lattice dimensions are defined by the geometry and type of cell simulated. Here these have been adapted to match the growth response of HCT116 cells, a commonly used human colorectal cancer cell line. We use a pixel size of 12 × 12μm on a 3000 × 3000 big lattice to model cells in two dimensional 6-well plates (well diameter 34.8 mm). For three dimensional models, a smaller grid of 200 × 200 × 200 voxel of the same size is used.

To enable the proliferation of the cells in the CA model we need to define specific rules: Here, each cell follows a four phase cell cycle (G1, S, G2, M phases). Depending on the level of detail to be simulated, cell cycle progression may be controlled by a set of ODEs modelling the temporal evolution of a number of cell cycle protein complexes as shown in Fig. 5.3. This approach, which was first proposed by Tyson and Novak (2001; Novak & Tyson, 2003), has the advantage of

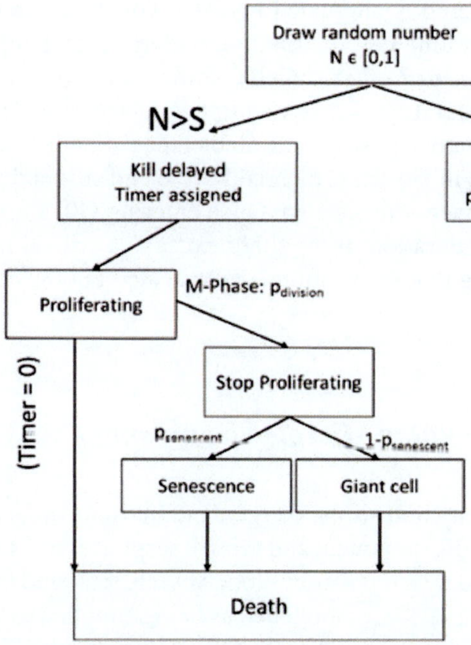

Fig. 5.3 Probability driven response cascade used to model the dynamic response of cells to radiation mimicking the processes of mitotic catastrophe. Upon irradiation, a random number N is drawn for each cell and compared to the respective surviving fraction S calculated according to Eq. (5.2). If the cell is drawn to die (N > S), it is assigned a time to cell death and continues to proliferate until for this time. Once the cell reaches M-phase, it exits the cell cycle at probability pdivison. These cells then either become senescent (at probability psenescent) or turn into a giant cell that increases in mass but does not divide. Once the delay to cell death is up, the cell and potential daughter cells are removed from the computational domain

closely modelling the biological reaction chain underlying the cell cycle, where different cycle stages are controlled by the level of cyclin complexes. These cyclin complex levels may be influenced by external factors such as the oxygenation level of the cell (here modelled by assuming the activation of the HIF pathway).

If detailed pathway dependence is not needed, a much simpler and computationally effective method for cell cycle implementation is a cell cycle timer that is assigned to each cell. This means that the cell cycle is characterized by cycle phase specific transition times to the following stage. Once the predefined doubling time is reached the cell divides. These parameters can either be measured experimentally, or estimated numbers can be used if validation data is not available. Typically, cell doubling times range around 24 h for tumor cells, but may be significantly longer for normal cells depending on the specific cell type. The computation time-step after which the cell cycle timers are incremented may therefore be in the order of several minutes to hours. Here we used a doubling time of 19.5 h, and assumed an infinite number of possible divisions for the cells.

Independent of the cell cycle model used cells progress through their cycle and will divide into two identical daughter cells at the end of M-phase. The position of the new cell is randomly selected from the neighborhood of its parental cell. The type of neighborhood considered is alternated between von Neumann and Moore neighborhoods to avoid generating cell distribution patterns matching the specific neighborhood (Powathil et al., 2012a, 2013). Cell colonies will thus grow in circular patterns and no further cell migration is assumed to take place. It is defined that, if there are no free neighboring lattice positions available, the cell will enter the reversible quiescent stage G0. Cells in G0 arrest their cell cycle and suspend their cellular activities until neighboring spaces are vacated. Once entering G0, a timer is started for the cell indicating the G0 duration. If the timer exceeds a critical limit, the cell is labeled as necrotic meaning that its position is not vacated but the cell is considered 'dead'.

5.3.2 *Modelling the Cellular Microenvironment*

The cellular microenvironment controls the oxygen and nutrient supply of the cells and therefore influences the cells' treatment and growth response. In a simple model, oxygen may either be supplied by a number of blood vessels scattered over the computational grid (Powathil et al., 2013), or modelled as being supplied externally from the surrounding medium, i.e. from outside a 3D cell aggregate or uniformly in the case of 2D cultures in a dish. The oxygen distribution $K(x,t)$ may then be calculated by iteratively solving the reaction-diffusion equation given in Fig. 5.2 using finite difference method. Here, the change in oxygen level at each position is calculated from contributions of oxygen diffusion, oxygen production from potential vessel locations, and oxygen consumption by the cells. The parameters used are the diffusion coefficient of oxygen in the respective tissue D_K, the oxygen production rate $r(x)$ at vessel locations of cross section $m(x)$, and the oxygen consumption rate of the cells, Φ, if a cell is present at position x (*cells(x,t)*). To iteratively solve the diffusion equation it is essential to use meaningful boundary conditions. These may be full oxygenation outside the cell aggregate (no vessels), or no-flux boundary conditions are imposed in the case of vessel distributions if an internal part of a tumor is simulated. An example of an oxygen distribution originating from a number of randomly distributed blood vessels is shown in Fig. 5.2. Gradients of other diffusive molecules, such as nutrients, drugs, or signaling molecules can be modelled accordingly.

5.3.3 Treatment Delivery and Response Modelling

Since the cell cycle timer increments are on the order of minutes to hours, irradiation using therapeutic dose rates of several Gray per minute may be simulated by assigning the total dose delivered to each cell within one time increment. For modelling cell survival of each individual cell, the probability of cell survival, S, at the specific dose level, d, is first calculated using the linear quadratic (LQ) cell survival model (Fowler, 1989):

$$S(d) = e^{-\left(\alpha d + \beta d^2\right)} \tag{5.1}$$

The LQ-model uses two cell line and treatment modality depended parameters, α and β. These parameters were referred to contributions of cell kill originating from double and single strand breaks in the original publication of the LQ model, but their interpretation has since then been debated. In a systems biology simulation, each cell is assigned specific values for α and β according to the intrinsic radio sensitivity of the cell type modelled. If a large number of cells from the same type are simulated, values displaying small differences from the known means may be assigned to each cell while maintaining the overall average for the population as a whole.

In addition to the intrinsic radio sensitivity of the cell type, the cells cycle stage also influences radio sensitivity due to differences in pre-activated DNA repair pathways (Lauber et al., 2012). One option to account for this is the inclusion of a cell cycle stage specific weighting factor γ in (5.1) (Powathil et al., 2012a).

$$S(d) = e^{-\gamma\left(\alpha d + \beta d^2\right)} \tag{5.2}$$

Microenvironmental factors, such as the oxygenation level, may further influence the radio sensitivity of the cells. This may be accounted for by multiplying the radiation dose with an oxygen level dependent weighting factor. This oxygen modification factor (OMF) ranges between 1 and 3 depending on the current partial oxygen pressure at the cells location. The respective equation is shown in the summary Fig. 5.2. The OMF introduces a higher probability of cell survival in the absence of oxygen and increases the cell-kill if the cell is well oxygenated (Powathil et al., 2012b).

The LQ-model generally describes clonogenic cell survival (Franken et al., 2006), meaning that it gives an estimation of the overall proportion of a treated cell population that are still able to successfully undergo division and may therefore be the origin of increasing tumor burden. The LQ-model does not inform the dynamics of the cell kill, or how quickly cells which are unable to form a colony will be unable to divide and/or die. To account for this problem, we are using a probability driven response cascade to model the dynamics of radiation induced mitotic catastrophe which is the major pathway involved in radiation-induced cell kill (Lauber et al., 2012) (see Fig. 5.3). In this cascade, a random number N between 0 and 1 is drawn for each cell in the first time step post irradiation. Whereas cell with $N \leq S$ continue to cycle normally after a short delay to allow damage repair, cells with a number drawn that exceeds the survival threshold are labelled as non clonogenic. Non clonogenic cells are assigned a delay time to cell death randomly sampled from an exponential distribution. These cells will continue to progress though their cell cycle, but in M-phase will not divide at a probability $p_{division}$, but become either senescent (not proliferating), or giant cell (increasing in size, but no division). Once the delay to cell death has been reached, the cell is removed from the computational grid if connected to empty space, or labeled as necrotic if at the center of a colony. It is important to note, that the threshold probabilities and delay times used, may be cell line and dose range dependent. This makes it essential to validate the simulation framework experimentally before making predictions.

5.3.4 *Simulation Validation*

Experimental validation is an essential part of any model to allow making meaningful predictions and to minimize the number of free parameters. In the case of systems biology simulations, validation may happen on any of the spatio-temporal scales used, resulting in a potentially very large number of datasets needed to fully validate every part of the model. Whereas simulations modelling in vitro scenarios may be relatively easy to validate, it is much more difficult in more complex simulations of realistic, in vivo-like cases. In vitro experiments used to calibrate a simulation to model the behavior of a particular cell type include cell size measurements, cell cycle analysis using Flow cytometry, cell survival evaluation with clonogenic and/or cell viability assays, as well as cellular growth curves of treated and untreated populations. 3D cell aggregates such as spheroids may be used to easily validate cellular behavior in 3D geometries and to include micro environmental aspects in the validation data. Histological analysis of spheroid sections may be used to validate oxygen distributions, as well as proportions of proliferating, cell-cycle arresting, and dead cells. Finally, tissue specimens from excised tumors can be stained to quantify vessel and oxygen distributions, proportions of tumor and normal cells and their viability.

Recently, Brüningk et al. (2018) presented a validated hybrid CA model for studying the effects of combinations of radiation and hyperthermia (heat) treatments. In this study, the response of a commonly used human colorectal cancer cell line (HCT116) was analyzed using clonogenic assays, and manually counted growth curves. For manual growth curve counts, samples were cultured in 6-well plates resulting in several hundreds of thousands to millions of cells per well. This means that even if experimental reference data is available, a direct model calibration is only possible if a large number of cells can be simulated which may be difficult in a CA model. In their model, Brüningk et al. first calibrated their framework to normal cellular growth by adapting cell size, doubling time, and cell cycle distribution. Then, growth curves of irradiated samples were used to adapt parameters in the response cascade used to model the radiation response dynamics. Finally, the simulation was successfully validated on another set of cellular growth curves for single and multiple treatment fractions without making further changes to the underlying model.

5.4 Modelling Indirect Effects: Radiation-Induced Bystander Effects

Recent advances in radiobiology expanded beyond studying the direct (targeted) effects of radiation and explored various factors associated with non- targeted effects of radiation, including the phenomena known as the "bystander effects", where the signals produced by irradiated cells influence the behavior of non- irradiated cells (Blyth & Sykes, 2011; Prise & O'Sullivan, 2009; Mothersill & Seymour, 2004; Morgan, 2003a, b). At low radiation doses, since the direct cell kill is relatively low, the non-targeted effects are shown to play a major role in cellular response to radiation (Mothersill & Seymour, 2004; Munro, 2009; Prise & O'Sullivan, 2009). This has major implications in several areas, especially, in clinical radiotherapy and radioecology, where cells or organisms are exposed to varying radiation levels, especially at low dose rates.

Although it is difficult to separate direct effects from bystander effects, there are several experimental studies that investigate these effects in various ways (Azzam et al., 2000; Fernandez-Palomo et al., 2015; Hu et al., 2006; Lorimore et al., 2005; Lyng et al., 2000; Mothersill & Seymour, 2002; Prise & O'Sullivan, 2009; Seymour & Mothersill, 2000; Smith et al., 2016). The cells that are in direct contact with each other are thought to support bystander signaling through the gap junctions (Prise & O'Sullivan, 2009). The experimental evidences also indicate that these effects are also mediated through the release of diffusive protein-like molecules, such as cytokines, from the cells that are irradiated or exposed to bystander signals. Recently, other factors such as exosomes containing RNA, UVA photons and NOS have been identified as potential candidates for bystander signals (Ahmad et al., 2013; Al-Mayah et al., 2012; Prise & O'Sullivan, 2009). The term "bystander signals" are generally used to denote multiple factors such as diffusive chemicals/factors (Fernandez-Palomo et al., 2015; Prise & O'Sullivan, 2009) or biophotons (Le et al., 2017), that contribute to the bystander effects in non-irradiated cells. Here, we will see a hybrid mathematical and computational modelling approach to study multiple effects of radiation and radiation-induced bystander effects on tumor and normal cells (Powathil et al., 2016). In this model, we consider bystander effects are produced only via bystander signals that diffuse through the medium/microenvironment (Fernandez-Palomo et al., 2015; Prise & O'Sullivan, 2009). In an in vitro study, it has been shown that these signals produced by the irradiated cells reach a maximum after 30 min of radiation and remain steady for at least 6 h after the radiation, transferring anywhere within the medium (Hu et al., 2006). The non-radiated cells that are exposed to these bystander signals respond with varying effects (Fig. 5.4).

The indirect radiation bystander effects are produced by radiation-induced signals sent by irradiated cells that are directly exposed to the radiation (Prise & O'Sullivan, 2009). The summary of the hybrid model is given in Fig. 5.2. Here, we are using a cellular potts model approach in a CompuCell3D framework to implement the hybrid multiscale model. The "bystander signals" are included in to the model as a field of bystander signal con- centration (Bs (x, t)) which by diffusing to nearby cells, produces bystander effects (Powathil et al., 2016). This concentration of bystander signals acts as a collection of multiple bystander diffusible signals that are observed experimentally. Following Powathil et al., 2016, the spatio-temporal evolution of these signals is modelled by a reaction-diffusion equation, given by:

$$\frac{\partial B_S(x,t)}{\partial t} = \underbrace{D_S \nabla^2 B_S(x,t)}_{\text{Diffusion}} + \underbrace{r_S cell_{Rad}(\Omega,t)}_{\text{Production}} - \underbrace{\eta_S B_S(x,t)}_{\text{Decay}}$$

Here, *Bs (x,t)* denotes the strength or concentration of the signal at position x and at time *t*, *Ds* is the diffusion coefficient of the signal (which is assumed to be constant),

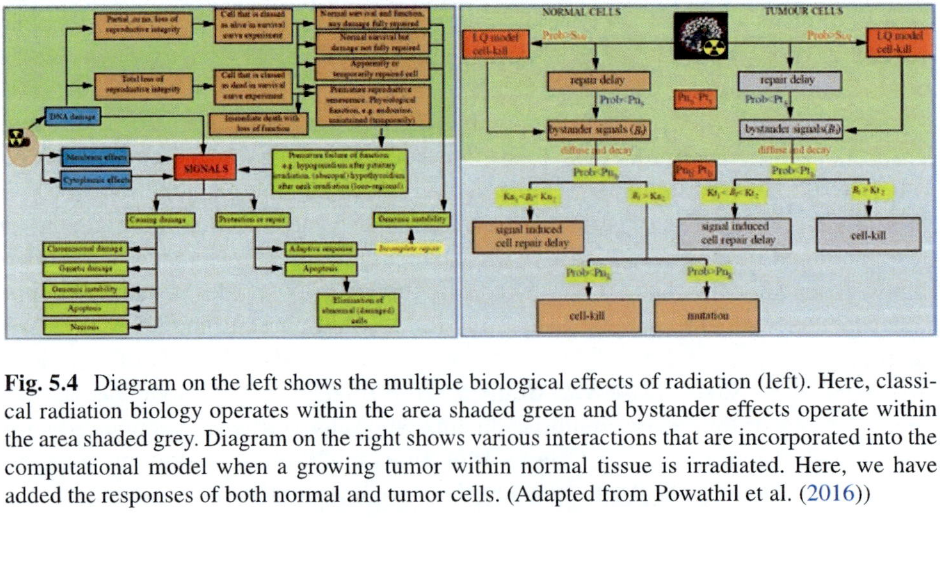

Fig. 5.4 Diagram on the left shows the multiple biological effects of radiation (left). Here, classical radiation biology operates within the area shaded green and bystander effects operate within the area shaded grey. Diagram on the right shows various interactions that are incorporated into the computational model when a growing tumor within normal tissue is irradiated. Here, we have added the responses of both normal and tumor cells. (Adapted from Powathil et al. (2016))

rs is the rate at which the signal is produced by an irradiated cell, *cellRad* (Ω, t) (*cellRad* $(\Omega, t) = 1$ if position $x \in \Omega$ is occupied by a signal-producing irradiated cell at time t and zero otherwise) and ηs is the decay rate of the signal (Powathil et al., 2016).

We assume a heterogeneous distribution of cells where the cancer cells are surrounded by normal cells. Moreover, cancer cells are assumed to proliferate continuously (based on their internal cell-cycle) while normal cells proliferate with the availability of free space (contact growth inhibition). To study the direct and indirect effects of radiation, we consider that the tumor cells are treated homogeneously to the prescribed dose per fraction and the surrounding normal cells receives lower doses with decreasing intensity from the tumor boundary. Two other scenarios of radiation exposure are given in Powathil et al. (2016).

The cells that are directly exposed to the radiation (including cells that undergo cell-death with LQ survival probability) are assumed to produce the bystander signals with some probability and these signals diffuse over the medium over time (4). Figure 5.5 shows an illustrative case of the spatio-temporal evolution of the cells (both normal and tumor cells) and corresponding signals when the cells are treated

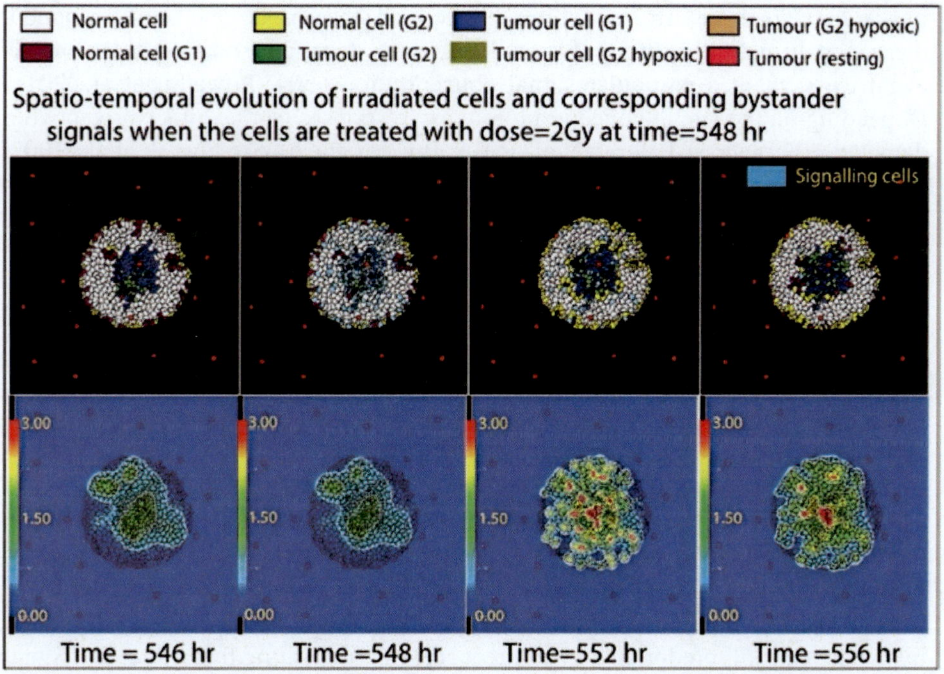

Fig. 5.5 Plots showing the spatio-temporal evolution of host-tumor dynamics with and without treatment. Plots show the changes in the spatial distribution of irradiated cells and bystander signals when the host-tumor system is irradiated at times = 548 h. Upper panel shows the changes in cell distribution as well as the signaling cells after irradiation and the lower panel shows the distribution of bystander signals with color map indicating various threshold values. (Adapted from Powathil et al. (2016))

with one of 5 doses of 2 Gy radiation dose at time = 548 h. The light blue labelled cells in the upper panel of the Fig. 5.5 show the bystander signal producing cells that are under repair delay. The cells that are exposed to higher gradient of given dose are likely to undergo cell-death due to direct effects and cells that are exposed to low-dose radiation are likely to turn into a signal producing cells, resulting a higher number of signaling cells at low dose regions. The lower panel plots show the scaled signal concentrations at different time points with maximum value 3, beyond which bystander responses are activated (Powathil et al., 2016). The plots also show that, even if the level of bystander signals are minimal after each fraction, the contributions of bystander signals from multiple fractions can leads to a concentration level that might trigger bystander responses in responding cells. The parameters of the model are chosen in accordance with the experimental observations that the bystander signals are active at least up to around 6–10 h after radiation exposure (Hu et al., 2006).

The potential direct and indirect biological effects of radiation, and the effects that are incorporated into the model are illustrated in the Fig. 5.4. Since not all cells that are exposed to bystander signals respond to these signals (Gomez-Millan et al., 2012; Prise et al., 2005), it is assumed that the bystander responses are triggered probabilistically with higher chances for tumor cells than normal cells. Once the bystander signals are diffused within the medium, the non-targeted cells respond to these signals in multiple ways (bystander effects), depending on probability of each event to occur at a particular signal concentration level (Powathil et al., 2016). Depending on the signal concentration, these responses can be either protective or damaging (Fernandez-Palomo et al., 2015; Prise et al., 2005; Shuryak et al., 2007). The model considers both of these responses up to some extent and incorporate signal induced repair delay, signal induced bystander cell-kill and possibility of mutagenesis of normal cells. It is indeed hard to predict exact probabilities and concentration thresholds that are used within the model (Mothersill & Seymour, 2006). However, using sensitivity analysis, we have shown that qualitative behavior of the model remained unaffected with a relative change in the parameters. The details of this probabilistic model and the parameters that are used are given in Powathil et al. (2016).

Figure 5.6 shows the direct and indirect effects of radiation when the cells are exposed to 5 fractions of varying radiation doses (Powathil et al., 2016). The total cell-kill due to the exposed radiation and contributions from targeted and non-targeted cell-kill are given in Fig. 5.6a. As expected, plots show an increase in cell-kill with an increase in the dose per fraction. However, at the lower doses (dose = 0.25 Gy and dose = 0.5 Gy) the contributions from the radiation-induced bystander cell-kill predominate the direct cell-kill, indicating that bystander signals might play a role in radiation damage at low dose levels. The survival fraction of the cells after radiation, with and without bystander cell-kill are shown in Fig. 5.6b. The plots show a region on high cell-kill at the doses 0.25 Gy and 0.5 Gy as compared with doses greater than 0.5 Gy. The region of hyper radio-sensitivity at low dose levels or inverse-dose effect is also observed in several experimental studies, as shown in Fig. 5.6c (Joiner et al., 2001; Marples & Joiner, 2000; Prise et al., 2005)

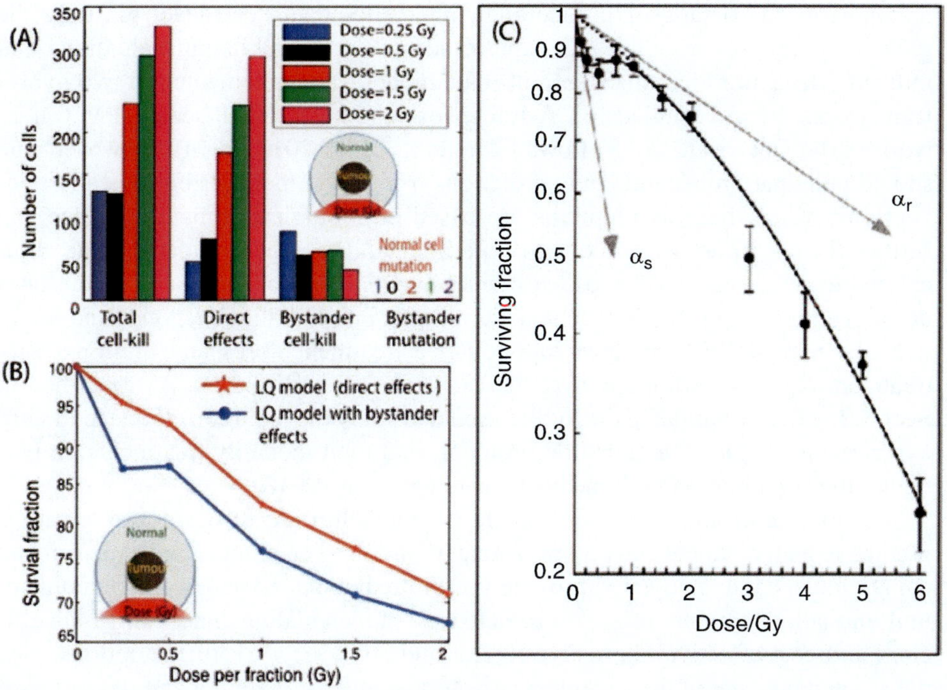

Fig. 5.6 Plots show (**a**) the number of cells killed under the direct effects and indirect effects of radiation and other bystander signal responses, (**b**) the differences in the survival fraction when bystander responses are considered and (**c**) Experimental result: survival of asynchronous T98G human glioma cells irradiated with 240 kVp X-rays, measured using the cell-sort protocol (Figure from Joiner et al. (2001), used with copyright permission). (Adapted from Powathil et al. (2016))

and is not predicted using the traditional LQ models. These in silico results were further supported by experimental study by Fernandez-Palomo et al. (2015), where they indicated that the bystander effects indeed play a role in hypersensitivity at low dose levels (Fernandez-Palomo et al., 2015; Powathil et al., 2016).

5.5 Conclusions and Overview

Over the past decade or so, with the rapid developments in acquisition of genetic, proteomic and other bio- chemical and biological data, systems biology has emerged as an important field in biomedical research. Systems biology aims to use interdisciplinary tools and skills to predict emergent behaviors in complex biomedical problems. Given the complex nature of most of the biomedical problems with inherent nonlinearities, included in cancer progression, systems biological approaches have been used recently to develop multiscale, multilevel predictive mathematical models to understand events that occur at multiple spatial and temporal level and resultant emergent behavior. Here, we discussed a hybrid multiscale approach that incorporates the details of intracellular, intercellular and cell-level to study the growth and progression of (cancer and/or normal) cells and its multiple direct and indirect effects to radiation. The multiscale model was implemented in two hybrid frameworks: hybrid multiscale CA framework and hybrid multiscale CPM framework (Brüningk et al., 2018; Powathil et al., 2012a; 2013, 2014). However, one should note that, in general, any predictions from (any) model is biologically relevant only when these assumptions are based on biological/clinical evidence and further, the results are validated with experimental data. Once validated using clinical and experimental data, these systems biological approaches, such as the ones described here, can provide mechanistic insights into radiation response and effectiveness, potentially helping to individualize treatment doses and multimodality treatment regimes (Brüningk et al., 2018; Powathil et al., 2013). As described in Sect. 5.3, we have validated the hybrid model to study the radiation effects in in vitro experiments to study fractionation protocols and multimodality treatments, in particular the combination with the heat (Brüningk et al., 2018).

The long-term aim of these modelling approaches are to develop qualitative, predictive mathematical and computational tools based on clinical and experimental data to understand, study, and provide useful predictions related to the outcome of multimodality treatment protocols used to treat human malignancies. This is an efficient and cost-effective alternative to test and study new treatment options, new dosage and efficacy of the treatments given, to allow the planning and designing of more adventurous treatments in silico, prior to beginning actual testing and long and costly clinical trials or experiments. As observed experimentally (Mothersill & Seymour, 2004; Munro, 2009; Prise & O'Sullivan, 2009), treatments such as radiation also induce indirect effects on the treated volume, that has a wider clinical implications in the treatment effectiveness and control, indicating the importance of studying and planning such protocols. One such indirect effect of the radiation is radiation induced bystander effect. Most of our knowledge about bystander effects is primarily from in vitro and in vivo studies, which may have less clinical applicability. Multiscale mathematical and computational tools that incorporate essential multi- layer complexities can be used as powerful tools to understand and identify the multiple parameters that are significant in radiation-induced bystander responses (Powathil et al., 2016). As described Sect. 5.4, such models can provide potential testable, qualitative hypotheses that may be very significant in exploring the role of bystander effects in a winder context. Our *in silico* findings about the role of bystander effects in mediating increased cell-kill at low doses were later supported by the experimental results by the Mothersill Group (Fernandez-Palomo et al., 2015).

Non-targeted and low dose effects of radiation have wider implications in an ecological perspective with a complex multilevel interaction among multiple species and their environment. Multiscale mathematical and computational modelling approaches can be extremely useful in studying the multilevel, long term effects/consequences of radiation (including the effects of historic doses) on a complex ecosystem in a qualitative approach, generating potentially testable hypothesis and inferences. The hybrid multiscale mathematical model outlined in this article to study the direct and indirect effects of radiation in the context of cancer progression serve as an example on how we can use mathematical modelling approaches to study complex biological problems. Moreover, it has many similarities in the multi-level spatial and temporal complexities involved in an ecosystem level. A step towards developing a predictive ecological model would be a continuous dialogue with the experts in radiation ecology to understand the relevant questions that needed to be studied and incorporated into the multiscale modelling approach.

References

Ahmad, S. B., McNeill, F. E., Byun, S. H., Prestwich, W. V., Mothersill, C., Seymour, C., Armstrong, A., & Fernandez, C. (2013). Ultra-violet light emission from HPV-G cells irradiated with low let radiation from (90)Y; consequences for radiation induced bystander effects. *Dose Response, 11*, 498–516.

Al-Mayah, A. H., Irons, S. L., Pink, R. C., Carter, D. R., & Kadhim, M. A. (2012). Possible role of exosomes containing RNA in mediating nontargeted effect of ionizing radiation. *Radiation Research, 177*(5), 539–545.

Andasari, V., Roper, R. T., Swat, M. H., & Chaplain, M. A. (2012). Integrating intracellular dynamics using CompuCell3D and Bionet- solver: Applications to multiscale modelling of cancer cell growth and invasion. *PLoS One, 7*(3), e33,726.

Anderson, A. R., & Chaplain, M. A. (1998). Continuous and discrete mathematical models of tumor-induced angiogenesis. *Bulletin of Mathematical Biology, 60*, 857–899.

Azzam, E. I., de Toledo, S. M., Waker, A. J., & Little, J. B. (2000). High and low fluences of alpha-particles induce a G1 checkpoint in human diploid fibroblasts. *Cancer Research, 60*(10), 2623–2631.

Blyth, B. J., & Sykes, P. J. (2011). Radiation-induced bystander effects: What are they, and how relevant are they to human radiation exposures? *Radiation Research, 176*(2), 139–157.

Brüningk, S., Powathil, G., Ziegenhein, P., Ijaz, J., Rivens, I., Nill, S., Chaplain, M., Oelfke, U., & ter Haar, G. (2018). Combining radiation with hyperthermia: a multiscale model informed by in vitro experiments. *Journal of the Royal Society Interface, 15*(138), 20170681

Fernandez-Palomo, C., Mothersill, C., Brauer-Krisch, E., Laissue, J., Seymour, C., & Schultke, E. (2015). H2AX as a marker for dose deposition in the brain of wistar rats after synchrotron microbeam radiation. *PLoS One, 10*(3), e0119,924.

Fowler, J. F. (1989). The linear-quadratic formula and progress in fractionated radiotherapy. *British Journal of Radiology, 62*(740), 679–694.

Franken, N. A. P., Rodermond, H. M., Stap, J., Haveman, J., & van Bree, C. (2006). Clonogenic assay of cells in vitro. *Nature Protocols, 1*(5), 2315–2319.

Glazier, J. A., Balter, A., & Popławski, N. J. (2007). Magnetization to morphogenesis: A brief history of the glazier-graner-hogeweg model. In *Single-cell-based models in biology and medicine* (pp. 79–106). Springer.

Gomez-Millan, J., Katz, I. S., Farias, V. D. E. A., Linares-Fernandez, J. L., Lopez-Penalver, J., Ortiz-Ferron, G., Ruiz-Ruiz, C., Oliver, F. J., & Ruiz de Almodovar, J. M. (2012). The importance of bystander effects in radiation therapy in melanoma skin-cancer cells and umbilical-cord stromal stem cells. *Radiotherapy and Oncology, 102*(3), 450–458.

Hu, B., Wu, L., Han, W., Zhang, L., Chen, S., Xu, A., Hei, T. K., & Yu, Z. (2006). The time and spatial effects of bystander response in mammalian cells induced by low dose radiation. *Carcinogenesis, 27*(2), 245–251.

Joiner, M. C., Marples, B., Lambin, P., Short, S. C., & Turesson, I. (2001). Low-dose hypersensitivity: Current status and possible mechanisms. *International Journal of Radiation Oncology Biology Physics, 49*(2), 379–389.

Kansal, A. R., Torquato, S., Harsh, G. R., IV, Chiocca, E. A., & Deisboeck, T. S. (2000). Cellular automaton of idealized brain tumor growth dynamics. *Biosystems, 55*, 119–127.

Lauber, K., Ernst, A., Orth, M., Herrmann, M., & Belka, C. (2012). Dying cell clearance and its impact on the outcome of tumor radiotherapy. *Frontiers in Oncology, 2*(9), 116.

Le, M., Fernandez-Palomo, C., McNeill, F. E., Seymour, C. B., Rainbow, A. J., & Mothersill, C. E. (2017). Exosomes are released by bystander cells exposed to radiation-induced biophoton signals: Reconciling the mechanisms mediating the bystander effect. *PLoS One, 12*(3), e0173, 685.

Lorimore, S. A., McIlrath, J. M., Coates, P. J., & Wright, E. G. (2005). Chromosomal instability in unirradiated hemopoietic cells resulting from a delayed in vivo bystander effect of gamma radiation. *Cancer Research, 65*(13), 5668–5673.

Lyng, F. M., Seymour, C. B., & Mothersill, C. (2000). Production of a signal by irradiated cells which leads to a response in unirradiated cells characteristic of initiation of apoptosis. *British Journal of Cancer, 83*(9), 1223–1230.

Marples, B., & Joiner, M. C. (2000). Modification of survival by DNA repair modifiers: A probable explanation for the phenomenon of increased radioresistance. *International Journal of Radiation Biology, 76*(3), 305–312.

Moreira, J., & Deutsch, A. (2002). Cellular automaton models of tumor development: A critical review. *Advances in Complex Systems (ACS), 5*(02), 247–267.

Morgan, W. F. (2003a). Non-targeted and delayed effects of exposure to ionizing radiation: I. Radiation-induced genomic instability and bystander effects in vitro. *Radiation Research, 159*(5), 567–580.

Morgan, W. F. (2003b). Non-targeted and delayed effects of exposure to ionizing radiation: II. Radiation-induced genomic instability and bystander effects in vivo, clastogenic factors and transgenerational effects. *Radiation Research, 159*(5), 581–596.

Mothersill, C., & Seymour, C. B. (2002). Bystander and delayed effects after fractionated radiation exposure. *Radiation Research, 158*(5), 626–633.

Mothersill, C., & Seymour, C. B. (2004). Radiation-induced bystander effects–implications for cancer. *Nature Reviews in Cancer, 4*(2), 158–164.

Mothersill, C., & Seymour, C. (2006). Radiation-induced bystander effects: Evidence for an adaptive response to low dose exposures? *Dose Response, 4*(4), 283–290.

Munro, A. J. (2009). Bystander effects and their implications for clinical radiotherapy. *Journal of Radiological Protection, 29*(2A), A133–A142.

Novak, B., & Tyson, J. J. (2003). Modelling the controls of the eukaryotic cell cycle. *Biochemical Society Transactions, 31*, 1526–1529.

Patel, A. A., Gawlinski, E. T., Lemieux, S. K., & Gatenby, R. A. (2001). A cellular automaton model of early tumor growth and invasion. *Journal of Theoretical Biology, 213*, 315–331.

Poleszczuk, J., & Enderling, H. (2014). A high-performance cellular automaton model of tumor growth with dynamically growing domains a high-performance cellular automaton model of tumor growth with dynamically growing domains. *Applied Mathematics, 5*(1), 144–152.

Powathil, G. G., Gordon, K. E., Hill, L. A., & Chaplain, M. A. (2012a). Modelling the effects of cell-cycle heterogeneity on the response of a solid tumour to chemotherapy: Biological insights from a hybrid multiscale cellular automaton model. *Journal of Theoretical Biology, 308*, 1–9.

Powathil, G., Kohandel, M., Milosevic, M., & Sivaloganathan, S. (2012b). Modeling the spatial distribution of chronic tumor hypoxia: Implications for experimental and clinical studies. *Computational Mathematical Methods in Medicine, 2012*, 410, 602.

Powathil, G. G., Adamson, D. J., & Chaplain, M. A. (2013). Towards predicting the response of a solid tumour to chemotherapy and radiotherapy treatments: Clinical insights from a computational model. *PLoS Computational Biology, 9*(7), e1003,120.

Powathil, G. G., Chaplain, M. A. J., & Swat, M. (2014). Investigating the development of chemotherapeutic drug resistance in cancer: A multiscale computational study. *arXiv*, 1407.0865 [q-bio.TO].

Powathil, G. G., Swat, M., & Chaplain, M. A. (2015). Systems oncology: Towards patient-specific treatment regimes informed by multiscale mathematical modelling. *Seminars in Cancer Biology, 30*, 13–20.

Powathil, G. G., Munro, A. J., Chaplain, M. A., & Swat, M. (2016). Bystander effects and their implications for clinical radiation therapy: Insights from multiscale in silico experiments. *Journal of Theoretical Biology, 401*, 1–14.

Prise, K. M., & O'Sullivan, J. M. (2009). Radiation-induced bystander signalling in cancer therapy. *Nature Reviews in Cancer, 9*(5), 351–360.

Prise, K. M., Schettino, G., Folkard, M., & Held, K. D. (2005). New insights on cell death from radiation exposure. *Lancet Oncology, 6*(7), 520–528.

Ribba, B., Colin, T., & Schnell, S. (2006). A multiscale mathematical model of cancer, and its use in analyzing irradiation therapies. *Theoretical Biology and Medicine Modelling, 3*, 7.

Richard, M., Kirkby, K., Webb, R., & Kirkby, N. (2007). A mathematical model of response of cells to radiation. *Nuclear Instruments and Methods in Physics Research Section B: Beam Interactions with Materials and Atoms, 255*(1), 18–22.

Satelli, A., Annono, P., Azzinin, I., Campolongo, F., Ratto, M., & Tarantola, S. (2010). Varianance based sensitivity analysis of model output. Design and estimator for the total sensitivity index. *Computer Physics Communications, 181*(2), 259–270.

Seymour, C. B., & Mothersill, C. (2000). Relative contribution of bystander and targeted cell killing to the low-dose region of the radiation dose-response curve. *Radiation Research, 153*(5 Pt 1), 508–511.

Shuryak, I., Sachs, R. K., & Brenner, D. J. (2007). Biophysical models of radiation bystander effects: 1. Spatial effects in three- dimensional tissues. *Radiation Research, 168*(6), 741–749.

Smith, R. W., Seymour, C. B., Moccia, R. D., & Mothersill, C. E. (2016). Irradiation of rainbow trout at early life stages results in trans-generational effects including the induction of a bystander effect in non-irradiated fish. *Environmental Research, 145*, 26–38.

Turner, S., & Sherratt, J. A. (2002). Intercellular adhesion and cancer invasion: A discrete simulation using the extended Potts model. *Journal of Theoretical Biology, 216*, 85–100.

Tyson, J. J., & Novak, B. (2001). Regulation of the eukaryotic cell cycle: Molecular antagonism, hysteresis, and irreversible transitions. *Journal of Theoretical Biology, 210*, 249–263.

Chapter 6
Immune Networks in the Context of Low Dose Ionizing Radiation

Dörthe Schaue, Keisuke S. Iwamoto, and William H. McBride

Abstract Irradiated tissues engage the immune system on many levels. The general assumption is that the initial damage alerts immune cells through universal danger sensing and signaling pathways that are pro-oxidant, pro-inflammatory at first, before morphing into an anti-oxidant, anti-inflammatory counter response. The perpetuating nature of the inflammatory forces that drive normal tissue toxicity seems to originate in part from cytosolic damaged DNA structures that directly and persistently activate innate immune cells, enforce senescence and feed back to the bone marrow, preferentially driving myelopoiesis and further immune activation. Many of the lasting radiation effects on the immune system resemble those seen in premature aging, and are seen in A-bomb survivor studies and amongst bone marrow transplant recipients. These include enhanced T cell senescence, a shrinking T cell repertoire, less T cell functionality, and overall a more pro-inflammatory immune outlook. The ultimate long-term effects also resemble failed attempts at regeneration with fibrosis, scarring, failure of tissue function, and possibly carcinogenesis and are a de facto multi-organ disease. Whether or not the dose response follows a linear, no-threshold low dose dependency, is less clear but it is reasonable to assume that different thresholds exist for different radiation-induced effects. For example, there is evidence that DNA damage that might be repaired at higher doses, at low doses may leave foci of unrepaired lesions on the balance sheet, which may serve as a nidus for a carcinogenic event or as a chronic low-dose stimulus. So, the potential for longer term, chronic low-grade responses may remain, with the possibility of its later amplification by a secondary event such as an infection or wounding. The status of immune activation at the time of radiation exposure, i.e. acute antigen exposure, immune cell activation, differentiation, metabolic and redox balance, proteasome status, and NFκB/Nrf2 activity will hugely affect the net-outcome of low dose radiation events. Understanding how the immune rheostat might be reprogrammed by low dose radiation, the importance of radiation dose, dose rate,

D. Schaue (✉) · K. S. Iwamoto · W. H. McBride
Division of Molecular and Cellular Oncology Department of Radiation Oncology, David Geffen School of Medicine, University of California at Los Angeles (UCA), Los Angeles, CA, USA
e-mail: dschaue@mednet.ucla.edu

M. D. Wood et al. (eds.), *Biomarkers of Radiation in the Environment*, NATO Science for Peace and Security Series A: Chemistry and Biology,
https://doi.org/10.1007/978-94-024-2101-9_6

and quality, as well as the impact of collateral signals in the context of danger or non-danger signaling is relevant to many life shortening and carcinogenetic events.

Keywords Inflammation · Immunity · T cells · Danger · ROS · Proteasome

6.1 Introduction

Within weeks of their discovery in 1895, Roentgen rays were shown to cause inflammation of the skin with severe burns and later cancer that led to hundreds of radiologists losing their hands and lives until protective means to avoid incidental exposure were implemented. Ironically, the minimal erythematous dose became the first useful measure of radiation delivery in patients, which minimized side effects and probably prompted the introduction of the concept of radiation fractionation into the clinic. The involvement of inflammation in the context of radiation damage was supported by early pathological observations of immune infiltration into irradiated sites that also noted microvasculature damage. In fact, for a long time vascular rather than parenchymal cell damage was stressed as the major cause of normal tissue effects, although the direct cytotoxic action of radiation was fully acknowledged by pioneering radiopathologists. The observation of intratumoral immune infiltrates also led to early intellectual debates as to whether the effects of radiation therapy (RT) were systemic or purely local (Shohan, 1916); discussions that have recently been resurrected in the form of bystander and abscopal effects, both of which were postulated over a century ago. It should be noted that, during its first 50 years, RT focused on both non-cancerous as well as cancerous diseases. Acne, mycobacterial infections, gas gangrene, lupus vulgaris, chronic inflammation, and many other, largely superficial conditions were prime diseases for treatment. It was only after World War II that fear of radiation-induced cancer and its associated litigation, as well as alternative treatments such as antibiotics, led to RT being limited largely to cancer treatment, at least in the USA. In Europe, benign conditions continue to be treated, and the evidence for success in treating chronic degenerative and inflammatory conditions with relatively low radiation doses is convincing and raises many questions regarding the effect of dose on immune responses.

6.2 Irradiated Tissues Respond with a *Bona fide* Inflammatory Response

It is now well established that radiation is a pro-inflammatory stimulus and, as such, has systemic consequences (McBride et al., 2004; Purbey et al., 2017; Schaue et al., 2015). Many side effects of cancer RT, most notably fatigue, are due to chemokines or cytokines released after RT rather than the physical-chemical events that immediately follow the deposition of the energy of ionization, which itself is quantitatively very low. The question of how much these and other indirect immune effects contribute to cell kill is still a matter of some controversy, especially in the context of low dose exposures. After higher doses, the appearance of circulating acute phase proteins made largely in the liver following exposure and the action of liberated cytokines acting on the hypothalamus to cause occasional somnolence and flu-like symptoms are obvious examples of systemic consequences of local radiation delivery (Hong et al., 1996; Schaue et al., 2015) but at low doses these are less obvious and may be limited.

Irradiation has long been known to cause microvascular leak (Mottram, 1933). Endothelial cells and platelets are activated leading to coagulation cascades, which involve some acute phase proteins, increased vascular permeability, release of chemotactic factors, and mobilization of myeloid phagocytic cells that extravasate into sites. The consequent cytokine cascades orchestrate the classic sequence of inflammatory cell changes that take place as part of an innate immune response. The first responders are neutrophils that cause further tissue damage in an attempt to mop up and to open up the site. These tend to die in large numbers and are engulfed by invading monocytes that become activated macrophages. More reactive oxygen and nitrogen species are generated but in time the pro-oxidant, pro-inflammatory damaging microenvironment morphs into an anti-oxidant, anti-inflammatory one as evidenced by angiogenesis, cell proliferation, tissue regeneration and remodeling; orchestrated again by cytokines – this time, anti-inflammatory and pro-healing in nature. These events are influenced by tissue-specific and genetic influences, but essentially are canonical events, triggered by many "danger" stimuli (McBride et al., 2004). Positive and negative feed-back control loops are inherent to the coordination of the response and appear as recurring waves of pro-inflammatory cytokines, reactive oxygen species (ROS) production, and DNA damage (Kim & McBride, 2010; Schaue et al., 2012a). The relationship between radiation dose and the inflammatory response is critical to the outcome. Different aspects of the inflammatory armament are triggered by low and high radiation doses as are different levels of feedback immune stimulation, ROS, and DNA damage. This lack of linearity in the radiation dose-response relationship with respect to the immune response is to be expected.

Radiation has, however, an additional layer of complexity compared to other sterile or infectious challenges, or oxidative stresses that may relate to the persistency of the immune response. It comes in the form of a relatively high frequency of DNA double strand breaks, as opposed to other forms of DNA damage. Latent DNA damage can persist in cells for a long time, and when cells divide micronuclei or other nuclear foci can occur that may be fatal or can persist in the cytoplasm for a long period of time (Ghandhi et al., 2018; Siddiqui et al., 2015; Udroiu et al., 2017; Werner et al., 2017); these can be recognized as a chronic immune stimulus (Harding et al., 2017; Mackenzie et al., 2017). In a sense, irradiated sites behave as wounds that do not heal, at least not properly, with fibrosis and loss of tissue function as common outcomes.

6.3 Danger & Inflammation Build the Bridge to Adaptive Immunity

Perhaps, one of the biggest fascinations for immunologists is how the innate and adaptive immune systems integrate so that rapid innate recognition of pathogens can segue into generating specific immunity, with the development of immunological memory as the ultimate prize. For a long time, the theory was that lymphocytes generated clonal diversity and that those T cell clones that responded to "self" molecules were deleted in the thymus leaving only those that responded to "non-self" (Hodgkin et al., 2007). This clean and simple concept was muddied by the discovery that tolerance to "self" was as much "peripheral" as it was "central", by the finding that T lymphocytes recognized self MHC molecules, and by the alarming discovery that removing some of the constraints on the immune system resulted in potentially fatal autoimmunity and cytokine "storms" (Lee et al., 2014). In fact, both innate and adaptive immunity have multiple, crucial negative control mechanisms in place including regulatory T cells, myeloid suppressor cells, anti-inflammatory cytokines, and inhibitory T cell signaling pathways, such as PD-1, etc. (Smith, 2012a, b). Shared ligands and receptors are employed to control the interactions within the immune systems and their interactions with other normal tissues so as to maintain tissue homeostasis. RT inevitably disturbs this homeostasis in many ways, most obviously when causing the death of small lymphocytes, whose extreme radiosensitivity was one of the first observations following the discovery of X-rays and hence relevant for the low dose discussion.

Another level of immune control is exerted through antigen presentation. Professional antigen-presenting dendritic cells (DC) are part of the innate immune system, but form the link to adaptive antigen-specific immunity. To do this DCs must undergo a maturation process under the influence of "danger" signals (Medzhitov, 2013; Seong & Matzinger, 2004). Without this crucial maturation process, DCs act to maintain peripheral tolerance. Importantly, radiation can directly and indirectly signal "danger" to the immune system (McBride et al., 2004), which leads to the notion that RT can build a bridge between innate and adaptive immunity and, by extension, present targets for modifying radiation responses.

"Danger" signaling following irradiation may take several forms (McBride et al., 2004; Ratikan et al., 2015). One of the primordial mechanisms is through release of adenosine 5'-triphosphate (ATP), which is one of the earliest responses to even low doses of radiation (Kojima et al., 2017; Mancuso et al., 2011; Ohshima et al., 2012). Active ATP release from cells has many putative conduits, including connexin hemichannels, P2X7 receptors, anion channels, and vesicles. Extracellular ATP may activate P2Y purine receptors, but it also is relatively quickly converted to adenosine monophosphate (AMP) via ectonucleoside triphosphate diphosphohydrolase 1 (CD39), and subsequently through the ecto-5′-nucleotidase (CD73) to adenosine (Sitkovsky et al., 2008). Through this cascade, nanomolar concentrations of extracellular ATP mediate a broad range of autocrine/paracrine signaling effects in many cell types. Adenosine signaling, for instance, can occur through four distinct adenosine receptors. One or more of these can be targeted for modulation of radiation responses, as well as immune responses, enhancing DNA repair and myeloid system recovery (Grant et al., 1976; Pospisil et al., 1995; Schaue & McBride, 2015). Clinically, adenosine is famous for its ability to counteract tachycardia, but most relevant to this discussion is its ability to dampen acute inflammation. Other molecules secreted after RT include the free radical scavenger glutathione that is also anti-inflammatory. In fact, low doses of radiation appear to affect many of the channels that maintain intracellular homeostasis, resulting in secondary production of reactive oxygen species (ROS) and metabolic re-programming (Dodson et al., 2013; Schaue et al., 2012b).

The most studied danger signaling pathways today are those activated through pattern recognition receptors (PRRs) and pro-inflammatory cytokine receptors (Medzhitov, 2013; O'Neill et al., 2013). Originally thought to recognize microbial products only, or so-called pathogen associated molecular patterns (PAMPs). PRRs – as it turns out – also bind endogenous, damage associated molecular patterns (DAMPs). The fact that PAMPS and DAMPS universally engage PRRs reveals yet another example of breakdown of the "self-versus-non-self" distinction in the immune system. PRRs are ubiquitously, yet variably expressed by non-immune and immune cells alike, so they have relevance at many levels of cell-cell communication, but the most prominent is to initiate innate immune responses. Several pathways may be activated downstream of PRRs depending on their quantity and cellular location, tissue type, and what other pathways are activated. Most common are type I interferon and pro-inflammatory cytokine production, although the anti-inflammatory cytokine IL-10 can be produced to minimize tissue damage by inhibiting pro-inflammatory cells and cytokines. What determines the balance has yet to be determined but the simultaneous concomitant presence of pro- and anti-inflammatory immune forces is a common, and probably necessary, feature of immunity (McBride et al., 2017).

Many DAMPS, including ATP and glutathione, are normally expressed intracellularly at high levels and actively secreted by damaged cells, or released upon cell death. For example, high mobility group protein 1 (HMGB1) is a nuclear protein with various functions, including roles in DNA repair, differentiation and development, but when released through pathways involving secretory lysosomes it generates IL-6, interferon-α, and TNF-α to cause inflammation and to mature DCs for antigen presentation (Schaue et al., 2015). The PRRs that HMGB1 generally activates are the cell surface Toll-like receptors (TLR) 4 and 2. However, it has multiple roles in immunity and tissue healing, which appear to involve it switching between oxidation states, binding partners, and receptors, including PRRs that bind advanced glycation end products (RAGE) involved in metabolic reactions. From a radiation perspective, inhibiting HMGB1 using the liquorice extract and anti-inflammatory agent glycyrrhizin, prolongs survival when given after potentially lethal doses of whole-body irradiation (WBI), hinting at the negative effects of immune engagement in the aftermath of normal tissue damage (Schaue et al., 2015). Radiation can also induce the expression of TLR genes suggesting another level of radiation-danger-immune intersection (Menendez et al., 2011).

6.3.1 The Sting of Radiation

While some PRRs are membrane bound receptors, others are endosomal or cytoplasmic. TLR 3, 7, 8, 9, and 13 and RIG-like PRRs, for instance, bind nucleic acids to sense the presence of intracellular infections in the cytoplasm. Of particular relevance to radiation is the cyclic GMP-AMP synthase (cGAS) that binds double-stranded DNA (self or non-self) to produce cGAMP, which binds and activates the adaptor protein STING ultimately driving the production of type I interferons and inflammatory cytokines through TBK1/IRF3 and IKK/NF-κB pathways, respectively (Harding et al., 2017; Purbey et al., 2017; Vanpouille-Box et al., 2017). Normally, self-nucleic acids do not gain access to the cytoplasm or endosome, but in pathological situations, including after RT, this may occur (Spektor et al., 2017). It has been known for many decades that nuclear lesions can persist in irradiated cells for a long time, most obviously in the form of chromosome damage. In fact, such damage in lymphocytes is often used to assess exposures to radiation many years after the event and is taken as evidence for the long-lived nature of recirculating T cells (McLean & Michie, 1995). Other sources of cytoplasmic DNA in an irradiated site may come from phagocytosis of dead cells by macrophages, including dead neutrophils. This may be a primitive defense mechanism. Neutrophils produce extracellular traps (NETs) that are networks of extracellular fibers, primarily composed of DNA, that can trap bacteria and other dangerous material (Brinkmann & Zychlinsky, 2007). These are cleared eventually by macrophages, and influence their behavior. Radiation-induced senescence and autophagy may also allow persistence of cytoplasmic DNA foci, be self-perpetuating, and feed recurring immune activation.

The mechanism of cell death is probably critical to the ability of DNA to stimulate the cGAS/STING pathway. While radiation can cause some cell types, such as small lymphocytes, to rapidly apoptose, it has been known since the time of Puck and Marcus (1956) that many irradiated cell types can divide before dying by mitotic catastrophy; the number of divisions depending upon the radiation dose up to about 8 Gy, above which division is prevented. The most obvious form of chromosome abnormality precipitated during mitosis in irradiated cells are micronuclei that can persist for several cell cycles. cGAS is localized in the cytoplasm of non-dividing cells but enters the nucleus and associates with chromatin DNA during mitosis so micronuclei and cytoplasmic foci of damaged DNA can activate cGAS (Yang et al., 2017). Radiation-induced senescence and autophagy may also activate cGAS; indeed, cGAS seems to be a necessary requirement for senescence in some cases (Yang et al., 2017), including the chronic production of inflammatory cytokines. The cGAS/STING pathway has also been shown to assist in the generation of radiation-induced tumor immunity and this has been associated with specific DC subsets, which are specialized in taking up material from dying cells for cross-presentation of cell-associated antigens (Deng et al., 2014; Vanpouille-Box et al., 2017). The DNA that activates this pathway is lost after very high radiation doses and there is evidence that this is due to DNAse III activation, although as noted above, lack of cell division may also be involved. The emerging concept is that radiation-induced self-nucleic acids can result in persisting DNA damage signals that will drive chronic inflammation and autoimmunity. For one, the type I IFN that ensues can persistently recruit inflammatory monocytes in a CCR2/CCL2-dependent manner that largely fail to mature, sustaining a chronic inflammatory loop (Groves et al., 2018; Lee et al., 2009). While the cGAS-STING pathway is likely involved in many of these cases, chronic radiation inflammation is exquisitely orchestrated and a much more highly nuanced process. A case in point is that activation of the cGAS-STING pathway does not end with cytokine production in all cell types and is not the only pathway that can result in interferon and pro-inflammatory cytokine production after RT (Purbey et al., 2017). As Campisi and colleagues have elegantly shown, abnormal foci in senescent cells can also take the form of DNA SCARS (segments with chromatin alterations reinforcing senescence) without engaging cGAS, and yield all of the hallmarks of senescence such the senescence-associated secretory phenotype (SASP) that includes chronic production of inflammatory cytokines, most notably IL-6 and IL-8 (Rodier et al., 2011). What is clear, is that nuclear material can persist for a long time in tissues after irradiation, some being generated due to tissue turnover, and this is a likely source, along with SASP, of recurring immune stimulation during the latent period and that feeds radiation-induced tissue damage, as has been reported in the lung (Beach et al., 2018; Ghandhi et al., 2018; Groves et al., 2018).

6.3.2 Radiation-Induced DNA-Damage – A Gift That Keeps on Giving

What this means in terms of late effects on the tissue level and the complexity thereof can be illustrated by how the same dose of radiation to the lung results primarily in pneumonitis in C3H mice but fibrosis in the C57Bl/6 strain. Over the first two months, pro-inflammatory cytokine levels in the lung increase, but generally are comparable in the two strains. Thereafter, this seems to spiral out of control in C3H mice while being more contained in the lungs of C57Bl/6 mice (Chiang et al., 2005; Hong et al., 1999; McBride et al., 2017; Rubin et al., 1995; Schaue et al., 2012a, 2015; Schaue & McBride, 2012). These differences are genetically determined and under control of immune mechanisms rather than driven by lung stem cells and/or their damage *per se*. How this links to DNA damage and persistence is not clear and multiple mechanisms are probably involved. It appears that even when pneumonitis is prevented, tissue pathology still proceeds to fibrosis at least in non-human primates pointing at the independence of these immune-driven events (Cline et al., 2018). One aspect which is unexplained, and this is true of most chronic inflammatory states, is the marked presence and timing of infiltrating lymphocytes. The function of these cells is still unknown, but it is hard to believe that they do not contribute to the radiation-induced alterations in tissue microenvironment. Here again are important nuances. The distinction between Th1 and Th2 cells developed several decades ago presented a clean concept to explain how the immune system could manifest so many different guises (Mosmann & Sad, 1996). Th1 cells through type II interferon production and macrophage activation mount cell-mediated responses against intracellular pathogens, while Th2 responses focus on antibody production and immunity to extracellular pathogens. Th1 are considered more pro-inflammatory, while Th2 are more anti-inflammatory. Today we know that this T helper differentiation is much more complex than a Th1/2 division. The discovery of regulatory T cells and their role in controlling cell mediated responses and peripheral tolerance for instance, has dramatically increased the options available for considering immune involvement in radiation responses as has the finding of IL-23 driven Th17 cells that play a major role in balancing chronic inflammation and autoimmunity through production of IL-17, IL-6, and TNF-a (Barbi et al., 2013; Chen & Oppenheim, 2014; Kimura & Kishimoto, 2010; Waite & Skokos, 2012). All are under genetic control, but a certain balance must be maintained for homeostasis.

Classic radiobiological theory dictates that the time to expression of radiation-induced tissue damage (latency) is dependent on the rate of tissue turnover and, as a result, quite predictable. Thus "late effects" of irradiation may take months or years to emerge. This is somewhat surprising, as molecular and cellular events, for example those associated with acute inflammation, are initiated very rapidly after tissue irradiation. Inflammation then appears to continue in "waves" up until loss of tissue function, at which point pro-inflammatory responses are at their highest, speaking to their involvement in loss of tissue function (Chiang et al., 1997; McBride et al., 2004; Schaue et al., 2015). Crises can be explained when cells die upon division because they carry latent, potentially lethal DNA lesions. These same waves occur, albeit at a lower level, at doses that are insufficient to result in overt loss of tissue function and are thought to be repeated attempts at recovery. That these cyclical responses are important can be judged by late consequences especially in the context of additional challenges such as infection or trauma. At the higher dose levels, even if mice recover from potentially lethal WBI they tend to die early, with mortality coming in waves (Travis et al., 1985) and few completing their normal life span. Inflammation and disturbed immune homeostasis are common in late radiation damage and are seen in various forms, in multiple organs, including heart, lung, liver, bone marrow.

6.4 Inflammaging – How Radiation Makes us Immunologically Older

Life shortening has long been known as a consequence of low and high dose WBI exposure (Kallman & Kohn, 1958; Mole, 1959; Yuhas, 1969). Reevaluation of the A-bomb survivor data has shown, additionally to cancers, significantly increased incidences of late non-cancerous conditions including thyroid and liver disease, cataracts, glaucoma, hypertension, and myocardial infarction, as well as persistently altered immune and inflammatory status with underlining T cells immunosenescence (Fujiwara et al., 2008; Kajimura et al., 2017; Kusunoki et al., 2010; Kusunoki & Hayashi, 2008; Preston et al., 2003; Shimizu et al., 2010; Shimura & Kojima, 2014). Survivors at the Mayak facility exposed to waterborne uranium fission products developed late chronic inflammatory hematological and late-onset non-hematological complications. This is similar to multiple organs disease syndrome (MODS) that is also associated with dysregulated inflammation (Aikawa, 1996; Williams & McBride, 2011). A clinical parallel can be drawn from childhood cancer survivors treated with WBI plus bone marrow transplantation who have a higher, but variable incidence of radiation pneumonitis, veno-occlusive disease, renal hypertension, cardiac damage, neurocognitive abnormalities, cataracts, endocrine dysfunction, and hormonal deficits (Arora et al., 2016).

In acute radiation syndromes, the parallel between loss of stem cells and mortality is inescapable, but it seems likely that for late effects after <u>lower</u> radiation doses immune phenomena play a larger role. Our experience with late effects after WBI is that the myeloid:lymphocyte balance in the immune system is greatly increased and this is associated with increased chronic inflammation that can affect all organ systems. In the bone marrow, this myeloid shift is associated with a limiting of the output of primitive bone marrow stem cells i.e. a shift in the organization of cellular subsets within the hematopoietic system. The hypothesis that emerges is chronic stimulation without resolution leads the immune system to recruit its most primitive mechanism – myeloid cell activation – in a last-ditch attempt to survive. This involves a "walling off" wound healing mechanism involving collagen production that is mediated by TGF-β production, which is immunosuppressive, anti-proliferative, pro-fibrotic, and pro-senescence. Additional systemic consequences are immunohematopoietic and immunological exhaustion.

It is worth thinking of radiation-induced life shortening as being similar in many ways to premature aging and increased frailty (Antoch et al., 2017; Hernandez et al., 2015; Lowe & Raj, 2014). In fact, many of the observed radiation-induced changes in the T cell compartment of A-bomb survivors resembles those of an older person, such as decrease in naïve T cell output, while the memory T cells pool expands, an increasingly limited TCR repertoire, suboptimal T cell function, all alongside a higher pro-inflammatory baseline, a.k.a. inflammaging (Denkinger et al., 2015; Franceschi et al., 2017; Frank & Caceres, 2015; Fulop et al., 2018; Kusunoki et al., 2010). Senescence is an obvious link, and the fact that both age and radiation increase the baseline micronuclei frequency in lymphocytes, the incidence of MODS, the shift in myeloid:lymphocyte balance, and the development of frailty indicates other strong parallels (Nefic & Handzic, 2013; Norman et al., 1984; Wojda et al., 2007). What drives the radiation-induced immune waves, their dose dependency, and the final outcome is therefore of general relevance even -or perhaps especially- at the lower dose end of the spectrum. Again, the A-bomb survivor story teaches us that damage to the immune system is proportional to dose and even if the hematopoietic system largely self-repairs, residual injury persists and likely re-emerges upon a subsequent challenge (Kusunoki & Hayashi, 2008; Rodrigues-Moreira et al., 2017).

6.5 Redox, Radiation-Induced Signaling Networks and Immune Engagement

Radiation pathways may be activated in many ways (Fig. 6.1). Our focus here has been largely on immune pathways, cognizant of the fact that the immune system interacts intimately with other cell types to maintain homeostasis. ROS play critical roles in so many cellular processes that they form the most obvious earliest point of intersection of biology with ionizing radiation-induced physico-chemical events (Spitz & Hauer-Jensen, 2014). Even very low doses of radiation can affect redox to completely alter both the intracellular content and extracellular environment of the cell. It should be remembered that after irradiation, ROS can be generated from many sources other than from ionization radiation events *per se*. Pro-inflammatory cytokines generate ROS and can cause DNA damage (Westbrook et al., 2012). Additional sources are through mitochondrial damage, NOX and other oxidases, and metabolic pathways including those leading to autophagy and senescence, and ER-stress. ROS are critical for many early radiation effects, in addition to the release of ATP and glutathione. A formidable array of signaling molecules are redox-sensitive, including some key players such as NF-κB/IκB, Nrf2/Keap1, thioredoxin, APE-1, p53, HIF-1α, PK-M2, ATM, PTEN and other phosphatases, and 26S proteasomes. These are specific targets for oxidative stresses and there must be a hierarchy of sensitivities, with location being a major contributing element, although it is not clear how these sensitivities coordinate to make meaningful responses. The fact that there is a major shutdown of 26S proteasomes that occurs after very low doses of radiation, causing a pile up of oxidized and polyubiquitinated proteins (Kim et al., 2008; Pajonk & McBride, 2001; Pervan et al., 2001, 2005) as well as major metabolic changes that can be seen as spikes in.

Autofluorescence (Schaue et al. 2012b) suggests that the reprogramming towards pro-oxidant status is an early event, likely feeding into protective mechanisms that promote cell survival. The Nrf2 pathway is a key regulator of this protective response aimed at restoring redox homeostasis as well as detoxification (Zenkov et al., 2017). It does so largely through affecting the expression of a battery antioxidant proteins, e.g. GST, γ-GCS, GPx, Trx, TrxR1, catalase, SOD1, SOD2, HO-1, NQO1, UGT, and ferritin. Upon oxidative stress, Keap1 releases Nrf2 from its grip which allows it to translocate to the nucleus where it can bind to the antioxidant response elements to initiate gene transcription. Increase in Nrf2 production can also overwhelm Keap1 restriction. Nrf2 also drives an increase in 20S proteasomes and their Pa28 (11S) regulators, which are superior at degrading oxidized proteins and hence stress resistance. Immunoproteasomes that are generated through IRF1, not Nrf2, seem less responsive. Of note, Nrf2 is responsible for much of the cellular glutathione

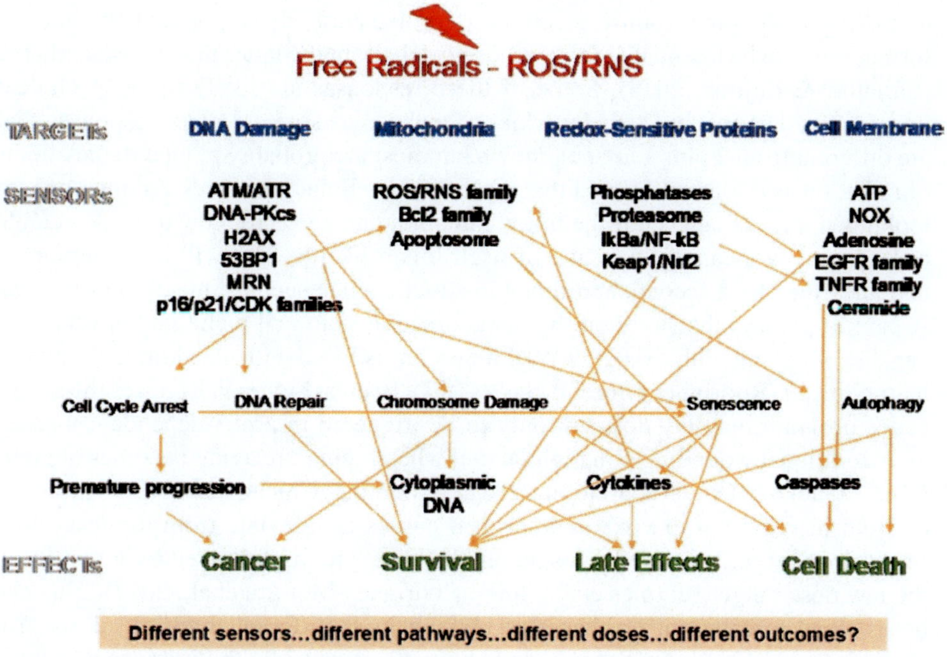

Fig. 6.1 Radiation-induced signaling networks

production, and has recently been shown to negate the initial ROS-mediated low dose radiation hypersensitivity of adult hematopoietic stem cells. Protection is however transient, and increased ROS levels reappear, dampening long-term HSC self-renewal (Rodrigues-Moreira et al., 2017). The link between redox and the immune system is further exemplified in that redox and Nrf2 sets the Th1:Th2 immune response balance (Rockwell et al., 2012).

Another poorly recognized nuance of the response to radiation that is especially important to immunity is that even low doses can affect the response to subsequent stimulation (Hong et al., 1996). If two doses of radiation are given, the change in the second response is referred to as an adaptive response. In fact, signals other than radiation can be used to modify subsequent exposure to radiation with the outcome being dependent on the sequencing and the system. For example, in vivo, lipopolysaccharide (LPS) radioprotects against lethality from WBI, but fails to do so if given after WBI. In contrast, in vitro, irradiation acts as a danger signal and primes macrophages to mount proinflammatory responses to LPS (Ratikan et al., 2015). It is hard to generalize when dealing with a system so complex as immunity, but activation of lymphocytes and other cell types tends to radioprotect them, which may be why antigens given even shortly before WBI can even enhance immune responses, blocking radiation's normal immunosuppressive cytotoxic action.

For example, radiation exposure tends to limit antibody formation when given prior to antigen exposure, much less so when given afterwards (Taliaferro et al., 1964). However, the end result can vary as is seen by comparing autoimmune disease with allergic responses. Both may involve antibody responses but while the former tends to be attenuated by low-dose irradiation, the latter may be exacerbated (Shimura & Kojima, 2014). Some of these responses are likely to be of clinical importance. The mechanisms by which signal 1 plus signal 2 gives responses that are different from 2 plus 1 are not known but most are probably related to the effects of radiation as a "danger" signal that reprograms cellular responses. Another way of looking at this is that even if the initial radiation damage is too low to be detectable or biologically meaningful in and of itself, in all likelihood it will be remembered and unearthed by a second, additional insult/challenge such as infection or trauma. A case in point is the study on the hematopoietic stem cell niche that is very well capable of coping with a 0.02Gy WBI hit until it is faced with an additional, inflammatory insult (Rodrigues-Moreira et al., 2017). Even at higher WBI doses the senescence program may lay dormant only to be triggered in a p53-dependent manner after additional wounding or a high fat diet when premature aging becomes obvious (Dr. A. Gudkov, Annual Radiation Research Meeting, Chicago 2018).

Both in vitro and in vivo, cell survival curves can deviate from the usual log-linear dose response curve by displaying hypersensitivity followed by a plateau in the low dose range (Marples et al., 2004; Rodrigues-Moreira et al., 2017). This can be ascribed to different mechanisms kicking in at different threshold doses, for example the Nrf2 response mentioned above. By extension, the immune system is likely engaged differently across a dose range (Fig. 6.2) (McDonald et al., 2010; Tsukimoto et al., 2010). Other possibilities are that DNA damage at very low doses may not trigger sufficient DNA repair, e.g. H2AX may not be phosphorylated sufficiently below 20–40 mGy (Elgart et al., 2015; Rothkamm & Lobrich, 2003) while at very high doses repair may be compromised; or diverse redox-sensitive pathways may be activated at different doses altering the probability of recovery (Brush et al., 2009; Lehnert & Iyer, 2002; Smith et al., 2012). Such considerations challenge the standard linear-no-threshold hypothesis that damage, and therefore risk, is always directly proportional to dose. In other words, a low dose response is not always a partial high dose response.

6.6 Summary

A concept that emerges from the arguments proposed above is that early pro-inflammatory responses to radiation sustain a redox imbalance, driving ROS production for the general purpose of microbial killing. In this way, the immune system becomes an integral player in radiation-induced ROS-mediated responses. As with all immune responses, pro-inflammatory responses are countered by anti-inflammatory responses, such as those mediated through Nrf2, and these opposing forces co-exist as long as the lesion persists. This leads to questions as to why the redox imbalance persists after irradiation, what is the role of the immune system in late radiation damage, and what are the dose and other requirements for this effect?

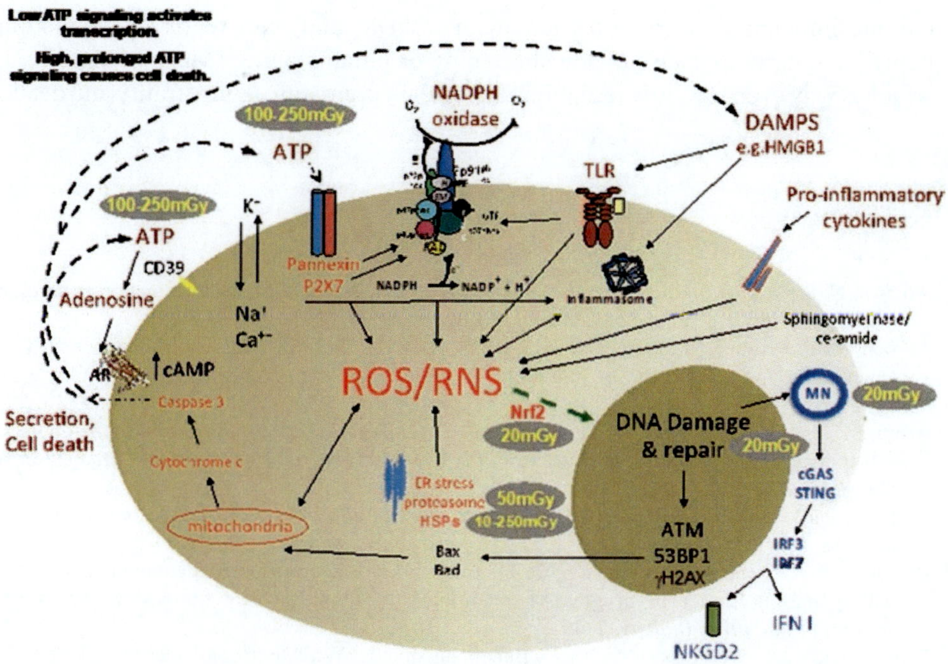

Fig. 6.2 Cellular responses to low radiation doses (putative threshold doses highlighted in yellow)

For tissue or tumor irradiation, the probability of a complete effect like lethality or tumor cure can be described by a steep deterministic dose-response curve fitted to a probit distribution. The overriding concept is that these curves reflect failure of a critical target cell population to replace functional cells, and that the time to response (latency) reflects the tissue turnover time (Withers & McBride, 1998). In contrast, after lower radiation doses it is likely that the immune system plays a larger role and that morbidity or mortality, if it occurs, is a more stochastic life shortening effect that is mechanistically distinct from direct radiation cytotoxic DNA damage and involves chronic inflammation. Radiation-induced lesions do not show the same consistent pathology at low as at high doses, vary for any given tissue, and may form at discrete inflammatory foci in different tissues within, or even outside, the field. The most likely source of continuing immune stimulation is residual DNA damage, ROS generation, inflammation, and further DNA and tissue damage, but the contribution of radiation-induced imbalance within the immune system acting through other pathways, and of vascular lesions, which may be precipitated immunologically, cannot be ignored. There are chronic inflammatory diseases that similarly have DNA as a source of stimulation, for example systemic lupus erythematosus, or persistent viral infections, or for that matter any situation where there is abnormal tissue turnover and cell death (Gao et al., 2015). One parallel is in the waves of responses seen in so many of these chronic conditions, including after irradiation. It is likely that similar waves occur after cancer chemotherapy and some forms of surgery. The persistent ROS, DNA and tissue damage associated with chronic inflammation are consistent with MODS, and with redox dysregulation being an underlying cause of life shortening or inflammaging (Meng et al., 2017), which is why even low dose radiation can result in premature aging, frailty and death.

References

Aikawa, N. (1996). Cytokine storm in the pathogenesis of multiple organ dysfunction syndrome associated with surgical insults. *Journal of Japan Surgical Society*, 97771–97777.

Antoch, M. P., Wrobel, M., Kuropatwinski, K. K., et al. (2017). Physiological frailty index (PFI): Quantitative in-life estimate of individual biological age in mice. *Aging (Albany NY), 9*(3), 615–626.

Arora, M., Sun, C. L., Ness, K. K., et al. (2016). Physiologic frailty in nonelderly hematopoietic cell transplantation patients: Results from the bone marrow transplant survivor study. *JAMA Oncology, 2*(10), 1277–1286.

Barbi, J., Pardoll, D., & Pan, F. (2013). Metabolic control of the Treg/Th17 axis. *Immunological Reviews, 252*(1), 52–77.

Beach, T. A., Groves, A. M., Johnston, C. J., et al. (2018). Recurrent DNA damage is associated with persistent injury in progressive radiation-induced pulmonary fibrosis. *International Journal of Radiation Biology*, 1–12.

Brinkmann, V., & Zychlinsky, A. (2007). Beneficial suicide: Why neutrophils die to make NETs. *Nature Reviews. Microbiology, 5*(8), 577–582.

Brush, J. M., Kim, K., Sayre, J. W., et al. (2009). Imaging of radiation effects on cellular 26S proteasome function in situ. *International Journal of Radiation Biology, 85*(6), 483–494.

Chen, X., & Oppenheim, J. J. (2014). Th17 cells and Tregs: Unlikely allies. *Journal of Leukocyte Biology*.

Chiang, C. S., Hong, J. H., Stalder, A., et al. (1997). Delayed molecular responses to brain irradiation. *International Journal of Radiation Biology, 72*(1), 45–53.

Chiang, C. S., Liu, W. C., Jung, S. M., et al. (2005). Compartmental responses after thoracic irradiation of mice: Strain differences. *International Journal of Radiation Oncology, Biology, Physics, 62*(3), 862–871.

Cline, J. M., Dugan, G., Bourland, J. D., et al. (2018). Post-irradiation treatment with a superoxide dismutase mimic, MnTnHex-2-PyP(5+), mitigates radiation injury in the lungs of non-human primates after whole-thorax exposure to ionizing radiation. *Antioxidants (Basel, Switzerland), 7*(3), 40.

Deng, L., Liang, H., Xu, M., et al. (2014). STING-dependent cytosolic DNA sensing promotes radiation-induced type I interferon-dependent antitumor immunity in immunogenic tumors. *Immunity, 41*(5), 843–852.

Denkinger, M. D., Leins, H., Schirmbeck, R., et al. (2015). HSC aging and senescent immune remodeling. *Trends in Immunology, 36*(12), 815–824.

Dodson, M., Darley-Usmar, V., & Zhang, J. (2013). Cellular metabolic and autophagic pathways: Traffic control by redox signaling. *Free Radical Biology & Medicine*, 63207–63221.

Elgart, S. R., Bostani, M., Mok, K. C., et al. (2015). Investigation of DNA damage dose-response kinetics after ionizing radiation schemes similar to CT protocols. *Radiation Research, 183*(6), 701–707.

Franceschi, C., Garagnani, P., Vitale, G., et al. (2017). Inflammaging and 'Garb-aging'. *Trends in Endocrinology & Metabolism, 28*(3), 199–212.

Frank, M. O., & Caceres, B. A. (2015). Inflammaging: A concept analysis. *The Journal for Nurse Practitioners, 11*(2), 258–261.

Fujiwara, S., Suyama, A., Cologne, J. B., et al. (2008). Prevalence of adult-onset multifactorial disease among offspring of atomic bomb survivors. *Radiation Research, 170*(4), 451–457.

Fulop, T., Larbi, A., Dupuis, G., et al. (2018). Immunosenescence and inflamm-aging as two sides of the same coin: Friends or foes? *Frontiers in Immunology*, 8(1960).

Gao, D., Li, T., Li, X.-D., et al. (2015). Activation of cyclic GMP-AMP synthase by self-DNA causes autoimmune diseases. *Proceedings of the National Academy of Sciences, 112*(42), E5699.

Ghandhi, S. A., Turner, H. C., Shuryak, I., et al. (2018). Whole thorax irradiation of non-human primates induces persistent nuclear damage and gene expression changes in peripheral blood cells. *PLoS One, 13*(1), e0191402.

Grant, G. A., Barlow, J. A., & Leach, K. E. (1976). Modification of survival of gamma irradiated mice by adenosine nucleotides. *Strahlentherapie, 152*(3), 285–291.

Groves, A. M., Johnston, C. J., Williams, J. P., et al. (2018). Role of infiltrating monocytes in the development of radiation-induced pulmonary fibrosis. *Radiation Research, 189*(3), 300–311.

Harding, S. M., Benci, J. L., Irianto, J., et al. (2017). Mitotic progression following DNA damage enables pattern recognition within micronuclei. *Nature, 548466.*

Hernandez, L., Terradas, M., Camps, J., et al. (2015). Aging and radiation: Bad companions. *Aging Cell, 14*(2), 153–161.

Hodgkin, P. D., Heath, W. R., & Baxter, A. G. (2007). The clonal selection theory: 50 years since the revolution. *Nature Immunology, 8*(10), 1019–1026.

Hong, J., Chiang, C., Campbell, I. L., et al. (1996). Induction of acute phase gene expression by brain irradiation. *International Journal of Radiation Oncology Biology Physics*, 33619–33626.

Hong, J.-H., Chiang, C.-S., Tsao, C.-Y., et al. (1999). Immediate-acute responses of cytokine gene expression in lung following thoracic irradiation. *Proceedings of the American Association for Cancer Research Annual Meeting, 40422.*

Kajimura, J., Lynch, H. E., Geyer, S., et al. (2017). Radiation- and age-associated changes in peripheral blood dendritic cell populations among aging atomic bomb survivors in Japan. *Radiation Research, 189*(1), 84–94.

Kallman, R. F., & Kohn, H. I. (1958). Life shortening by whole- and partial-body x-irradiation in mice. *Science, 128*(3319), 301–302.

Kim, K., Brush, J. M., Iwamoto, K. S., et al. (2008). EGFRvIII expression in U87 glioblastoma cells alters their proteasome composition, function, and response to irradiation. *Molecular Cancer Research, 6*(3), 426–434.

Kim, K., & McBride, W. H. (2010). Modifying radiation damage. *Current Drug Targets, 11*(11), 1352–1365.

Kimura, A., & Kishimoto, T. (2010). IL-6: Regulator of Treg/Th17 balance. *European Journal of Immunology, 40*(7), 1830–1835.

Kojima, S., Ohshima, Y., Nakatsukasa, H., et al. (2017). Role of ATP as a key signaling molecule mediating radiation-induced biological effects. *Dose-Response, 15*(1), 1559325817690638.

Kusunoki, Y., & Hayashi, T. (2008). Long-lasting alterations of the immune system by ionizing radiation exposure: Implications for disease development among atomic bomb survivors. *International Journal of Radiation Biology, 84*(1), 1–14

Kusunoki, Y., Yamaoka, M., Kubo, Y., et al. (2010). T-cell immunosenescence and inflammatory response in atomic bomb survivors. *Radiation Research, 174*(6), 870–876.

Lee, D. W., Gardner, R., Porter, D. L., et al. (2014). Current concepts in the diagnosis and management of cytokine release syndrome. *Blood, 124*(2), 188–195.

Lee, P. Y., Li, Y., Kumagai, Y., et al. (2009). Type I interferon modulates monocyte recruitment and maturation in chronic inflammation. *The American Journal of Pathology, 175*(5), 2023–2033.

Lehnert, B. E., & Iyer, R. (2002). Exposure to low-level chemicals and ionizing radiation: Reactive oxygen species and cellular pathways. *Human & Experimental Toxicology, 21*(2), 65–69.

Lowe, D., & Raj, K. (2014). Premature aging induced by radiation exhibits pro-atherosclerotic effects mediated by epigenetic activation of CD44 expression. *Aging Cell, 13*(5), 900–910.

Mackenzie, K. J., Carroll, P., Martin, C. A., et al. (2017). cGAS surveillance of micronuclei links genome instability to innate immunity. *Nature.*

Mancuso, M., Pasquali, E., Leonardi, S., et al. (2011). Role of connexin43 and ATP in long-range bystander radiation damage and oncogenesis in vivo. *Oncogene, 30*(45), 4601–4608.

Marples, B., Wouters, B. G., Collis, S. J., et al. (2004). Low-dose hyper-radiosensitivity: A consequence of ineffective cell cycle arrest of radiation-damaged G2-phase cells. *Radiation Research, 161*(3), 247–255.

McBride, W. H., Chiang, C. S., Olson, J. L., et al. (2004). A sense of danger from radiation. *Radiation Research, 162*(1), 1–19.

McBride, W. H., Ganapathy, E., Lee, M. H., et al. (2017). A perspective on the impact of radiation therapy on the immune rheostat. *The British Journal of Radiology, 90*(1078), 20170272.

McDonald, J. T., Kim, K., Norris, A. J., et al. (2010). Ionizing radiation activates the Nrf2 antioxidant response. *Cancer Research, 70*(21), 8886–8895.

McLean, A. R., & Michie, C. A. (1995). In vivo estimates of division and death rates of human T lymphocytes. *Proceedings of the National Academy of Sciences of the United States of America, 92*(9), 3707–3711.

Medzhitov, R. (2013). Pattern recognition theory and the launch of modern innate immunity. *Journal of Immunology, 191*(9), 4473–4474.

Menendez, D., Shatz, M., Azzam, K., et al. (2011). The toll-like receptor gene family is integrated into human DNA damage and p53 networks. *PLoS Genetics, 7*(3), e1001360.

Meng, J., Lv, Z., Qiao, X., et al. (2017). The decay of Redox-stress Response Capacity is a substantive characteristic of aging: Revising the redox theory of aging. *Redox Biology*, 11365–11374.

Mole, R. H. (1959). Life shortening by multiple doses of irradiation. *Gerontologia*, 3159–3160.

Mosmann, T. R., & Sad, S. (1996). The expanding universe of T-cell subsets: Th1, Th2 and more. *Immunology Today*, 17138–17146.

Mottram, P. (1933). A reaction in the skin occurring during the latent period following X-radiation. *Nature, 132317.*

Nefic, H., & Handzic, I. (2013). The effect of age, sex, and lifestyle factors on micronucleus frequency in peripheral blood lymphocytes of the Bosnian population. *Mutation Research, 753*(1), 1–11.

Norman, A., Cochran, S., Bass, D., et al. (1984). Effects of age, sex and diagnostic X-rays on chromosome damage. *International Journal of Radiation Biology and Related Studies in Physics, Chemistry and Medicine, 46*(3), 317–321.

Ohshima, Y., Tsukimoto, M., Harada, H., et al. (2012). Involvement of connexin43 hemichannel in ATP release after gamma-irradiation. *Journal of Radiation Research, 53*(4), 551–557.

O'Neill, L. A., Golenbock, D., & Bowie, A. G. (2013). The history of Toll-like receptors – redefining innate immunity. *Nature Reviews. Immunology, 13*(6), 453–460.

Pajonk, F., & McBride, W. H. (2001). Ionizing radiation affects 26s proteasome function and associated molecular responses, even at low doses. *Radiotherapy and Oncology, 59*(2), 203–212.

Pervan, M., Iwamoto, K. S., & McBride, W. H. (2005). Proteasome structures affected by ionizing radiation. *Molecular Cancer Research, 3*(7), 381–390.

Pervan, M., Pajonk, F., Sun, J. R., et al. (2001). Molecular pathways that modify tumor radiation response. *American Journal of Clinical Oncology, 24*(5), 481–485.

Pospisil, M., Hofer, M., Znojil, V., et al. (1995). Radioprotection of mouse hemopoiesis by dipyridamole and adenosine monophosphate in fractionated treatment. *Radiation Research, 142*(1), 16–22.

Preston, D. L., Shimizu, Y., Pierce, D. A., et al. (2003). Studies of mortality of atomic bomb survivors. Report 13: Solid cancer and noncancer disease mortality: 1950–1997. *Radiation Research, 160*(4), 381–407.

Puck, T. T., & Marcus, P. I. (1956). Action of x-rays on mammalian cells. *The Journal of Experimental Medicine, 103*(5), 653–666.

Purbey, P. K., Scumpia, P. O., Kim, P. J., et al. (2017). Defined sensing mechanisms and signaling pathways contribute to the global inflammatory gene expression output elicited by ionizing radiation. *Immunity, 47*(3), 421–34 e3.

Ratikan, J. A., Micewicz, E. D., Xie, M. W., et al. (2015). Radiation takes its Toll. *Cancer Letters, 368*(2), 238–245.

Rockwell, C. E., Zhang, M., Fields, P. E., et al. (2012). Th2 skewing by activation of Nrf2 in CD4(+) T cells. *Journal of Immunology, 188*(4), 1630–1637.

Rodier, F., Munoz, D. P., Teachenor, R., et al. (2011). DNA-SCARS: Distinct nuclear structures that sustain damage-induced senescence growth arrest and inflammatory cytokine secretion. *Journal of Cell Science, 124*(Pt 1), 68–81.

Rodrigues-Moreira, S., Moreno, S. G., Ghinatti, G., et al. (2017). Low-dose irradiation promotes persistent oxidative stress and decreases self-renewal in hematopoietic stem cells. *Cell Reports, 20*(13), 3199–3211.

Rothkamm, K., & Lobrich, M. (2003). Evidence for a lack of DNA double-strand break repair in human cells exposed to very low x-ray doses. *Proceedings of the National Academy of Sciences of the United States of America, 100*(9), 5057–5062.

Rubin, P., Johnston, C. J., Williams, J. P., et al. (1995). A perpetual cascade of cytokines postirradiation leads to pulmonary fibrosis. *International Journal of Radiation Oncology, Biology, Physics, 33*(1), 99–109.

Schaue, D., Kachikwu, E. L., & McBride, W. H. (2012a). Cytokines in radiobiological responses: A review. *Radiation Research, 178*(6), 505–523.

Schaue, D., & McBride, W. H. (2015). Opportunities and challenges of radiotherapy for treating cancer. *Nature Reviews. Clinical Oncology, 12*(9), 527–540.

Schaue, D., & McBride, W. H. (2012). T lymphocytes and normal tissue responses to radiation. *Frontiers in Oncology, 2119.*

Schaue, D., Micewicz, E. D., Ratikan, J. A., et al. (2015). Radiation and inflammation. *Seminars in Radiation Oncology, 25*(1), 4–10.

Schaue, D., Ratikan, J. A., & Iwamoto, K. S. (2012b). Cellular autofluorescence following ionizing radiation. *PLoS One, 7*(2), e32062.

Seong, S. Y., & Matzinger, P. (2004). Hydrophobicity: An ancient damage-associated molecular pattern that initiates innate immune responses. *Nature Reviews. Immunology, 4*(6), 469–478.

Shimizu, Y., Kodama, K., Nishi, N., et al. (2010). Radiation exposure and circulatory disease risk: Hiroshima and Nagasaki atomic bomb survivor data, 1950–2003. *BMJ*, 340b5349.

Shimura, N., & Kojima, S. (2014). Effects of low-dose-gamma rays on the immune system of different animal models of disease. *Dose-response: A Publication of International Hormesis Society, 12*(3), 429–465.

Shohan, J. (1916). Some theoretical considerations on the present status of roentgen therapy. *The Boston Medical and Surgical Journal*, 175321–175327.

Siddiqui, M. S., Francois, M., Fenech, M. F., et al. (2015). Persistent gammaH2AX: A promising molecular marker of DNA damage and aging. *Mutation Research, Reviews in Mutation Research*, 7661–7619.

Sitkovsky, M., Lukashev, D., Deaglio, S., et al. (2008). Adenosine A2A receptor antagonists: Blockade of adenosinergic effects and T regulatory cells. *British Journal of Pharmacology, 153*(Suppl), 1S457–1S464.

Smith, J. T., Willey, N. J., & Hancock, J. T. (2012). Low dose ionizing radiation produces too few reactive oxygen species to directly affect antioxidant concentrations in cells. *Biology Letters, 8*(4), 594–597.

Smith, K. A. (2012b). Toward a molecular understanding of adaptive immunity: A chronology – Part II. *Frontiers in Immunology, 3364.*

Smith, K. A. (2012a). Toward a molecular understanding of adaptive immunity: A chronology, part I. *Frontiers in Immunology, 3369.*

Spektor, A., Umbreit, N. T., & Pellman, D. (2017). Cell biology: When your own chromosomes act like foreign DNA. *Current Biology, 27*(22), R1228–R1R31.

Spitz, D. R., & Hauer-Jensen, M. (2014). Ionizing radiation-induced responses: Where free radical chemistry meets redox biology and medicine. *Antioxidants & Redox Signaling, 20*(9), 1407–1409.

Taliaferro, W. H., Taliaferro, L. G., & Jaroslow, B. N. (1964). In F. Fremont-Smith (Ed.), *Radiation and immune mechanism* (Monograph series on radiation biology). Academic Press.

Travis, E. L., Peters, L. J., McNeill, J., et al. (1985). Effect of dose-rate on total body irradiation: Lethality and pathologic findings. *Radiotherapy and Oncology, 4*(4), 341–351.

Tsukimoto, M., Tamaishi, N., Homma, T., et al. (2010). Low-dose gamma-ray irradiation induces translocation of Nrf2 into nuclear in mouse macrophage RAW264.7 cells. *Journal of Radiation Research, 51*(3), 349–353.

Udroiu, I., Antoccia, A., & Sgura, A. (2017). Long-term genotoxic effects in the hematopoietic system of prenatally X-irradiated mice. *International Journal of Radiation Biology, 93*(3), 261–269.

Vanpouille-Box, C., Formenti, S. C., & Demaria, S. (2017). TREX1 dictates the immune fate of irradiated cancer cells. *Oncoimmunology, 6*(9), e1339857.

Waite, J. C., & Skokos, D. (2012). Th17 response and inflammatory autoimmune diseases. *International Journal of Inflammation*, 2012819467.

Werner, E., Wang, Y., & Doetsch, P. W. (2017). A single exposure to low- or high-LET radiation induces persistent genomic damage in mouse epithelial cells in vitro and in lung tissue. *Radiation Research, 188*(4), 373–380.

Westbrook, A. M., Wei, B., Hacke, K., et al. (2012). The role of tumour necrosis factor-alpha and tumour necrosis factor receptor signalling in inflammation-associated systemic genotoxicity. *Mutagenesis, 27*(1), 77–86.

Williams, J. P., & McBride, W. H. (2011). After the bomb drops: A new look at radiation-induced multiple organ dysfunction syndrome (MODS). *International Journal of Radiation Biology, 87*(8), 851–868.

Withers, H. R., & McBride, W. H. (1998). Biologic basis of radiation therapy. In C. A. Perez & L. W. Brady (Eds.), *Principles and practiee of radiation oneology* (pp. 79–118).

Wojda, A., Zietkiewicz, E., & Witt, M. (2007). Effects of age and gender on micronucleus and chromosome nondisjunction frequencies in centenarians and younger subjects. *Mutagenesis, 22*(3), 195–200.

Yang, H., Wang, H., Ren, J., et al. (2017). cGAS is essential for cellular senescence. *Proceedings of the National Academy of Sciences of the United States of America, 114*(23), E4612–E4E20.

Yuhas, J. M. (1969). The dose response curve for radiation-induced life shortening. *Journal of Gerontology, 24*(4), 451–456.

Zenkov, N. K., Kozhin, P. M., Chechushkov, A. V., et al. (2017). Mazes of Nrf2 regulation. *Biochemistry (Moscow), 82*(5), 556–564.

Chapter 7
Learning from NATO Biomarker Research for Humans

A. Giovanetti, R. Marconi, C. Bartoleschi, M. C. Pardini, R. Turchi, N. Awad, G. Sanguineti, P. Pinnarò, F. Tortolici, S. Rufini, M. Canfora, and L. Strigari

Abstract Early knowledge of the absorbed dose is of paramount importance in radiation emergencies in order to focus the available resources only on those subjects who need urgent medical treatment. Several biomarkers have been established for measuring dose, the count of dicentrics or micronuclei are the main utilized techniques but give results only from 2 to 3 days after. It is therefore highly needed a tool to sort subjects un-exposed or exposed to very low doses, from those patients likely to develop health consequences, requiring medical evaluation and intervention. With the aim of validating a novel tool of early biomarkers for assessing the absorbed range of dose early after the exposure to radiation recently NATO SPS has

A. Giovanetti (✉) · C. Bartoleschi · M. C. Pardini
ENEA, Division of Health Protection Technologies, Rome, Italy
e-mail: anna.giovanetti@enea.it

R. Marconi
Scientific Direction, National Institute for Infectious Diseases "Lazzaro Spallanzani" IRCCS, Rome, Italy

R. Turchi · F. Tortolici · S. Rufini
Department Biology, University of Rome Tor Vergata, via della Ricerca Scientifica, Rome, Italy

N. Awad
Epidemiology Department, High Institute of Public Health, Alexandria University, Alexandria, Egypt

G. Sanguineti · P. Pinnarò
Departments of Radiation Oncology, IRCCS Regina Elena National Cancer Institute, Rome, Italy

M. Canfora
Clinical Trial Center, Biostatistics and Bioinformatics, IRCCS Regina Elena National Cancer Institute, Rome, Italy

L. Strigari
Scientific Direction, National Institute for Infectious Diseases "Lazzaro Spallanzani" IRCCS, Rome, Italy

Medical Physics Unit, "S. Orsola-Malpighi" Hospital, Bologna, Italy

M. D. Wood et al. (eds.), *Biomarkers of Radiation in the Environment*, NATO Science for Peace and Security Series A: Chemistry and Biology,
https://doi.org/10.1007/978-94-024-2101-9_7

funded an Italian-Egyptian collaborative project, still running. The selected biomarkers include blood count, radio-inducible proteins, Comet assay, H2AX foci and for validation also the micronuclei count. Patients recruited in Rome and in Alexandria are sampled before and 3 h after the first exposure to 250; 340 or > 500 cGy of X rays for therapeutic purposes. Recently in the frame of a bilateral collaborative research call, launched by the Italian Ministry of Foreign Affairs and International Cooperation and the Egyptian STD we have received a further funding for integrating the NATO SPS project, in particular for what concerns increasing training, the addition of a further sampling time points (24 h after RT), the analysis of immune-response and the characterization of the circulating exosomes as potential biomarkers. Preliminary results are presented.

Keywords Radiation dose · Comet assay · Micronuclei · Exosomes · Dosimetry

7.1 Biological Dosimetry for Radiation-Exposure in Humans

In radiation emergencies early knowledge about the absorbed dose is of paramount importance for prompt intervention and risk assessment. Physical dosimeters are not likely to be available thus dose is quantified by using biological dosimetry. Biomarkers do not measure exposure in real time but the changes induced by radiation on selected biological parameters. Biomarkers may assess the dose of radiation exposure or its effect. In case of exposure to high doses the severity of the effects increases with the dose as for acute radiation syndrome clinic (ARS), characterized by skin lesions and hematopoietic, gastrointestinal, and cerebrovascular systems damages: deterministic effects. For low doses is the probability of occurrence, not the severity, which increases with the dose, the effects are defined as stochastic and the biomarkers used for measuring the dose not necessarily imply a clear detriment of health. However an increase in the frequency of these indicators was often associated with an increased risk of radiation-induced cancer and may be indicative of radio-sensitivity to take into account in occupational and medical situation.

A suitable biomarker must have the following characteristics:

- Should be measured on tissues or fluids easily obtainable;
- Changes must be specific of radiation;
- Response should vary depending on the dose and measures also chronic or repeated exposure;
- Should it be possible to measure retrospectively exposure also after years;
- The technical procedures must be simple, fast or automated.

The most validated and utilized biodosimetry techniques for individual dose assessment are the count of dicentric chromosomes (the "gold standard"), and the micronucleus assay. Other methods used to measure human exposure to ionizing radiation are blood count, Comet Assay, pH2AX foci, FISH technique, measurement of radio-inducible proteins, gene expression.

In the last years the field of biodosimetry has advanced significantly beyond this original initiative, with expansion into the fields of genomics, proteomics, metabolomics and transcriptomics. In addition traditional cytogenetic assessment methods have been greatly strengthened through the use of automated platforms and the development of laboratory surge capacity networks. The major part of biomarkers are usually measured in blood, plasma, urine or saliva, sample types that are readily obtainable in a field setting, although some markers have also been explored in samples taken from both tooth enamel and nail and hair clippings.

7.2 A Panel of Biomarkers as Novel Tool for Early Detection of Radiation-Exposure (G4815 NATO SPS-Funded Project)

7.2.1 Introduction

A radiological or nuclear (R/N) emergency may cause dozens or hundreds of mass casualties. As previously reported the medical facilities in charge would likely be overwhelmed by number of casualties too large to manage. It is therefore highly needed a tool to sort the unaffected, or "worried-well" subjects, that can be sent home at the moment, from those patients likely to develop health consequences, requiring medical evaluation and intervention. The early assessment of absorbed dose may also aid in predicting the severity of later health outcomes and to put in place early and effective medical countermeasures and treatments. Early screening may also predict the likelihood of stochastic long-term effects (as cancer) in order to undertake countermeasures, such as antioxidant treatments, proven to minimize radio-induced genomic instability, chemoprevention and other appropriate risk mitigation strategies; and to inform long term planning for cancer health services based on expected need. Currently, to measure the absorbed dose the reference methods are the dicentrics or MN count in peripheral lymphocytes. They conform to several parameters but results become available after at least 50 h and 74 h, making hard in this way an appropriate management of the first phase of the R/N emergency. Due to the urgency of triage decisions new tools for sorting, depending on dose, subjects exposed to radiation within the first hours after exposure are strongly needed. Biomarkers capable of measuring radiation exposure in a short time are currently available but used alone are not considered reliable because they lack specificity or sensitivity.

With the aim of validating biomarkers for assessing the absorbed radiation dose soon after exposure, at the ENEA Casaccia Center (IT) in collaboration with the clinical IRE-IFO National Cancer Institute (IT) and with the HIPH of Alexandria

University (ET) we are carrying out the NATO SPS funded research project titled: "*A Panel of Biomarkers as Novel Tool for Early Detection of Radiation Exposure*".

The selected biomarkers provide results in a few hours, a dose-response relationship was reported and the experimental protocols do not need very experienced staff and complex and expensive equipment. They could hence be used in structures or in countries, which do not have the possibility to invest considerable resources for R/N emergencies. These biomarkers are based on different biological mechanisms, thus we expect, integrating all the results, to achieve, the same degree of specificity and sensitivity of the established cytogenetic assays.

Other objectives are the training of Egyptian young scientists, the establishment of the cell culture and cell biology laboratories in Alexandria, the building of a web platform with an open accessible e-training space with protocols, videos, articles and a restricted areas with the patients' data and results.

The utilized biomarkers include:

Blood Count The lympho-hematopoietic elements are among the most highly replicated tissues in mammals and as such are among the most radiosensitive, as consequence the peripheral blood count may be used as a biological indicator of damage. The "normal" range for absolute lymphocyte count can vary, even in a healthy adult population. Samples collected from small number of "healthy" workers at AFRRI suggest a "normal" range from about $1.35–3.5 \times 10^9$ cells/liter. The kinetics of the lymphocyte response to radiation is likely to be at least as important as the absolute lymphocyte count.

Blood count is a good indicator of the hematopoietic syndrome (absorbed dose 2–3 Gy and 8 Gy), while mitotically active hematopoietic progenitors are unable to divide after a whole-body exposure >2–3 Gy, which results in a hematologic crisis in the ensuing weeks. At lower doses (<2 Gy) radiation induces mild cytopenias without significant bone marrow damage. Peripheral blood lymphopenia may develop within the first 6–24 h after a moderate – to high-dose exposure (López & Martín, 2011).

Oxidative Stress The short and long-term effects induced by exposure to ionizing radiation have been demonstrated to be mostly caused by the alteration of the oxidative status and free radical damage/oxidative stress is a very early event leading to DNA damage and other genetic disorders (Islam, 2017) and oxidative stress due to increased free radicals and reactive oxygen metabolite production can be used as diagnostic tool for quantify the X dose.

Cytogenetic Techniques Cytogenetic analysis has been the most utilized biological methods adopted for dosimetry purposes. As proposed by Hutchinson (1966) the main target for mutagens, in particular ionizing radiation, is DNA which, along with proteins, constitutes the chromosomes, localized in the nucleus of every eukaryotic cell. During mitosis or meiosis, the chromosomes condense and become identifiable. In human have been identified as 23 pairs chromosomes. Initially, the pairs of chromosomes were classified according to size and shape, later on the use of band-ing techniques and probes allowed to more accurately identify individual chromosomes and chromosomal regions (Terzoudi & Pantelias, 2006).

Micronuclei Count Micronuclei are small round bodies present in cytoplasm outside the main nucleus arising from acentric fragments that fail, because of the lack of a centromere, to incorporate into the daughter nuclei during cell division. The problem over the fate of micronuclei in cells that have divided more than once, was solved by Fenech and Morley (1985) blocking cytokinesis after 44 h of culturing by adding cytochalasin B, which results in the formation of binucleated cells. Micronuclei are then scored only in these cells. In vitro studies on radiation-induced micronuclei revealed dose-response relationship, which indicates that micronuclei can be utilized as biological indicator of radiation exposures. Specific application for radiation dosimetry is reported in the IAEA report (, 2011). Compared to dicentrics count, analysis of micronuclei does not require highly trained personnel and is much faster but culture time is 72 h. This permits an easy analysis of many thousands of cells in a very short time.

Detection of Phosphorylated H2AX foci Thanks to the advances in molecular biology, new methods to measure exposure to radiation have been developed not requiring cell culturing and stimulation, that can be applied to all cell types at any stage of the cycle as the count of phosphorylated histones foci (pH2AX). One of the earliest steps in the cellular response to DNA double strand breaks (DSBs) is the phosphorylation of serine 139 of H2AX, a subclass of eukaryotic histone proteins that are part of the nucleoprotein structure called chromatin. Discrete nuclear foci, either induced by exogenous agents such as IR or generated endogenously during programmed DNA rearrangements, can be visualized at sites of DSBs by using a fluorescent antibody specific for the phosphorylated form of pH2AX. Their detection and quantification can be achieved by light microscopy, flow cytometry or western blotting. The count of H2AX foci is an early and sensitive biodosimeter to be used in case of emergencies (Rothkamm et al., 2015). However it is not stable during the time because their number rapidly decreased.

Comet Assay The Comet Assay or Single Cell Electrophoresis is an early, sensitive, and fast method to measure DNA damage in individual cells (Tice & Strauss, 1995). This assay is based on the ability of negatively charged fragments of DNA to be drawn through an agarose gel in response to an electric field. Cells, embedded in agarose are submitted to lysis and the to electrophoresis. DNA moves towards the anode and, if present, fragments migrate faster forming the tail of the comet. The relative amount of DNA present in the Tail depends directly on the degree of DNA fragmentation (Giovanetti et al., 2008).

Unlike the cytogenetic test described above, for Comet Assay cells don't have to be cultured, the test can be applied at any stage of the cell cycle and to all cell types. With this technique is possible to measure the breakage of single (SSB) -or double (DSB)-stranded DNA, alkali-labile sites and apoptotic index.

Protein Biomarkers Organisms respond to irradiation by altering the expression and/or the post-translational modifications of some proteins in cells, tissues and/or organic fluids, as serum or urine and protein expression profiling can be used to measure radiation exposure or to differentiate between detrimental and harmless upcoming injuries (Guipaud & Benderitter, 2009). Proteins are easily obtained using non-invasive (urine) or semi-invasive (blood) collection methods. Their quantifications using immune-detection techniques in biological fluids (urine, serum), cells (circulating lymphocytes) or tissue (biopsies) are fast and reliable. Two proteins have been proposed as bioindicators for radiation exposure effects: the amylase, an indicator of radiation damage to the parotid gland, and the Flt3-ligand (Flt3-L), an indicator of bone marrow damage, while the amino acid citrulline has been used as a physiologic marker for epithelial radiation induced small bowel damage.

The rise of serum amylase that results from the irradiation of salivary tissue reflects death of serous salivary cells and has also been proposed as biochemical indicators of salivary gland injury during iodine-131 therapy of patients and for exposure to cosmic radiation during prolonged space travel (Becciolini et al., 2001). Elevations in serum amylase activity must be measured early, i.e. 0.5 to 2 days, after suspected radiation exposure to serve as a biochemical indicator triage tool for identifying individuals with potentially severe radiation injury.

Flt3-ligand is a hematopoietic cytokine structurally homologous to the stem cell factor (SCF) and the colony stimulating factor 1 (CSF-1). The Flt3-L concentration is increased in the blood of patients with aplastic anaemia and plasma Flt3-L concentration during the first 5 days after radiation therapy directly correlated with the radiation dose in a non-human primate model. The variations in plasma Flt3-L concentration has been shown to directly reflect the radiation-induced bone marrow damage during fractionated local radiation therapy and can be use also to measure radiation dose in case of emergency (Bertho et al., 2008).

Citrulline is an amino acid specifically produced by enterocytes and its concentration is correlated with the enterocyte mass in some pathological situations such as small bowel disease) and after irradiation). Recently, the citrulline concentration was assessed in patients accidentally irradiated (Bertho et al., 2008).

7.2.2 Experimental Approach

Patients The study is conducted on patients treated with increasing therapeutic doses of X rays, recruited at the oncology medical services of IRE-IFO (Rome, Italy) and HIPH (Alexandria, Egypt). Patients are sampled before and 3 h after radio-treatment. The number of patients was calculated assuming a difference of about 20% in the reference biomarker. In each country was planned to recruit:

50 patients exposed to a single dose of 200–250 cGy of X rays

15 patients exposed to a single dose of 340 cGy of X rays
5 patients exposed to a single dose ≥500 cGy of X rays

In reality at IRE-IFO Hospital, to obtain statistically relevant data the total number of patients enrolled was 80 (51 patients treated with 200–250 cGy; 22 with340 cGy; 7 > 500 cGy).

Sampling and Processing Whole Blood The day of the radio-treatment 12 ml of blood are withdrawn from patients before and 3 h after the first therapeutic treatment with X-rays. Blood is distributed in five vacutainers: 1 with Na-EDTA as anticoagulant for blood count; 1 with Na-EDTA as anticoagulant for exosomes analysis in plasma (see below); 1 with lithium heparin as anticoagulant for Comet assay, micronuclei, plasma lymphocytes collection; 1 with coagulation factor for collecting serum for amylase analysis.

Whole blood for white cells count and serum for amylase are processed in the hospital analysis Unit. Lymphocytes and plasma are immediately separated using Histopaque plasma is separated by centrifuging at 3000 g x 15 min, all plasma are distributed in 300 ul aliquots in cryovials and frozen at −20 °C, transported in the ENEA Casaccia laboratory where are stored at −80 °C; the vacutainer with Li heparin is brought in the ENEA laboratory.

In the ENEA laboratory 100 ul of whole blood are processed for Comet assay and the remaining 1.5 ml are seeded in 3 flasks with 5 ml of completed medium for micronuclei count.

For Lymphocytes and Plasma Separation Whole blood is centrifuged 8 min at 1600 RPM (460 g), plasma is carefully removed, distributed in cryovials (200 μl in each one) and stored at −80 °C, to the pellet it is added the same volume of the initial blood (3–3.5 ml) of PBS 1X/1 mM EDTA (pH 7.4), and the same volume of Histopaque and the total solution is centrifuged 30 min at 1600 RPM (460 g). Lymphocytes, present as cellular ring between PBS-EDTA and Hystopaque are removed using a 2 ml syringe with normal needle, then resuspended in 4 ml PBS1X/1 mM EDTA and centrifuged for 10 min at 1600 RPM (460 g). Surnatant is removed and resuspended in 1 ml Ethanol 70% (30% distilled water) maintained at 4 °C. Then the fixed cells are transferred in 2 cryovials, hold for 10 min at 4 °C and stored at −20 °C for the further H2AX foci analysis.

For exosomes purification. Microvesicles are isolated from blood plasma by ultracentrifugation and filtration as described (Baranyai et al., 2015). Purity of the preparation has been confirmed by the identification of exosomal markers (i.e. Alix and CD81) and by visualization using Transmission Electron Microscopy. Mean average diameter of the vesicles has been also studied with Dynamic Light Scattering while the apparent z-potential has been determined as previously described (Cerreto et al., 2017).

The Alkaline Comet Assay This analysis is performed as described by Giovanetti et al. (2008). Cells contained in 20 μl of blood, will be embedded in agarose on a microscope slide and lysed with detergent and high salt to form nucleoids containing supercoiled loops of DNA. After lysis and washing the slides will be submitted to electrophoresis at high pH, fragments, if present will migrate faster forming a tail of structures resembling comets; the intensity of the comet tail relative to the head reflects the number of DNA breaks. This will be followed by visual analysis with staining of DNA and calculating fluorescence to determine the extent of DNA damage. Results obtained with visual scoring were compared to those obtained with IAS dedicated software.

Micronuclei Count This test is realized on lymphocytes cultured for 72 h in complete medium supplemented with mitogen and cytochalasin B for obtaining binucleated cells. After fixation and centrifugation the pellet will be dropped an slides and stained with ethidium bromide. For each experimental point, a minimum of 1000 binucleated cells will be analysed on coded slides for MN induction.

Sample Conservation Plasma, serum aliquots are isolated and stored at −80 °C and analysed for oxidative stress, radio-inducible proteins, inflammation markers.

Biostatistics and bio-mathematical approaches have been developed for data analyses to be shared among partners participating to the project.

7.2.3 Preliminary Results

The first results we analyzed on genetic toxicity measured with Comet assay and micronuclei count showed an increase of damage 3 h after the first fraction of radiotherapy when compared with the basal (pre-radiotherapy) values.

Comet Assay As shown in Fig. 7.1, DNA breaks are increasing after radiation; differences are not significant due to the high inter-individual variability in basal values of DNA breaks and in radiation sensitivity.

Micronuclei Also the % of micronuclei increased in all the patients after radiotreatment, this increase is similar in the 250 and 340 cGy groups and higher in the >500 cGy group (Fig. 7.2). Also for the micronuclei a high inter-individual variability regarding the basal level of damage and the response to radiation was observed, and because of this variability the differences between the patient groups are not significant.

The Pearson correlation test (Pearson, 1896) was used to test for a possible relationship between radiation dose and the modification of biochemical and hematological parameters, i.e. for investigating blood related variables and nominal dose to the target and micronuclei number (Figs. 7.3, 7.4, 7.5, 7.6, 7.7, 7.8 and 7.9). In particular, the absolute and relative variations were calculated, as follows: the net variation i.e. the increased absolute number was obtained as the difference of values after the first fraction of radiotherapy and the basal ones (and indicated using the symbol Δ), while the percentage variation was determined as the difference between the value after the first fraction of radiotherapy minus one before, divided by the value before the first fraction of radiotherapy (and indicated using the symbol δ and expressed in percentage).

The net numbers of WBC (whole blood cells) and of eosinophils (ES) decreased when the dose per fraction increased (Figs. 7.3 and 7.5, respectively). The increment of the percentage numbers of MN (Fig. 7.4) and eosinophils (Fig. 7.6) showed a dose-dependent decrease. This is likely due to the increase of radiation-induced damage with the dose per fraction. In most of the analyzed patients the net numbers of MN increase with the WBC (Fig. 7.7), eosinophils (Fig. 7.8) and comet assay (Fig. 7.9) net values 3 h after first radiotherapy fraction.

More in details, a statistical significant correlation has been found between the nominal dose to the target and the increase of micronuclei number from the basal values to the values at 3 h after first radiotherapy fraction (p-value = 0.025) and between the nominal dose to the target and the increase of WBC (p-value = 0.021).

There was a statistical significant correlation between the net variation of MN between the values 3 h after first radiotherapy fraction and the basal ones and the variation of both WBC (p-value = 0.005) and comet assay (p-value = 0.005); while there was a trend between the net variation of MN and the net variation of eosinophils (p-value = 0.087). In addition the percentage numbers of MN and eosinophils showed a dose-response correlation.

Patients dosimetry obtained from treatment planning have been used for the evaluation of biological effective integral dose. Additional dosimetric models are under investigation to take into consideration the blood circulation during the irradiation.

As future prospective the effective dose to the whole body is under calculation and will be correlated with the modification of biochemical and hematological parameters when the results of the biomarkers will be completed.

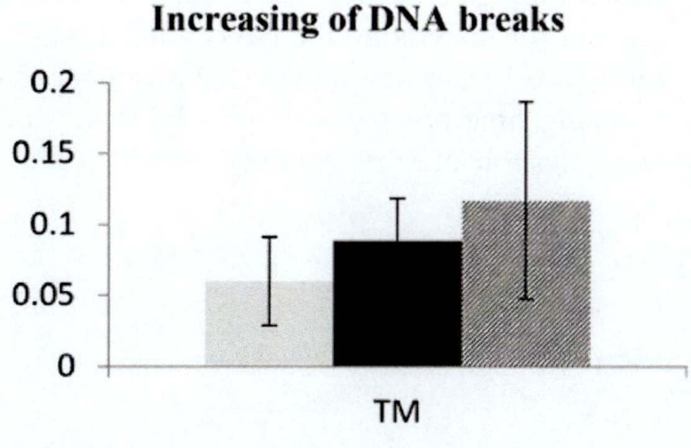

Fig. 7.1 The increasing of DNA breaks measured with IAS software 3 h after radio-treatment in patients treated with 250 cGy (grey) 350 cGy (black) and > 500 cGy (striped)

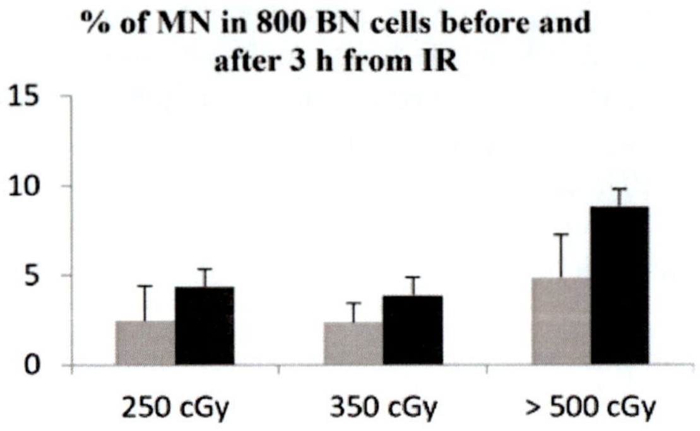

Fig. 7.2 The % of micronuclei (MN) in patients treated with 250, 350 and > 500 cGy, before (grey) and 3 h after the radio-treatment before (black)

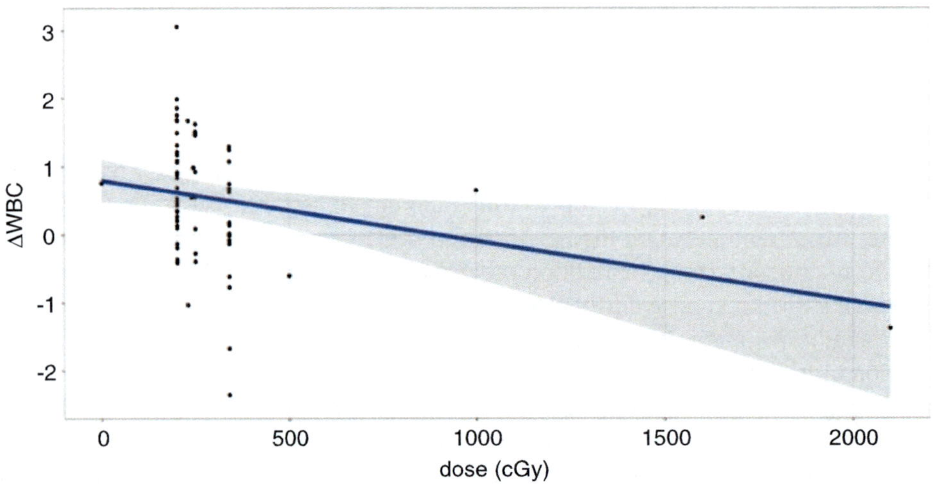

Fig. 7.3 Correlation between the absolute increase of WBC and the nominal prescribed dose $t = -2.3696$, $df = 62$, p-value = 0.021

7.3 Implementation of a Novel Set of Early Biomarkers for Measuring Radiation Dose in Case of Emergency

Recently in the frame of a bilateral collaborative research call, launched by the Italian Foreign Affairs Minister and the STD (Egypt) we have received a further funding for integrating the NATO SPS G4815 project, in particular for what concerns travel expenses, a major involvement of end-users, the addition of a further sampling time points (24 h after RT), and the analysis of immune-response and exosomes role after radiation.

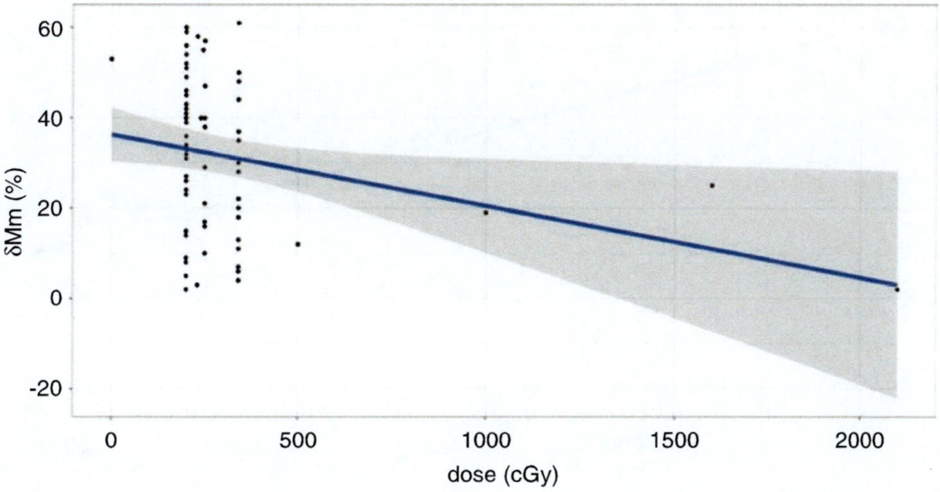

Fig. 7.4 Correlation between the percentage variation of MN breaks and the nominal prescribed dose t = −2.2895, df = 62, p-value = 0.025

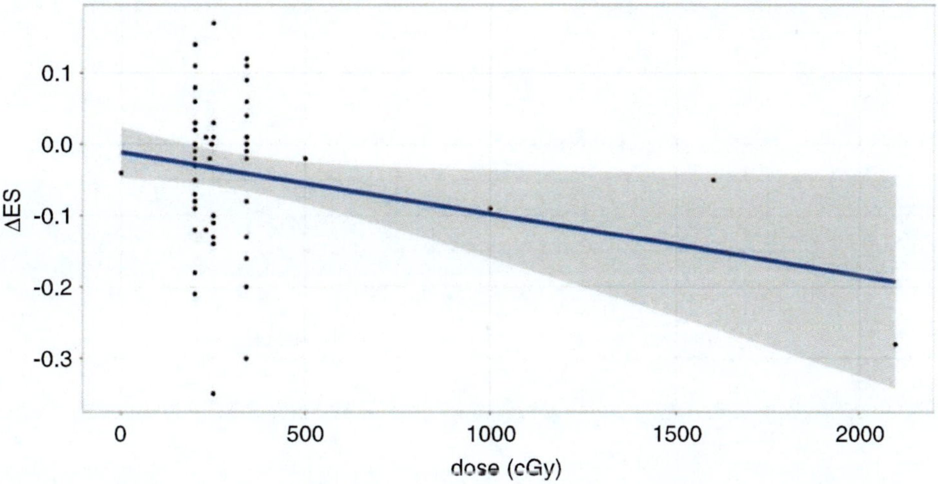

Fig. 7.5 Correlation between the absolute increase of eosinophils (ES) and the nominal prescribed dose t = −2.0951, df = 62, p-value = 0.040

7.3.1 Exosomes as Biomarkers

Exosomes are small, secreted vesicles with a diameter of approximately 40–100 nm, released from cells both constitutively and following stimulation. They represent a sophisticated cellular double-layered product carrying complex signals such as proteins, mRNA, and miRNA. Their unique "packaging" provides both protection of water-soluble cargo and membrane molecular machineries to sites even very far from their origin districts. Although the discovery of these vesicles dates back to

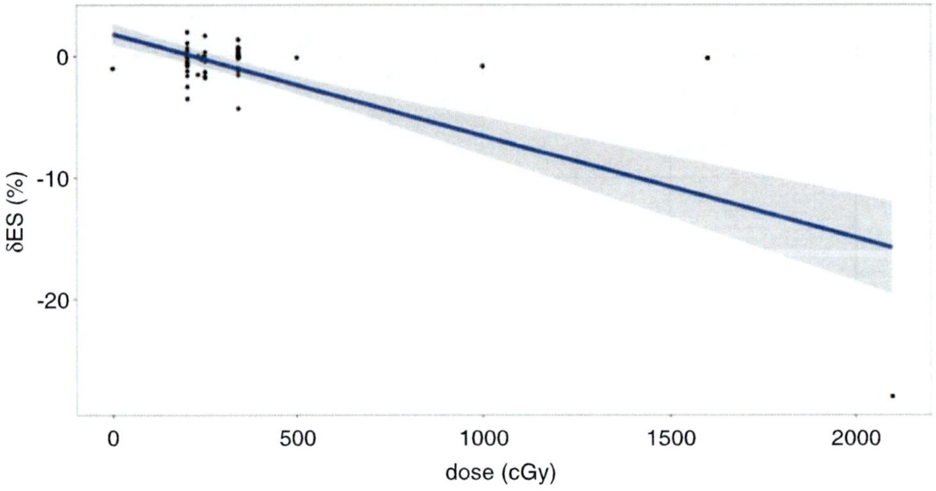

Fig. 7.6 Correlation between the relative variation of eosinophils (ES) and the nominal prescribed dose $t = -7.9631$, $df = 62$, $p\text{-value} < 0.001$

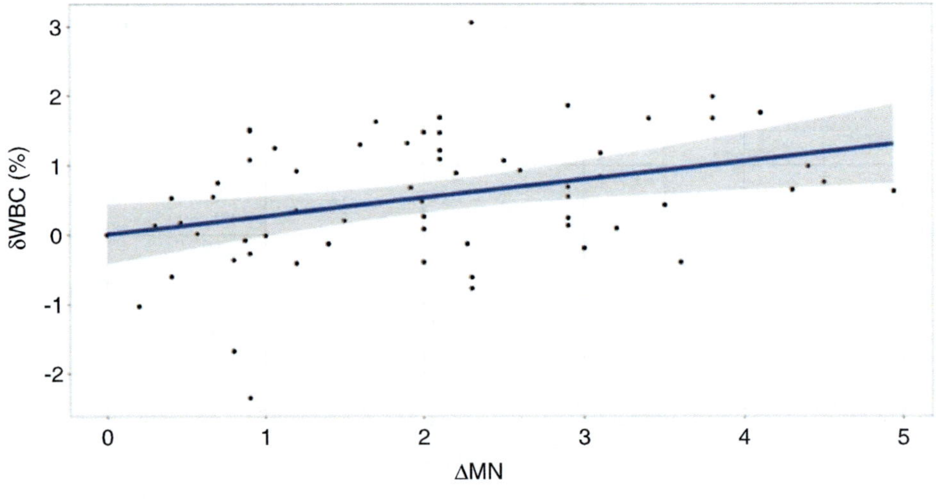

Fig. 7.7 Correlation between the relative increased number of WBC and the absolute increased number of MN breaks $t = 2.8869$, $df = 61$, $p\text{-value} = 0.005$

more than 30 years ago, their physiological function is still not clearly understood. The modification of secreted exosomes is suggested to be associated with unwanted radiotherapy effects. A major challenge is to establish the correlation between physical property and messages carried by micro-vesicles, for a potential use of exosomes as biomarkers for radiation exposure. This research, still ongoing, includes *in vitro* and *ex vivo* studies. As we have previously described, the irradiated SH-SY5Y neuroblastoma cells release exosomes with different physico-chemical characteristics compared with those issued by the non-irradiated ones and able to increase cell

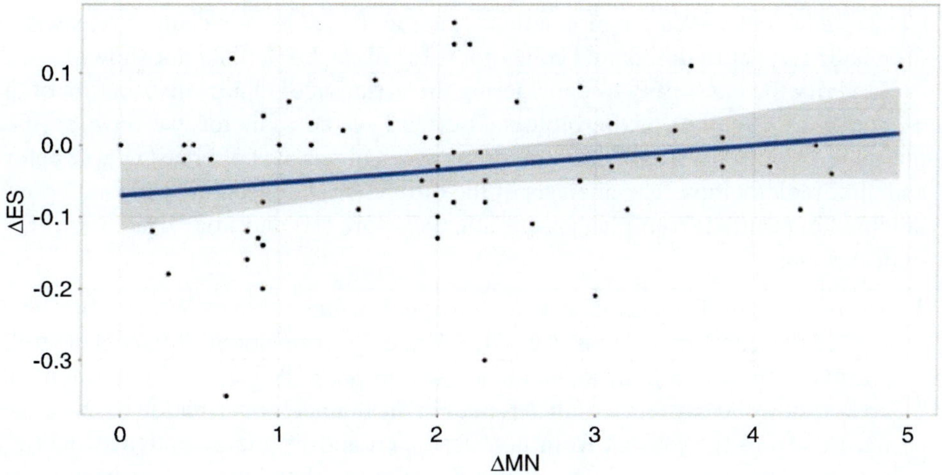

Fig. 7.8 Correlation between the absolute increased number of eosinophils (ES) and of MN breaks t = 1.741, df = 61, p-value = 0.087

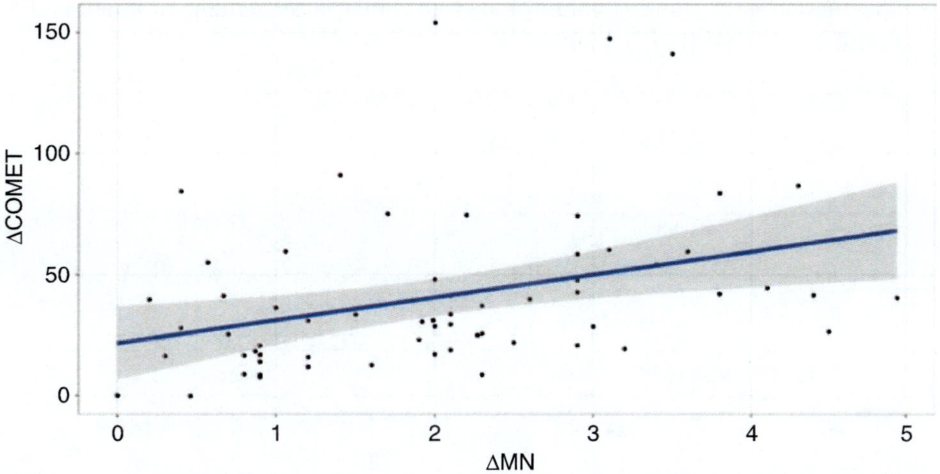

Fig. 7.9 Correlation between the absolute increased number of values determined by Comet assay and of MN breaks t = 2.9322, df = 61, p-value = 0.005

survival (Cerreto et al., 2017). Analysis by Light Scattering (DLS) shows that SH-SY5Y cells release at least two different populations of vesicles sharing the same diameter (Figure 7.10a) but different ζ-potential values i.e. −52 and −24 mV (Figure 7.10b). After cells irradiation (500 cGy), the diameter distribution of vesicles released by the cells do not change respect to the unirradiated (Figure 7.10a'), while the ζ-potential of the more negative population of vesicles increases from −52 to −44 mV (Figure 7.10b'). Furthermore, quantification of two marker proteins of

CD 81 and Alix by Western blot demonstrates an increase of exosomes released by inradiated respect to unradiated cells of a value of 38 ± 5% (data not shown).

To verify the possibility of considering the variation of charge distribution of the exosomes as a marker of the biological effects of the radiation we have purified plasma exosomes from patients undergoing surgical removal of breast cancer before and after radiotherapy. The analysis of the differences between the exosomal populations, i.e. purified from plasma of patients before (E) and after irradiation (E*), indicates that:

1. The quantity of the exosomes circulating per ml, quantified by Western blot analysis of two marker proteins (i.e. CD 81 and Alix), is about 70% higher in the plasma of the subjects irradiated than the unirradiated. The DLS measurements do not show the existence of differences in the diameter distribution of exosomes purified from the plasma from non-irradiated and irradiated patients (data not shown). These data go in the same direction as that observed in the *in vitro* experiments above described.
2. The analysis of the apparent ζ-potential distribution of the exosomes extracted from the plasma of unirradiated patients (E) reveals the presence of a single population with a mean value of −14 mV that, after patient irradiation (E*), shifts a – 10 mV (Fig. 7.11).

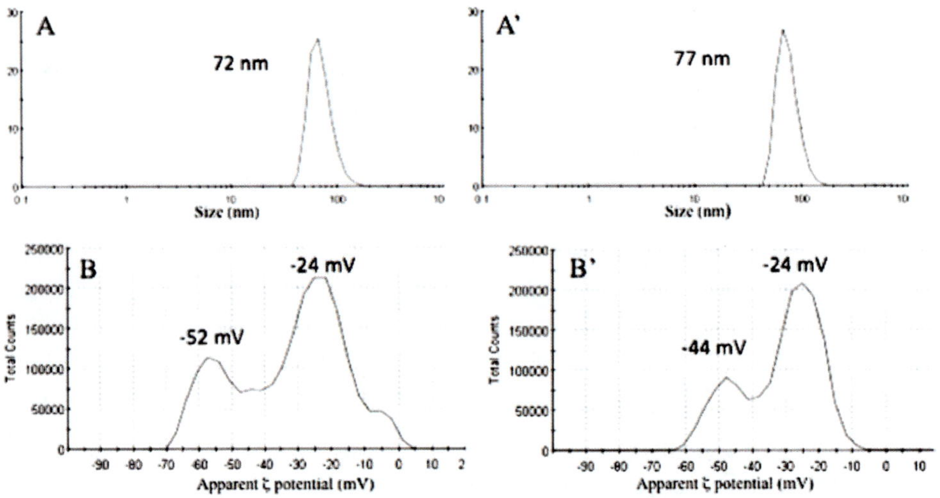

Fig. 7.10 Diameter distribution of exosomes secreted by unirradiated (A) and irradiated (A') SH-SY5Y cells. ζ potential distributions of exosomes secreted by unirradiated (B) and irradiated (B′) SH-SY5Y cells, showing two coexisting groups of charged particles, which mean values are −52 mV and −24.9 mV (B) and −44.4 mV and −24.2 mV (B′)

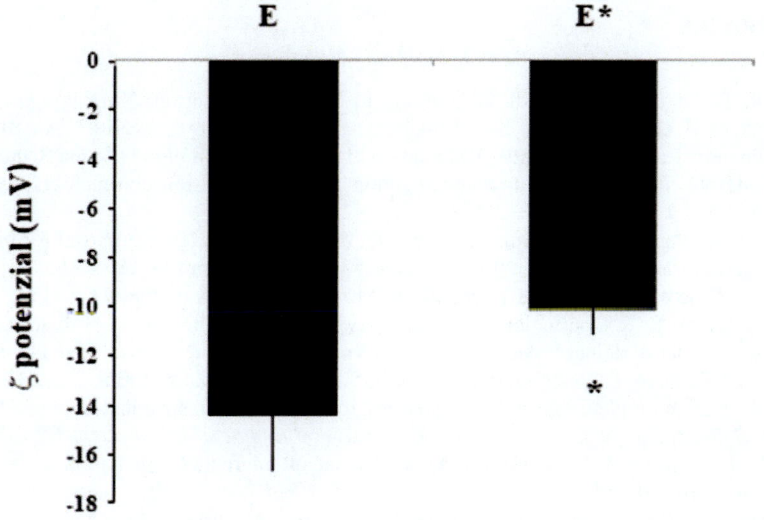

Fig. 7.11 Mean ± SD of ζ-potential values of exosomes purified from the plasma of 8 breast cancer patients subjected to tumor removal, before (E) and after (E *) irradiation with a dose of 2 Gy. The differences are statistically significant ($p < 0.001$)

7.4 What Can Be Transferred from Biomarkers Research on Humans for Dose Assessment in Environmental Radiological Protection?

It is still too early to draw conclusions from the current study while experiments and statistical analyses are still ongoing. The results obtained so far show that genotoxicity tests as Comet assay and MN show a net increase after radiation and could potentially be applied in case of environmental contamination in reference organisms where inter-individual variability is lower comparing to humans. In a previous study utilizing Comet assay, together with other parameters such as mortality, changes in animals' weight, mitochondrial damage assessed by neutral red retention time assay, as well as bioaccumulation of uranium in the earthworm *Eisenia fetida* we demonstrated the higher toxicity of natural uranium comparing to depleted uranium (Giovanetti et al., 2010).

While in reference organisms it is expected to be a lower inter-individual variability than in humans, several models are available to calculate the absorbed dose. This field is still to be developed for reference organisms that include animal and vegetable species at various levels of biological organization. Further research is needed to establish whether and how dosimetry should account for differences in tissue physiology, organism life stages and seasonal variability.

Acknowledgments This research was supported by NATO Science for Peace and Security (SPS) Programme (grant G4815) and by the Italian Ministry of Foreign Affairs and International Cooperation (grant PGR00782).

References

Baranyai, T., Herczeg, K., Onódi, Z., Voszka, I., Módos, K., Marton, N., Nagy, G., Mäger, I., Wood, M. J., El Andaloussi, S., Pálinkás, Z., Kumar, V., Nagy, P., Kittel, Á., Buzás, E. I., Ferdinandy, P., & Giricz, Z. (2015). Isolation of exosomes from blood plasma: Qualitative and quantitative comparison of ultracentrifugation and size exclusion chromatography methods. *PLoS One, 10*, e0145686.

Becciolini, A., Porciani, S., Lanini, A., Balzi, M., & Faraoni, P. (2001). Proposal for biochemical dosimeter for prolonged space flights. *Physica Medica, 17*(Suppl 1), 185–186.

Bertho, J. M., Roy, L., Souidi, M., Benderitter, M., Gueguen, Y., Lataillade, J. J., Prat, M., Fagot, T., De Revel, T., & Gourmelon, P. (2008). New biological indicators to evaluate and monitor radiation-induced damage: An accident case report. *Radiation Research, 169*, 543–550.

Cerreto, M., Sennato, S., Tortolici, F., Casciardi, S., Giovanetti, A., & Rufini, S. (2017). Effect of the irradiation on Neuroblastoma-derived microvesicles: A physical and biological investigation. *Colloids and Surfaces, A: Physicochemical and Engineering Aspects, 532*, 195–202.

Fenech, M., & Morley, A. A. (1985). Measurement of micronuclei in lymphocytes. *Mutation Research, 147*, 29–36.

Hutchinson, F. (1966). The molecular basis for radiation effects on cells. *Cancer Research, 26*, 2045–2052.

Giovanetti, A., Deshpande, T., & Basso, E. (2008). Persistence of genetic damage in mice exposed to low dose of X rays. *International Journal of Radiation Biology, 84*, 227–235.

Giovanetti, A., Fesenko, S., Cozzella, M. L., Asencio, L. D., & Sansone, U. (2010). Bioaccumulation and biological effects in the earthworm Eisenia fetida exposed to natural and depleted uranium. *Journal of Environmental Radioactivity, 101*, 509–516.

Guipaud, O., & Benderitter, M. (2009). Protein biomarkers for radiation exposure: Towards a proteomic approach as a new investigation tool. *Annali dell'Istituto Superiore di Sanità, 45*, 278–286.

IAEA. (2011). *Cytogenetic dosimetry: Applications in preparedness for and response to radiation emergencies* (247 p). EPR Biodosimetry.

Islam, M. T. (2017). Radiation interactions with biological systems. *International Journal of Radiation Biology, 93*, 487–493.

López, M., & Martín, M. (2011). Medical management of the acute radiation syndrome. *Reports of Practical Oncology and Radiotherapy, 16*, 138–146.

Pearson, K. (1896). Mathematical contributions to the theory of evolution. III. Regression, heredity, and panmixia. *Philosophical Transactions of the Royal Society of London, Series A, 187*, 253–318. Reprinted in Karl Pearson's Early Statistical Papers, 113–178. Cambridge: Cambridge University Press, 1948.

Rothkamm, K., Barnard, S., Moquet, J., Ellender, M., Rana, Z., & Burdak-Rothkamm, S. (2015). DNA damage foci: Meaning and significance. *Environmental and Molecular Mutagenesis, 56*, 491–504.

Terzoudi, G. I., & Pantelias, G. E. (2006). Cytogenetic methods for biodosimetry and risk individualisation after exposure to ionising radiation. *Radiation Protection Dosimetry, 122*, 513–520.

Tice, R. R., & Strauss, G. H. (1995). The single cell gel electrophoresis/comet assay: A potential tool for detecting radiation-induced DNA damage in humans. *Stem Cells, 13*(Suppl 1), 207–214.

Part III
Biomarkers for Risk Evaluation

Chapter 8
Exosomes as Radiation Biomarkers

Kishore Kumar Jella

Abstract Exosomes are now considered as important mediators of intercellular communication. The exosome cargo contains proteins; mRNA, microRNA (miRNA) and DNA that delivers the information between one cell to other cells can play an important role in identifying the pathophysiological conditions of any cell. Increased understanding of mechanism about exosome release and its communication could provide a novel strategy for the development of biomarkers in various health conditions. Recent shreds of evidence revealed that radiation induces the secretion and alters the composition of exosomes released from radiated cells. The exosomes released from radiated cells alters the signaling pathways in recipient cells. Unraveling the mechanisms related to radiation and exosomes would shed light on the unknown factors that are involved in radiation-induced non-targeted effects. This can provide pavement for the development of biomarkers in radiation emergency situations.

Keywords Exosomes · Radiation · miRNA · Genomic instability

8.1 Introduction

Exosomes are extracellular vesicles with diverse sizes and shape secreted by both prokaryotic and eukaryotic cells. Exosomes are formed by double-layered lipid membrane with proteins, RNA, DNA, glycans and lipids. The size of exosomes ranges from 30–150 nm in diameter and the buoyant density ranges from 1.13–1.19 g/ml (Kahlert et al., 2014; Kalluri, 2016; Patel et al., 2016; Raposo et al. 2003). Exosomes have been observed in variety of biofluids and were reported in various pathophysiological process (Azmi et al., 2013). Exosomes in the process of circulation within the body interacts with platelets and various cells *in vivo* (Vickers & Remaley, 2012). Exosomes secreted by one kind of cell interacts with other cells

K. K. Jella (✉)
Department of Radiation Oncology, Emory University, Atlanta, USA
e-mail: kjella@emory.edu

M. D. Wood et al. (eds.), *Biomarkers of Radiation in the Environment*, NATO Science for Peace and Security Series A: Chemistry and Biology,
https://doi.org/10.1007/978-94-024-2101-9_8

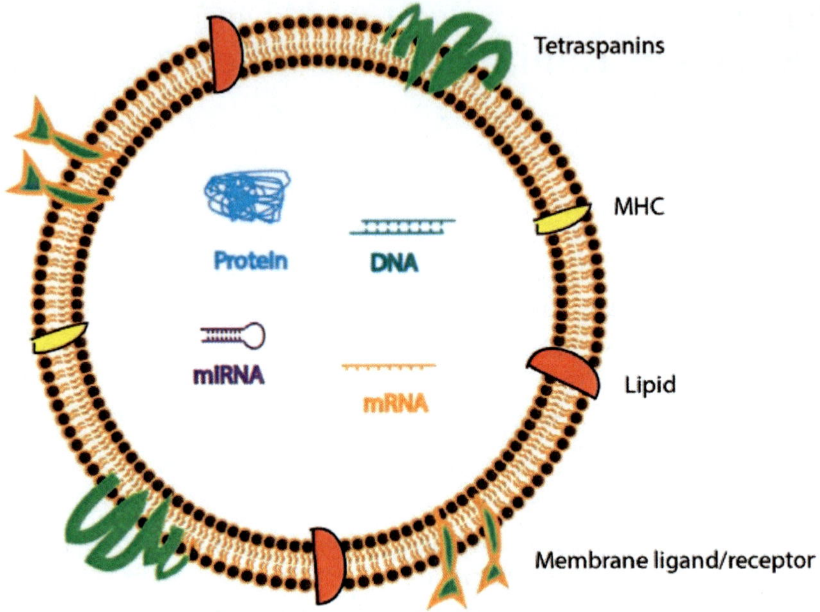

Fig. 8.1 Representative image of basic structure of exosome

and play a significant role in the development of diseases (Gong et al., 2012; Jaiswal et al., 2012). The basic structure of exosome was provided in Fig. 8.1.

Exosomes are double-layered membrane vesicles, the released exosomal lipid layer matches with its parental cell. The lipid layer of exosomes play an important role in cell-to-cell communication and it involves in the regulation of various signaling pathways that play an important role in various diseases. According to Exocarta database (Version 3.1; http://exocarta.org), till date, 41,860 proteins, 4986 mRNAs, 2838 microRNAs (miRNAs) and 1116 lipids have been reported from wide variety of tissues. The basic structure of exosome has been depicted in the Fig. 8.1. Exosomes communicate between cells through horizontal transfer of its cargo contents that are biologically active. The cargo also can shuttle to distant sites and promote the function of cells (Lasser, 2012). Here in this chapter, we discuss few details of exosomes in the field of radiation biology.

8.1.1 Exosome Biogenesis and Release

Two different pathways are involved in the production of exosomes and microvesicles. Both microvesicles and exosomes are formed by initial formation of endocytic vesicles which later fuse together to form early endosome further they convert into late endosomes (Brinton et al., 2015; Lee et al., 2012; Raposo & Stoorvogel, 2013). Endosomes contain many small vesicles so-called multivesicular bodies (MVBs) often called as intraluminal endosomal vesicles (Lee et al., 2012). Some MVBs follow degradation pathway by uniting with lysosomes and others fuse with the plasma membrane to release as exosomes into extracellular space (Lee et al., 2012). Exosome biogenesis occurs with the involvement of endosomal sorting complex required for transport (ESCRT) complexes that involve in the sorting of proteins within endosomes into MVBs. Interactions of ESCRT-1, 11 and 111 with receptor tyrosine kinase of mammalian substrates and Vps27 helps in the transport of proteins into MVBs (Babst et al., 2002; Bang & Thum, 2012).

The clear mechanism of biogenesis and secretion of exosomes are still unknown but the reports suggest the involvement of syndecan heparin sulphate proteoglycans and syntenin in the regulation of exosomes (Baietti et al., 2012; Roucourt et al., 2015). Thus formed exosomes within the cells are secreted to outer extracellular space are regulated by Rab GTPases pathway (Ostrowski et al., 2010; Stenmark, 2009). The intracellular pH, Ca^{+2} channels and ESCRT complexes are also involved in the secretion and delivery of released exosomes to their recipient cells (Bobrie et al., 2011; Iero et al., 2008; Michelet et al., 2010; Parolini et al., 2009; Ramachandran & Palanisamy, 2012; Savina et al., 2003). The released exosomes are taken up the local or distant cells through fusion mechanism or receptor-mediated endocytosis or through rapid endocytosis (Zhang et al., 2015). Microvesicles are generated through a shredding mechanism called exocytosis, a less complex mechanism than compared with exosomes (Zomer et al., 2010). The secreted exosomes involve in the modulation of gene expression of recipient cells. The genomic and proteomic content of exosomes in recipient cells involves in the alteration of the morphology of recipient cells by upon interfering with signaling mechanisms.

8.1.2 Communication Between Donor and Recipient Cells

Exosomes are involved in communication between near and distant neighboring cells but the mechanism is not yet clear (Lasser et al., 2012). Several *in vitro* studies and based on other shreds of evidence it has been proposed that exosomes after releasing into extracellular space, binds to the surface of recipient cells and fuses via adhesion molecules present on the surface of exosomes (Fig. 8.2). The attached exosomes were taken up by the recipient cells through phagocytosis or through receptor mediated endocytosis process. The interaction between the exosomes and recipient cells take place via transfer of ligands or receptors present on the surface of exosomes. This interaction results in the transfer of cargo between one to other cells resulting in communication between the cells via exosomes (Valadi et al., 2007). In biological fluids such as serum, plasma, cerebrospinal fluid the amount of cargo content released into each exosome is very less but they carry important genetic information to each recipient cells upon uptake (Chevillet et al., 2014; Mittelbrunn et al., 2011).

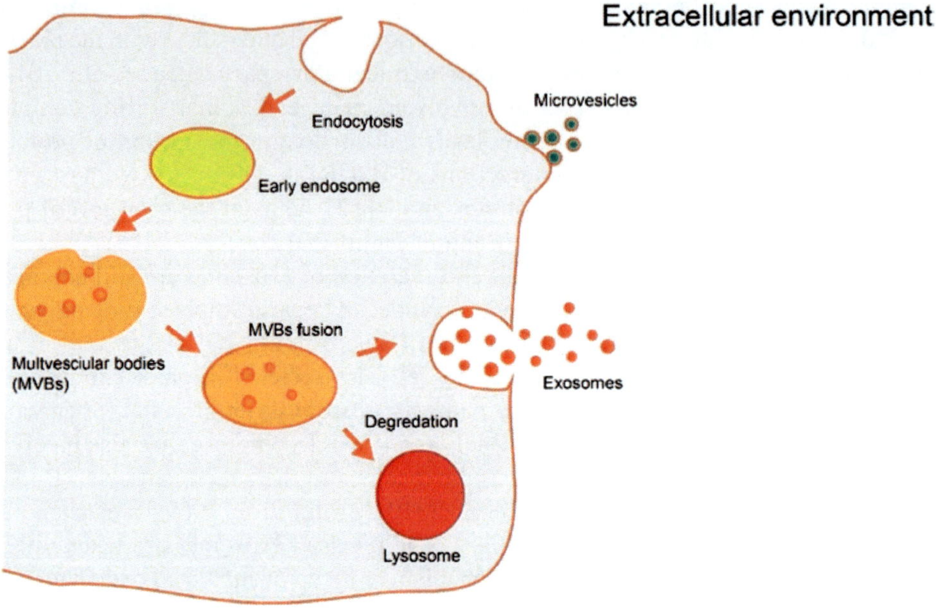

Fig. 8.2 Formation and release of exosomes to the extracellular environment

8.2 Exosome Isolation and Its Biophysical Characterization

Isolation of pure exosomes is a major set back in the field, needs to be purified other contaminating sources such as apoptotic bodies and microvesicles (Kowal et al., 2016; Lotvall et al., 2014; Witwer et al., 2013). Ultracentrifugation is considered as a gold standard method till now in the field of exosome biology and other methods such as density gradient purification, affinity column method, and precipitation by various types of resins (Lane et al., 2017). In the exosome purification process, cell debris will be removed at 230Xg, at 10,000Xg speed microvesicles can be removed/ collected and exosomes can be separated and washed at 100,000Xg using ultracentrifugation process (Dang et al., 2017; Jella et al., 2016).

The major hurdle in isolating exosomes using ultracentrifugation process is tedious and time-consuming and difficult to remove contaminants such as ribosomes and other protein complexes from exosomes (Alvarez et al., 2012; Thery et al., 2006). Using a discontinuous gradient method such as sucrose gradient purification/Opti-prep solution further purifies exosomes; the major drawback of this method is the loss of exosomes. Isolation of exosomes at low-speed centrifugation using polyethylene glycols (PEG) offered by many manufacturers. Isolation of exosomes at low-speed centrifugation process using PEG method offered by various manufacturers provides good yield but has the limitations in downstream processing which can overcome by various columns (Rekker et al., 2014). Exosomes purified by affinity column methods by using various surface antigens of exosomes as targeting agents e.g. CD9, CD81, flotillin-1 (Greening et al., 2015; Mathivanan et al.,

2010; Thery et al., 2006). Exosome purification by affinity-based methods can reduce contamination by providing pure exosomes (Bobrie et al., 2012; Mathivanan et al., 2010; Vaidyanathan et al., 2014). Exosomes are characterized by Immunoblot method to identify markers such as Tsg101, flotillin-1, flotillin-2, CD9, CD63 and CD81.

Transmission electron microscopy helps to identify the cup-shaped morphology by the negative-staining method. In electron microscopy method, if the sample were not pure, it would be difficult to distinguish between exosomes, proteins and other vesicles. Nanoparticle tracking analysis such as NanoSight method measures the size and distribution of exosomes in a particular volume. The direct measurement of size and distribution of exosomes can be performed using a Tunable resistive pulse sensing instrument (qNano system) (Maas et al., 2014). In summary, exosomes are complex molecules and it requires a sensitive method to optimize quality and quantity of exosomes.

8.3 Exosomal Cargo as a Diagnostic Biomarker

Exosomes are released by both normal and cancerous cells and play an important role in identifying the pathological conditions. They cargo of exosomes carrying protein and miRNA can be crucial for the identification of pathological conditions of various diseases. Numerous studies have been reported exosomal cargo can directly relate to cancer, infectious, metabolic and various neurodegenerative diseases.

Exosomal tetraspanins are highly enriched and they present on the surface of exosomes and they are considered as typical exosomal markers. In 2012, Welker et al. (2012) reported that CD81 is found to be elevated in patients suffering from chronic hepatitis C, providing clues that CD81 can be a potential marker for the diagnosis and treatment of hepatitis C. A various number of exosomal proteins are considered as potential tools for the diagnosis of various dreadful diseases. Exosomes isolated from the serum of patients suffering from glioblastoma contain specific epidermal growth factor receptor vIII that can serve as a biomarker (Welker et al., 2012). Exosomes isolated from cerebrospinal fluid (CSF) of Alzheimer's disease (AD) patients revealed the phosphorylation of tau protein at Thr-181, interestingly these patients have mild symptoms of disease (Saman et al., 2012). These findings could potentially suggest the early stages of an AD. Exosomal proteins have greater diagnostic values that have potential clinical applications. Isolation of exosomes from urine has profound clinical applications, especially in patients suffering from urinary tract diseases. Urinary exosomal protein, fetuin-A has been identified in patients with acute kidney injury and it is a clinical biomarker for identifying the patients (Zhou et al., 2006). The same group identified the activating transcription factor 3 in patients of AKI but not in normal patients (Zhou et al., 2008). In patients with prostate cancer, both PCA-3 and TMPRSS2:ERG was identified in exosomes isolated from prostate cancer patients urine samples (Nilsson et al., 2009). Urinary exosomes can be isolated from a variety of cancer patients and they are potential for the identification of biomarkers in clinical settings.

Small miRNA's are non-coding RNAs that involves the expression of various genes. These miRNA's are synthesized within the cells and are released into their target cells via exosomes. They are also transported in various body fluids such as blood, urine, and CSF and can be used as potential biomarkers to monitor various diseases (Table 8.1). MiRNA's of exosomes have been recently tested for hepatocellular carcinoma (Sugimachi et al., 2015), colorectal cancer (Matsumura et al., 2015; Ogata-Kawata et al., 2014), and acute myeloid leukemia (Hornick et al., 2015) for Alzheimer's disease (Cheng et al., 2015; Lugli et al., 2015).

Earlier studies have shown that exosomal miRNA is being safely circulated in biofluids and being protected from RNase-dependent degradation present in the biofluids (Hunter et al., 2008; Mitchell et al., 2008; Taylor & Gercel-Taylor, 2008). In ovarian cancer patients, 8 miRNA has been identified in the serum exosomes and these miRNAs are totally absent in normal controls, suggesting that they can be used as diagnostic biomarkers in identifying ovarian cancer patients (Taylor & Gercel-Taylor, 2008). In the case of lung adenocarcinoma patients, the group identified a similar pattern of exosomal miRNA both in circulating exosomes and in tumor biopsies and they are totally different from healthier patient samples (Rabinowits et al., 2009). In patients suffering from prostate cancer, miR-141 has been found to be in elevated levels than compared with normal healthy controls (Mitchell et al., 2008). In serum of prostate cancer patients they have identified miRNA in association with tumor progression, hence miR-141 and miR-375 can be a valuable marker for the identification of disease progression (Brase et al., 2011). It was reported that the patients suffering from esophageal squamous cell cancer have elevated levels of miR-21 in serum and this miRNA is in correlation with

Table 8.1 The table represents exosomal miRNA isolated from plasma and urine samples

Pathological condition	Biological sample	Exosomal miRNA	References
Prostate cancer	Plasma	miR-34b, miR-92a, miR-92b, miR-107, miR-197, miR-485, miR-328, miR-485, miR-486-5p, miR-640, miR-766 and miR-885-5p	Wittmann and Jack (2010)
Breast cancer	Plasma	miR-141 and miR-195	Corcoran et al. (2011)
Ovarian cancer	Plasma	miR-21, miR-141, miR-203, miR-200a, miR-200b, miR-200c, miR-205,	Keller et al. (2009) and Taylor and Gercel-Taylor (2008)
Esophageal squamous cell cancer (ESCC)	Plasma	miR-21, miR-1246	Takechita et al. (2013) and Tanaka et al. (2013)
Lung cancer	Plasma	miR-17, miR-3p, miR-21, miR-20b, miR-21, miR-223, miR-301 and let-7f	Brase et al. (2011) and Silva et al. (2011)
Renal fibrosis	Urine	miR-29c	Lv et al. (2014)

aggressiveness and tumor progression, especially in patients with the benign disease without any systemic inflammation (Tanaka et al., 2013). MiRNA of exosomes has demonstrated as biomarkers of clinical diagnosis in renal fibrosis and various cardiovascular diseases (Hong et al., 2009; Lv et al., 2014).

8.4 Radiation Treatments and Its Impact on Exosome Biogenesis

Exosomal content mainly depends on the type and state of a cell and it influences recipient cells in pathophysiological conditions. Radiation is one of the major contributors to environmental changes inducing cellular stress and alters gene expression. When cells were exposed with radiation, it not only impacts on radiated cells, but also bystander and distant un-irradiated cells termed as radiation-induced non-targeted effects. They include genomic instability, low-dose hypersensitivity, adaptive response and distant abscopal effects. They mainly cause biological effects on non-targeted cells and tissues such as changes in gene expressions, DNA damage, nitric oxide production, reactive oxygen species (ROS) production, cytokine production and activation of cell death pathways (Jella et al., 2013; Li et al., 2015; Lyng et al., 2012). The communication between radiation and non-irradiated cells occurs by gap junction intercellular communication (Azzam et al., 2001) or by releasing factors into the surrounding environment. Whereas radiation-induced abscopal effect occurs at remote sites and they are very well documented in both clinical and preclinical studies as systemic effects of local radiotherapy. The studies have shown inhibition of distant tumors with local irradiation and its impacts on distant tumor growth (Formenti & Demaria, 2009). Researchers have reported the role of p53 and immune signaling in mediating exosomes in radiation-induced abscopal effects (Camphausen et al., 2003; Demaria et al., 2004). As exosomes or other vesicles can travel to distant locations via blood (Kalluri, 2016), the contents of these vesicles are actively investigated to identify the potential role in abscopal responses. In coculture experiments conducted in a transwell system, TK6 cells have undergone changes in miRNA expression upon coculture with cells that are irradiated with 2 Gy X-rays (Chaudhry & Omaruddin, 2012). There is an increase in ROS production and p53 53BP-1 foci in human NSCLC cell line H1299, upon exposure with conditioned media harvested from 5 Gy X-ray irradiated from H1299 cells (Jiang et al., 2014).

Exosomes were shown to involve in radiation-induced genomic instability in MCF7 cell line upon exposure with conditioned media obtained from 2 Gy irradiated cells, they have identified RNA component of these exosomes as a critical mediator in transferring these effects (Al-Mayah et al., 2012, 2015). Along with these, Jella et al. (2014), showed that exosomes released from human keratinocyte cell lines, HaCaTs, were found to involve in exosome-mediated cell death, ROS production and calcium signaling in unirradiated cells upon incubation with exosomes obtained after γ-irradiated HaCaTs. Exosomes isolated from irradiated glioblastoma cells enhance migration in recipient cells altering their signaling pathways related to cell migration/invasion through activation of TrkA and FAK signaling (Arscott et al., 2013). A study on head and neck cancer cell models reported that irradiation has induced increased levels of proteins in exosomes and the proteins are related to basic cellular mechanisms that could potentially impact recipient cells (Jelonek et al., 2016). Exosomes isolated from γ-irradiated bronchial epithelial cells have shown differential expression of miRNA content, of these miR-7-5-p found to play a role in autophagy in recipient cells (Song et al., 2016). The Wang group reported that conditioned media obtained from MRC-5 fibroblasts on irradiation with 2 Gy X-rays showed a decrease in clonogenic survival, increase in 53 BP1 foci and occurrence of micronuclei (Xu et al., 2014). Irradiated conditioned media obtained from MRC-5 cells, was enriched with miR-21, upon transfer of these conditioned media to bystander cells have shown to repress BCL2 gene. In similar cell line, upon transfection with miR-21 have shown similar effects, suggesting the role of miR-21 in transferring radiation-induced bystander effects from donor to recipient cells (Xu et al., 2015).

There is increasing evidence suggesting that radiation increase with an increase in dose and time and this could be due to the activation of transmembrane protein tumor suppressor, stimulated by p53 via the activation of DNA damage (Yu et al., 2006). It was also confirmed in HaCaT cell lines (Jella et al., 2014), Glioblastoma cells (Arscott ct al., 2013), MCF-7 cells (Al-Mayah et al., 2015) and in prostate cancer cell lines (Lehmann et al., 2008). A mixture of both exosomes and microvesicles was found to be superior in biological activity when compared with either exosomes or microvesicles alone. Murine hematopoietic cell line upon exposure to five Gy irradiation showed a reversal of growth inhibition, DNA damage and apoptosis upon exposure to murine or human mesenchymal stem cells (MSC)-extracellular vesicles. The group identified that radiation damage to bone marrow stem cells can be reversed with extracellular vesicles originated from MSC extracellular vesicles (Wen et al., 2016). Kulkarni *et al,* isolated exosomes from both urine and serum upon whole body irradiation of mice and performed proteomic analysis. They have identified 47 differentially expressed proteins in urinary and serum exosomes isolated after whole body irradiation of mice (Kulkarni et al., 2016).

8.5 Conclusion

Research on exosome biogenesis, communication and its functions in recipient cells has increased exponentially over the past 7 years. Identification of exosomes in biological fluids has potential applications in clinical diagnosis. Exosome as biomarkers has higher sensitivity and specificity in biological fluids with greater clinical application. In addition, exosomes miRNA has valuable information that can be applied to clinical diagnostics. Till now, there are no specific radiation biomarkers;

exosomes have potential applications to expand as radiation biomarkers. The impact of radiation on exosomes has to be fully explored to develop exosomes as potential radiation biomarkers.

Acknowledgments I would like to thank the chair and organizing committee of the BRITE NATO Advanced Research Workshop for inviting and allowing me to produce this chapter. A special thanks to Addie Byrd for scientific corrections.

References

Al-Mayah, A., et al. (2015). The non-targeted effects of radiation are perpetuated by exosomes. *Mutation Research, 772*, 38.

Al-Mayah, A. H., Irons, S. L., Pink, R. C., Carter, D. R., & Kadhim, M. A. (2012). Possible role of exosomes containing RNA in mediating nontargeted effect of ionizing radiation. *Radiation Research, 177*, 539.

Alvarez, M. L., Khosroheidari, M., Ravi, R. K., & DiStefano, J. K. (2012). Comparison of protein, microRNA, and mRNA yields using different methods of urinary exosome isolation for the discovery of kidney disease biomarkers. *Kidney International, 82*, 1024–1032.

Arscott, W. T., et al. (2013). Ionizing radiation and glioblastoma exosomes: Implications in tumor biology and cell migration. *Translational Oncology, 6*, 638–648.

Azmi, A. S., Bao, B., & Sarkar, F. H. (2013). Exosomes in cancer development, metastasis, and drug resistance: A comprehensive review. *Cancer Metastasis Reviews, 32*, 623–642.

Azzam, E. I., de Toledo, S. M., & Little, J. B. (2001). Direct evidence for the participation of gap junction-mediated intercellular communication in the transmission of damage signals from alpha -particle irradiated to nonirradiated cells. *Proceedings of the National Academy of Sciences of the United States of America, 98*, 473.

Babst, M., Katzmann, D. J., Estepa-Sabal, E. J., Meerloo, T., & Emr, S. D. (2002). Escrt-III: An endosome-associated heterooligomeric protein complex required for mvb sorting. *Developmental Cell, 3*, 271–282.

Baietti, M. F., et al. (2012). Syndecan-syntenin-ALIX regulates the biogenesis of exosomes. *Nature Cell Biology, 14*, 677–685.

Bang, C., & Thum, T. (2012). Exosomes: New players in cell-cell communication. *The International Journal of Biochemistry & Cell Biology, 44*, 2060–2064.

Bobrie, A., Colombo, M., Raposo, G., & Thery, C. (2011). Exosome secretion: Molecular mechanisms and roles in immune responses. *Traffic, 12*, 1659–1668.

Bobrie, A., Colombo, M., Krumeich, S., Raposo, G., & Thery, C. (2012). Diverse subpopulations of vesicles secreted by different intracellular mechanisms are present in exosome preparations obtained by differential ultracentrifugation. *Journal of Extracellular Vesicles, 1*.

Brase, J. C., et al. (2011). Circulating miRNAs are correlated with tumor progression in prostate cancer. *International Journal of Cancer, 128*, 608–616.

Brinton, L. T., Sloane, H. S., Kester, M., & Kelly, K. A. (2015). Formation and role of exosomes in cancer. *Cellular and Molecular Life Sciences, 72*, 659–671.

Camphausen, K., et al. (2003). Radiation abscopal antitumor effect is mediated through p53. *Cancer Research, 63*, 1990.

Chaudhry, M. A., & Omaruddin, R. A. (2012). Differential regulation of microRNA expression in irradiated and bystander cells. *Molekuliarnaia Biologiia (Mosk), 46*, 634.

Cheng, L., et al. (2015). Prognostic serum miRNA biomarkers associated with Alzheimer's disease shows concordance with neuropsychological and neuroimaging assessment. *Molecular Psychiatry, 20*, 1188–1196.

Chevillet, J. R., et al. (2014). Quantitative and stoichiometric analysis of the microRNA content of exosomes. *Proceedings of the National Academy of Sciences of the United States of America, 111*, 14888–14893.

Corcoran, C., Friel, A. M., Duffy, M. J., Crown, J., & O'Driscoll, L. (2011). Intracellular and extracellular microRNAs in breast cancer. *Clinical Chemistry, 57*, 18–32.

Dang, V. D., Jella, K. K., Ragheb, R. R. T., Denslow, N. D., & Alli, A. A. (2017). Lipidomic and proteomic analysis of exosomes from mouse cortical collecting duct cells. *The FASEB Journal, 31*, 5399–5408.

Demaria, S., et al. (2004). Ionizing radiation inhibition of distant untreated tumors (abscopal effect) is immune mediated. *International Journal of Radiation Oncology, Biology, Physics, 58*, 862.

Formenti, S. C., & Demaria, S. (2009). Systemic effects of local radiotherapy. *The Lancet Oncology, 10*, 718.

Gong, J., et al. (2012). Microparticles and their emerging role in cancer multidrug resistance. *Cancer Treatment Reviews, 38*, 226–234.

Greening, D. W., Xu, R., Ji, H., Tauro, B. J., & Simpson, R. J. (2015). A protocol for exosome isolation and characterization: Evaluation of ultracentrifugation, density-gradient separation, and immunoaffinity capture methods. *Methods in Molecular Biology, 1295*, 179–209.

Hong, B. S., et al. (2009). Colorectal cancer cell-derived microvesicles are enriched in cell cycle-related mRNAs that promote proliferation of endothelial cells. *BMC Genomics, 10*, 556.

Hornick, N. I., et al. (2015). Serum exosome microRNA as a minimally-invasive early biomarker of AML. *Scientific Reports, 5*, 11295.

Hunter, M. P., et al. (2008). Detection of microRNA expression in human peripheral blood microvesicles. *PLoS One, 3*, e3694.

Iero, M., et al. (2008). Tumour-released exosomes and their implications in cancer immunity. *Cell Death and Differentiation, 15*, 80–88.

Jaiswal, R., et al. (2012). Microparticle conferred microRNA profiles--implications in the transfer and dominance of cancer traits. *Molecular Cancer, 11*, 37.

Jella, K. K., et al. (2016). Exosomal GAPDH from proximal tubule cells regulate ENaC activity. *PLoS One, 11*, e0165763.

Jella, K. K., et al. (2014). Exosomes are involved in mediating radiation induced bystander signaling in human keratinocyte cells. *Radiation Research, 181*, 138.

Jella, K. K., Garcia, A., McClean, B., Byrne, H. J., & Lyng, F. M. (2013). Cell death pathways in directly irradiated cells and cells exposed to medium from irradiated cells. *International Journal of Radiation Biology, 89*, 182–190.

Jelonek K., P. Widlak, M. Pietrowska. The influence of ionizing radiation on exosome composition, secretion and intercellular Communication. Protein and Peptide Letters 23, 656–663 (2016).

Jiang, Y., et al. (2014). The role of TGF-beta1-miR-21-ROS pathway in bystander responses induced by irradiated non-small-cell lung cancer cells. *British Journal of Cancer, 111*, 772.

Kahlert, C., et al. (2014). Identification of double-stranded genomic DNA spanning all chromosomes with mutated KRAS and p53 DNA in the serum exosomes of patients with pancreatic cancer. *The Journal of Biological Chemistry, 289*, 3869–3875.

Kalluri, R. (2016). The biology and function of exosomes in cancer. *The Journal of Clinical Investigation, 126*, 1208.

Keller, S., et al. (2009). Systemic presence and tumor-growth promoting effect of ovarian carcinoma released exosomes. *Cancer Letters, 278*, 73–81.

Kowal, J., et al. (2016). Proteomic comparison defines novel markers to characterize heterogeneous populations of extracellular vesicle subtypes. *Proceedings of the National Academy of Sciences of the United States of America, 113*, E968–E977.

Kulkarni, S., et al. (2016). Identifying urinary and serum exosome biomarkers for radiation exposure using a data dependent acquisition and SWATH-MS combined workflow. *International Journal of Radiation Oncology, Biology, Physics, 96*, 566–577.

Lane, R. E., Korbie, D., Trau, M., & Hill, M. M. (2017). Purification protocols for extracellular vesicles. *Methods in Molecular Biology, 1660*, 111–130.

Lasser, C. (2012). Exosomal RNA as biomarkers and the therapeutic potential of exosome vectors. *Expert Opinion on Biological Therapy, 12*(Suppl 1), S189–S197.

Lee, Y., El Andaloussi, S., & Wood, M. J. (2012). Exosomes and microvesicles: Extracellular vesicles for genetic information transfer and gene therapy. *Human Molecular Genetics, 21*, R125–R134.

Lehmann, B. D., et al. (2008). Senescence-associated exosome release from human prostate cancer cells. *Cancer Research, 68*, 7864–7871.

Li, Z., et al. (2015). Co-culturing with high-charge and energy particle irradiated cells increases mutagenic joining of enzymatically induced DNA double-strand breaks in nonirradiated cells. *Radiation Research, 184*, 249–258.

Lotvall, J., et al. (2014). Minimal experimental requirements for definition of extracellular vesicles and their functions: A position statement from the International Society for Extracellular Vesicles. *Journal of Extracellular Vesicles, 3*, 26913.

Lugli, G., et al. (2015). Plasma exosomal miRNAs in persons with and without Alzheimer disease: Altered expression and prospects for biomarkers. *PLoS One, 10*, e0139233.

Lv, L. L., et al. (2014). CD2AP mRNA in urinary exosome as biomarker of kidney disease. *Clinica Chimica Acta, 428*, 26–31.

Lyng, F. M., Desplanques, M., Jella, K. K., Garcia, A., & McClean, B. (2012). The importance of serum serotonin levels in the measurement of radiation-induced bystander cell death in HaCaT cells. *International Journal of Radiation Biology, 88*, 770–772.

Maas, S. L., De Vrij, J., & Broekman, M. L. (2014). Quantification and size-profiling of extracellular vesicles using tunable resistive pulse sensing. *Journal of Visualized Experiments, e51623.*

Mathivanan, S., et al. (2010). Proteomics analysis of A33 immunoaffinity-purified exosomes released from the human colon tumor cell line LIM1215 reveals a tissue-specific protein signature. *Molecular & Cellular Proteomics, 9*, 197–208.

Matsumura, T., et al. (2015). Exosomal microRNA in serum is a novel biomarker of recurrence in human colorectal cancer. *British Journal of Cancer, 113*, 275–281.

Michelet, X., Djeddi, A., & Legouis, R. (2010). Developmental and cellular functions of the ESCRT machinery in pluricellular organisms. *Biology of the Cell, 102*, 191–202.

Mitchell, P. S., et al. (2008). Circulating microRNAs as stable blood-based markers for cancer detection. *Proceedings of the National Academy of Sciences of the United States of America, 105*, 10513–10518.

Mittelbrunn, M., et al. (2011). Unidirectional transfer of microRNA-loaded exosomes from T cells to antigen-presenting cells. *Nature Communications, 2*, 282.

Nilsson, J., et al. (2009). Prostate cancer-derived urine exosomes: A novel approach to biomarkers for prostate cancer. *British Journal of Cancer, 100*, 1603–1607.

Ogata-Kawata, H., et al. (2014). Circulating exosomal microRNAs as biomarkers of colon cancer. *PLoS One, 9*, e92921.

Ostrowski, M., et al. (2010). Rab27a and Rab27b control different steps of the exosome secretion pathway. *Nature Cell Biology, 12*, 19–30.; ; sup, pp 11–13.

Parolini, I., et al. (2009). Microenvironmental pH is a key factor for exosome traffic in tumor cells. *The Journal of Biological Chemistry, 284*, 34211–34222.

Patel, G. K., Patton, M. C., Singh, S., Khushman, M., & Singh, A. P. (2016). Pancreatic cancer exosomes: Shedding off for a meaningful journey. *Pancreatic Disorders & Therapy, 6*, e148.

Rabinowits, G., Gercel-Taylor, C., Day, J. M., Taylor, D. D., & Kloecker, G. H. (2009). Exosomal microRNA: A diagnostic marker for lung cancer. *Clinical Lung Cancer, 10*, 42–46.

Ramachandran, S., & Palanisamy, V. (2012). Horizontal transfer of RNAs: Exosomes as mediators of intercellular communication. *Wiley Interdiscip Rev RNA, 3*, 286–293.

Raposo, G., & Stoorvogel, W. (2013). Extracellular vesicles: Exosomes, microvesicles, and friends. *The Journal of Cell Biology, 200*, 373–383.

Rekker, K., et al. (2014). Comparison of serum exosome isolation methods for microRNA profiling. *Clinical Biochemistry, 47*, 135–138.

Roucourt, B., Meeussen, S., Bao, J., Zimmermann, P., & David, G. (2015). Heparanase activates the syndecan-syntenin-ALIX exosome pathway. *Cell Research, 25*, 412–428.

Saman, S., et al. (2012). Exosome-associated tau is secreted in tauopathy models and is selectively phosphorylated in cerebrospinal fluid in early Alzheimer disease. *The Journal of Biological Chemistry, 287*, 3842–3849.
Savina, A., Furlan, M., Vidal, M., & Colombo, M. I. (2003). Exosome release is regulated by a calcium-dependent mechanism in K562 cells. *The Journal of Biological Chemistry, 278*, 20083–20090.
Silva, J., et al. (2011). Vesicle-related microRNAs in plasma of nonsmall cell lung cancer patients and correlation with survival. *The European Respiratory Journal, 37*, 617–623.
Song, M., et al. (2016). Bystander autophagy mediated by radiation-induced exosomal miR-7-5p in non-targeted human bronchial epithelial cells. *Scientific Reports, 6*, 30165.
Stenmark, H. (2009). Rab GTPases as coordinators of vesicle traffic. *Nature Reviews. Molecular Cell Biology, 10*, 513–525.
Sugimachi, K., et al. (2015). Identification of a bona fide microRNA biomarker in serum exosomes that predicts hepatocellular carcinoma recurrence after liver transplantation. *British Journal of Cancer, 112*, 532–538.
Takeshita, N., et al. (2013). Serum microRNA expression profile: miR-1246 as a novel diagnostic and prognostic biomarker for oesophageal squamous cell carcinoma. *British Journal of Cancer, 108*, 644–652.
Tanaka, Y., et al. (2013). Clinical impact of serum exosomal microRNA-21 as a clinical biomarker in human esophageal squamous cell carcinoma. *Cancer, 119*, 1159–1167.
Taylor, D. D., & Gercel-Taylor, C. (2008). MicroRNA signatures of tumor-derived exosomes as diagnostic biomarkers of ovarian cancer. *Gynecologic Oncology, 110*, 13–21.
Thery, C., Amigorena, S., Raposo, G., & Clayton, A. (2006). Isolation and characterization of exosomes from cell culture supernatants and biological fluids. *Curr Protoc Cell Biol.*, Chapter 3, Unit 3 22.
Vaidyanathan, R., et al. (2014). Detecting exosomes specifically: A multiplexed device based on alternating current electrohydrodynamic induced nanoshearing. *Analytical Chemistry, 86*, 11125–11132.
Valadi, H., et al. (2007). Exosome-mediated transfer of mRNAs and microRNAs is a novel mechanism of genetic exchange between cells. *Nature Cell Biology, 9*, 654–659.
Vickers, K. C., & Remaley, A. T. (2012). Lipid-based carriers of microRNAs and intercellular communication. *Current Opinion in Lipidology, 23*, 91–97.
Welker, M. W., et al. (2012). Soluble serum CD81 is elevated in patients with chronic hepatitis C and correlates with alanine aminotransferase serum activity. *PLoS One, 7*, e30796.
Wen, S., et al. (2016). Mesenchymal stromal cell-derived extracellular vesicles rescue radiation damage to murine marrow hematopoietic cells. *Leukemia, 30*, 2221–2231.
Wittmann, J., & Jack, H. M. (2010). Serum microRNAs as powerful cancer biomarkers. *Biochimica et Biophysica Acta, 1806*, 200–207.
Witwer, K. W., et al. (2013). Standardization of sample collection, isolation and analysis methods in extracellular vesicle research. *Journal of Extracellular Vesicles, 2*.
Xu, S., et al. (2015). Exosome-mediated microRNA transfer plays a role in radiation-induced bystander effect. *RNA Biology, 12*, 1355.
Xu, S., et al. (2014). MiR-21 is involved in radiation-induced bystander effects. *RNA Biology, 11*, 1161.
Yu, X., Harris, S. L., & Levine, A. J. (2006). The regulation of exosome secretion: A novel function of the p53 protein. *Cancer Research, 66*, 4795–4801.
Zhang, X., et al. (2015). Exosomes in cancer: Small particle, big player. *Journal of Hematology & Oncology, 8*, 83.
Zhou, H., et al. (2006). Exosomal Fetuin-A identified by proteomics: A novel urinary biomarker for detecting acute kidney injury. *Kidney International, 70*, 1847–1857.
Zhou, H., et al. (2008). Urinary exosomal transcription factors, a new class of biomarkers for renal disease. *Kidney International, 74*, 613–621.
Zomer, A., et al. (2010). Exosomes: Fit to deliver small RNA. *Communicative & Integrative Biology, 3*, 447–450.

Chapter 9
Monitoring Very Low Dose Radiation Damage in DNA Using "Field-Friendly" Biomarkers

Kai Rothkamm and Susanne Burdak-Rothkamm

Abstract The retrospective assessment of radiation exposures using quantitative biomarkers related to dose is well established for the purpose of biological dosimetry. Among the range of candidate biomarkers that could be used to estimate the dose received by a person, DNA damage-associated markers play a prominent role. One of the hallmarks of ionising radiation, which distinguishes it from many other environmentally relevant agents, is its ability to induce DNA double-strand breaks (DSB) at sites of ionisation clusters along the particle track. These frequently give rise to gross chromosomal rearrangements such as translocations, dicentrics and acentric fragments, the latter of which may form micronuclei in daughter cells. These chromosomal aberrations, and especially dicentrics, are regarded as the "gold standard" for radiation biodosimetry, and a lot of effort has gone into the development of ISO standards, automation and networking to enable robust and efficient dose assessments using these endpoints. However, alternative candidate markers of exposure related to DNA damage have also been investigated in recent years, including protein biomarkers of DSB and gene expression signatures associated with DNA damage signalling. Here the usefulness and applicability of DNA damage-associated biomarkers in environmental exposure assessment is discussed and their adaptability for a "field" setting explored. DNA damage-associated biomarkers of radiation exposure have proven their value in human biodosimetry applications, but all have certain restrictions, especially in the low dose region and in scenarios potentially including other pollutants. Their individual applicability and usefulness very much depends on the exposure scenario in question.

K. Rothkamm (✉) · S. Burdak-Rothkamm
Department of Radiotherapy and Radiation Oncology, University Medical Center Hamburg-Eppendorf, Hamburg, Germany
e-mail: k.rothkamm@uke.de

M. D. Wood et al. (eds.), *Biomarkers of Radiation in the Environment*, NATO Science for Peace and Security Series A: Chemistry and Biology,
https://doi.org/10.1007/978-94-024-2101-9_9

9.1 Ionising Radiation-Induced DNA and Chromosome Damage

Ionising radiation deposits its energy through ionisations along the tracks of charged particles. For electromagnetic radiation, these are secondary electrons ejected from their atomic shell following their initial interaction with an energetic X- or gamma-ray photon, while neutron irradiation results in the ejection of protons from atomic nuclei. Cellular chromosomal DNA is commonly regarded as the main biological target for the deleterious effects of ionising radiation. Radiation damage to DNA frequently involves altered bases as well as breaks in the sugar-phosphate backbone of the DNA. It is one of the hallmarks of ionising radiation that these lesions are not completely randomly distributed within the cell nucleus but frequently coincide in a localised cluster, forming locally multiply damaged sites and resulting in the formation of DNA double-strand breaks and other clustered DNA lesions. Overall, yields of these DNA lesions increase linearly with radiation dose and absolute numbers also depend on the DNA content of the irradiated cells. One gray of sparsely ionising radiation induces of the order of a thousand single-strand breaks in a human diploid cell and a similar yield of base lesions, but only a few dozen double-strand breaks.

While single-stranded lesions can be efficiently repaired using the sequence information on the intact complementary DNA strand, double-strand breaks or other clustered DNA lesions affecting both strands are more difficult to repair. As a consequence, they persist for longer and may give rise to mutations and chromosomal aberrations, if not correctly repaired. If two double-strand breaks coincide in close proximity, break ends originating from distinct breaks may accidentally be misrejoined to form gross structural chromosome aberrations such as translocations, dicentrics, inversions and deletions (Rothkamm & Lobrich, 2002). At low doses, these rearrangements originate from lesions induced along the same particle track (intra-track interaction), resulting in a linear dose response that is independent of dose-rate. The yield of such intra-track events is much higher for densely ionising particles than for X- or gamma-rays. At higher doses there is, at least in the case of sparsely ionising radiation, in addition an increasing chance for double-strand breaks formed at separate particle tracks to coincide sufficiently close to each other to allow their interaction. Accordingly, yields of exchange-type aberrations increase quadratically with dose towards high doses for sparsely, but not for densely ionising radiation, where – at relevant doses – tracks are too distant from each other to allow any significant interaction.

9.2 DNA Damage-Associated Biomarkers of Ionising Radiation Exposure

Any type of radiation-induced change that can be measured can in principle be used as a marker of radiation exposure (Rothkamm & Lloyd, 2014). Once a calibration curve is established, which links the amount of signal to the dose for a given type of radiation, this signal can be used for estimating the radiation dose. Radiation specificity and sensitivity as well as temporal stability and reproducibility of the signal are among the most critical parameters that determine the suitability of a radiation-induced signal as a biomarker of radiation exposure. Today the best established biomarkers of ionising radiation exposure can be divided into (i) electron spin resonance spectroscopy-based quantification of radicals in solid, crystalline biomaterials such as hydroxyapatite, which is found in tooth enamel and, to a lesser extent, bones, and (ii) assays detecting chromosomal DNA damage. The focus of this chapter is entirely on the latter, which can be divided into cytogenetic markers, reflecting aberrant DNA repair events, and those of unrepaired DNA damage, especially DNA (double-)strand breaks.

9.3 Cytogenetic Markers

9.3.1 The Dicentric Assay

Today, the gold standard biomarker of ionising radiation exposure is the frequency of dicentric chromosomes observed in metaphases of peripheral blood lymphocytes obtained from a potentially exposed person. This method was established in the 1960s and is to date the most widely used and accepted biological dosimetry method. A detailed description of the methodology, protocols and guidance for interpretation of results is available in an excellent IAEA publication (IAEA, 2011).

Dicentrics form when centric chromosome fragments originating from different chromosomes are erroneously joined together, leaving behind two acentric chromosome pieces which are also joined to form an acentric fragment. The formation of a dicentric and accompanying acentric fragment therefore requires the presence of two DNA double-strand breaks in close proximity, affecting two different chromosomes. The main double-strand break repair pathway, non-homologous end joining, is known to frequently join wrong break ends in situations where multiple breaks coincide (Rothkamm et al., 2001; Rothkamm & Lobrich, 2002). It is the requirement for at least two double-strand breaks clustering in close spatial and temporal proximity that makes this assay quite radiation-specific, as hardly any other genotoxic agent is capable of that.

International standards have been developed for normal (ISO 19238:2014) and triage mode (ISO 21243:2008) use of this assay in biological dosimetry. The dicentric assay forms the main methodological basis for a number of biodosimetry assistance networks such as the World Health Organization's BioDoseNet, the IAEA's RANET, as well as the regional South-American, Canadian, Japanese and European networks.

In brief, the procedure requires the stimulation of lymphocytes with e.g. phytohaemagglutinin, subsequent incubation for 48 h and mitotic blockage using a spindle poison such as colcemid in order to obtain first division metaphases. With consistent background levels in unirradiated individuals of 0–2 dicentrics per 1000 lymphcyte metaphases it is currently possible to detect an ionising radiation exposure at doses down to approximately 100 mGy of low linear energy transfer (LET) radiation using this assay. There are a number of challenges and caveats, though:

- Firstly, reliable identification of dicentrics in Giemsa-stained metaphase spreads is time-consuming and requires extensive training in order to achieve consistent results. Semi-automated scoring techniques are now available, which help reduce staff effort but require expensive equipment (Romm et al., 2013, 2014b). With their help, it may be feasible to improve sensitivity towards several tens of milligray. The introduction of fluorescent *in situ* hybridisation probes for centromere and telomere detection may help simplify automated as well as manual assessments (M'Kacher et al., 2014). Low cost alternatives to improve throughput include a rapid manual scoring strategy in triage mode (Flegal et al., 2010, 2012) and web-based scoring via internet-based sharing of electronic metaphase images (Romm et al., 2014a, 2017).
- Secondly, exposure to a few non-radiation agents may result in the formation of so-called derived dicentrics. It is therefore good practice to score only dicentrics that are accompanied by an acentric fragment, to ensure they were induced by radiation.
- Then, as lymphocytes containing dicentrics are replaced over time by new lymphocytes from undamaged precursor cells, the yield of dicentrics is not entirely stable over time but declines over a period of months to years. The loss depends on the severity of the exposure but is also influenced by other events involving the T cell response, such as infections, and can therefore not be estimated on an individual basis. As a result, dose estimations become less reliable the more time has passed since the exposure.
- Importantly, dicentrics yields vary with dose-rate (at least at moderate to high doses) and radiation quality, requiring separate calibration curves for different types of radiation, and necessitating some prior information on the nature of the exposure.
- One technical caveat is the need to capture a sufficiently large number of cells whilst they are undergoing their first mitotic division, in order to be able to analyse metaphases. Also, chromosomes would ideally have to be metacentric and reasonably large to enable efficient analysis of metaphases. Dicentrics are much harder to spot in mouse cell metaphases, for example, due to the acrocentric nature of murine chromosomes.

9.3.2 The Micronucleus Assay

Micronuclei are much more widely used than the dicentrics assay to assess the genotoxicity and clastogenicity of substances and agents. This is partly because micronuclei are technically easier to score, but it also reflects their broader response: whereas dicentrics are formed by the interaction of two DNA double-strand breaks which happened to form in close proximity – a situation that is almost unique for ionising radiation – micronuclei may form from a wider range of DNA lesions induced by many genotoxic agents. In consequence, this means that micronuclei may reflect ionising radiation-induced DNA damage, but other agents may just as well have contributed. The now commonly approved method originally developed by Fenech and Morley (1986) involves a cytochalasin B block to arrest lymphocytes at their first cytokinesis following stimulation. Subsequently, any binucleated cells would be scored. For human lymphocytes, this results in an incubation period of three days, i.e. one day longer than needed for the dicentric assay.

One putatively large contributor to micronucleus formation is the mis-segregation of chromosomes. This happens comparatively frequently to the inactive X-chromosome in females, resulting in a higher background frequency of micronuclei in women. Any interference with the proper function of the spindle apparatus may also increase the number of mis-segregated chromosome. In order to distinguish these "numerical" chromosome aberrations (which are not associated with radiation exposure) from structural ones (which do reflect radiation exposure), the micronucleus assay can be combined with a hybridisation step using fluorescent centromere probes. With this additional step, the micronucleus assay may detect radiation exposures as low as 50–100 mGy, with simpler morphology and well established automation procedures for scoring helping to make this feasible in a routine setting (Baeyens et al., 2011; Thierens et al., 1999, 2000, 2014; Vral et al., 1997, 2016).

The micronucleus assay shares several of the characteristics and challenges described above for the dicentric assay: an ISO standard has been established, defining performance criteria and providing a standard procedure for processing, scoring, data analysis and reporting of dose estimates (ISO 17099:2014). Several of the above mentioned assistance networks have embraced it. The signal diminishes with similar kinetics to the dicentric, due to the same processes, and radiation-induced yields depend on dose-rate and radiation quality. Whilst the overdispersed (wider than random) distribution of dicentrics among the scored population of cells may be used to detect partial body exposures, and even estimate the irradiated fraction and its dose, this is not possible for micronuclei. This is mainly because micronuclei distributions are overdispersed even following uniform exposures, as e.g. reported in an intercomparison exercise during the recent EU Multibiodose project (Ainsbury et al., 2014a).

9.3.3 The Premature Chromosome Condensation (PCC) Assay

Premature chromosome condensation (PCC) solves some of the limitations caused by the requirement for cells to reach mitosis for the dicentric assay or cytokinesis for the micronucleus assay. In brief, PCC can be achieved in G0, G1 and G2 phase cells by fusing the cells of interest with mitotic cells using polyethylene glycol (Pantelias & Maillie, 1984). Alternatively, G2 phase cells can be forced into PCC using phosphatase inhibitors such as calyculin A or ocadaic acid (Gotoh et al., 2005). Both approaches help overcome any DNA damage-induced G2/M cell cycle arrest and are therefore most useful in exposure scenarios involving very high doses. In such circumstances, e.g. in the course of the Tokai-mura criticality accident, the scoring of Giemsa-stained PCC ring chromosomes turned out to deliver good dose estimates (Hayata et al., 2001).

As rings form at a frequency an order of magnitude below that for dicentrics, they are not suitable for assessing low or moderate exposures. In those situations, excess PCC fragments are typically scored. Mathematical analysis of the distribution of such excess fragments among the scored cells also enables non-uniform exposures to be assessed using the same methodology employed for the dicentric assay (Darroudi, 1998). Alternatively, PCC dicentrics or translocations may be scored, using centromere and telomere or whole chromosome FISH painting (Terzoudi et al., 2017).

9.3.4 Translocation Analysis

Unstable chromosomal aberrations such as dicentrics, acentrics and micronuclei are lost from the lymphocyte pool as lymphocytes get replaced. Half times between several months and a few years have been reported for dicentrics; they vary a lot and may be drastically reduced in individuals who received a large radiation dose or had severe infections. The underlying issue is that haematopoietic stem cells carrying 'unstable' chromosome aberrations due to the radiation exposure fail to produce mature lymphocytes. This is because of the lethal effect that aberrations such as dicentric chromosomes and acentric fragments (which form micronuclei) have during cell division. Therefore, new lymphocytes would only be produced from undamaged stem cells, resulting in a progressive dilution of unstable chromosome aberrations. In contrast, stable aberrations, such as reciprocal translocations, can be passed on to daughter cells, as there are no chromosome segregation issues and no loss of chromosome material during mitosis. Consequently, haematopoietic stem cells containing reciprocal translocations may produce progenitor cells and subsequently mature lymphocytes and thus pass on this marker of a previous radiation exposure (Ainsbury et al., 2014b). Indeed it has been shown in a number of studies that, while the yield of dicentrics, acentrics and micronuclei declines over time, the number of translocations observed in "stable cells", i.e. those whose metaphases are free of any unstable aberrations, remains more or less constant over many years. The same would most likely be true for other radiation-induced stable aberrations, say inversions; however, these are much more difficult to detect using conventional cytogenetic methods.

Reciprocal translocations can easily be detected and reliably scored using chromosome painting probes for fluorescence in situ hybridisation (FISH) of metaphase spreads. Single colour staining of three large chromosomes (encompassing ca. 20% of the human genome, with the remainders just counterstained with a fluorescent DNA dye) already enables the detection of approximately 33% of all gross exchange-type aberrations. A formula developed by Lucas is used to calculate the whole genome equivalents of detected translocations. The percentage of detectable events can be further increased by using additional colours and chromosomes, up to mFISH, which distinguishes all 24 human chromosomes based on different fluorescent colour combinations. Whilst being extremely valuable for mechanistic studies and for resolving complex aberrations, such as those resulting from exposure to densely ionising radiations, the lower cost and simpler to analyse one-to-three colour option delivers similar results for most biological dosimetry applications, although more cells are required in order to obtain similar whole genome equivalents.

An earlier international collaborative effort showed that spontaneous background levels of translocations in unirradiated individuals increase with age, are influenced by smoking, and vary considerably between laboratories/countries (Sigurdson et al., 2008b). As a consequence, it is crucial to take age into account when estimating exposures based on translocation yields. Even then, uncertainties grow with age, resulting in fairly poor detection limits of the order of several hundred milligray for elderly individuals (Tucker & Luckinbill, 2011). Also, as for the dicentric and the micronucleus assays, laboratories require their own calibration curves to convert aberrations into dose estimates. As translocations were found to be induced at very similar yields to dicentrics, laboratories frequently use existing dicentrics curves for translocation-based dose estimations.

A number of large studies of exposed populations such as radiologic technologists (Bhatti et al., 2007; Sigurdson et al., 2008a), airline pilots (Yong et al., 2009), nuclear weapons test veterans (Wahab et al., 2008), former workers at the MAYAK nuclear plant (Sotnik et al., 2015) and residents of the upper reaches of the Techa river (Degteva et al., 2015; Vozilova et al., 2012, 2014) have been performed using FISH-based translocation analysis. They indicate that, while individual exposure assessments are limited to high doses, the effects of low doses may be detectable at the cohort level.

9.4 Markers of DNA Damage

Cytogenetic dose estimation is well established and validated for humans, based on the detection and scoring of structural chromosomal alterations in peripheral blood lymphocytes. However, this approach cannot always be easily adopted to dose assessments in the environment. Limitations include the need to obtain mitotic cells with condensed chromosomes (unless cell fusion-based premature chromosome condensation is employed) and, especially for the efficient detection of dicentrics, the availability of large, metacentric chromosomes. Assays for DNA damage are, in comparison, much easier to undertake in a variety of organisms. However, while chromosome aberrations typically signify the final consequence of (erroneous) repair events and persist for long periods, at least in non-dividing cells, radiation dose-dependent signals of DNA damage tend to be much more short-lived. This is because DNA repair mechanisms typically operate over time scales of minutes to hours, thus causing levels of unrepaired DNA damage to decline quickly post exposure. For this reason, markers of unrepaired DNA damage following ionising radiation exposure are less useful for estimating exposures long after the event or when the exact timing of the exposure is not known. Instead, they work fine for planned exposure situations such as computed tomography scans (Rothkamm et al., 2007) or radiotherapy treatments (Sak et al., 2007). In the case of protracted exposures, e.g. through a contaminated habitat, they only reflect the most recent damage inflicted by radiation, whereas cytogenetic markers indicate cumulative dose. Qualitatively, however, DNA damage assays such as single cell gel electrophoresis or gamma-H2AX foci may indicate persistent or recurrent damage resulting from e.g. an earlier low dose radiation exposure (Rothkamm et al., 2003) or even damage passed on trans-generationally from irradiated parents (Barber et al., 2006).

9.4.1 *Single Cell Gel Electrophoresis*

Single cell gel electrophoresis, also called the comet assay, can detect and quantify DNA damage in individual cells. To this end, cells are embedded in low gelling temperature agarose which is then spread out as a thin layer on a microscope slide. After cell lysis, the comet slides are subjected to electrophoresis in order to mobilise DNA fragments and stained with a DNA dye for subsequent microscopic analysis.

While DNA from intact, unirradiated cells is too large to migrate in the electric field and therefore remains in a sphere, DNA fragments resulting from radiation-induced strand breaks migrate towards the anode, forming a comet-shaped tail released from the "head" representing the cell nucleus. The fraction of DNA in the tail or the fraction multiplied with tail length (the 'tail moment') are used to quantify the amount of damage for each cell using digital image analysis methods. Measurement of total DNA signal allows simultaneous assessment of cell cycle position (e.g. Bauerschmidt et al., 2010; Burdak-Rothkamm et al., 2009). Alternatively, comets can be classified by eye, based on their appearance.

Single cell gel electrophoresis can be performed either in an alkaline buffer, causing DNA to denature into single strands and enabling the detection of single- as well as double-strand breaks, or in a neutral buffer, leaving DNA double-stranded. In this latter case, comets are believed to be a surrogate marker for double-strand breaks, though there are a number of other factors that influence comet formation.

Utilisation of enzymes that convert base damage to strand breaks, i.e. glycosylases/AP endonucleases, broadens the applicability of single-cell gel electrophoresis to other types of lesions. In combination with neutral conditions this could also be used as an approach for detecting radiation-induced non-double-strand break clustered lesions (Georgakilas, 2011). The comet assay may not only be used with cell suspensions. It can also be performed using thin tissue layers, e.g. in intact, isolated intestinal crypts (Brooks & Winton, 1996). As it can detect a wide range of DNA lesions it is frequently employed in chemical genotoxicity testing and biomonitoring of both humans and the environment (Glei et al., 2016). However, due to the rapid kinetics of DNA strand break repair and the large inter-experimental signal variability, it is not commonly used as a quantitative biological dosemeter for radiation exposure. Nonetheless, it may be a useful qualitative marker of ongoing genomic instability (Barber et al., 2006).

9.4.2 DNA Damage Foci

The immunocytochemical detection of proteins or protein modifications that accumulate in the vicinity of a DNA-double-strand break to form 'foci' has become a standard technique for detecting and quantifying the induction and repair of DNA double-strand breaks in individual cells and tissues following exposure to ionising radiation and other clastogenic agents (Rothkamm et al., 2015). While several different proteins form foci at sites of DNA double-strand breaks, antibodies targeting the phosphorylated histone variant gamma-H2AX and the 'p53-binding protein' 53BP1 are most commonly utilised as surrogate markers of DNA breaks and radiation exposure (Rothkamm & Horn, 2009). Enumeration of individual foci may be done by manual scoring at the fluorescence microscope or by automated scoring of digital microscopy images.

With kinetics of foci resolution closely following those for DNA double-strand break repair (Rothkamm et al., 2003) these markers of radiation exposure are fairly short-lived. However, as a single focus is believed to highlight the presence of an individual break, the methodology is very sensitive and has, in fact, been used to detect the DNA damaging effects of doses as low as a few milligray following in vitro (Rothkamm & Lobrich, 2003) as well as in vivo exposure to X-rays (Lobrich et al., 2005; Rothkamm et al., 2007).

DNA damage foci assays have been tested as rapid radiation biodosimeters in several studies and inter-comparison exercises (Ainsbury et al., 2014a; Barnard et al., 2015; Horn et al., 2011; Moquet et al., 2017; Rothkamm et al., 2013a, b, c). These studies showed that overall foci-based methods tend to be faster but less accurate than the gold standard dicentric assay. The biggest caveat is that, due to the rapid signal loss, only recent, acute radiation exposures can be reliably assessed. Ongoing protracted exposures may be qualitatively detectable using DNA damage foci, but without any information on the cumulative dose.

Thanks to the high level of evolutionary conservation observed for the phosphorylation site of H2A(X), a serine four residues from the C-terminus, this marker flags up DNA double-strand breaks across a wide range of species including e.g. yeasts (Nakamura et al., 2004) and plants (Charbonnel et al., 2011). Siddiqui et al. (2013) studied the phosphorylation of the H2AX homologue H2avB in fruit flies following exposure to ionising radiation. This study demonstrates clearly and convincingly the concept of phospho-H2avB as a marker of radiation exposure in this organism, which is of significant environmental interest.

Apart from the highly short-lived nature of DNA damage foci, reflecting the dynamics of DNA damage induction and repair, the other important caveat for these assays is the lack of specificity for radiation. A wide range of environmental substances, pollutants, pathogens etc. may cause DNA damage in any exposed organisms. As DNA damage foci may not only form at sites of prompt DNA double-strand breaks but also whenever replication or even transcription processes encounter single-stranded DNA lesions, it is not always possible to conclusively link increases in DNA damage foci or, in fact most other biomarkers, to ionising radiation exposure in real world situations, i.e. outside controlled laboratory scenarios. Some random examples of "interfering" substances include benzo[a]pyrene, which was reported to induce mitotic chromosomal abnormalities, gamma-H2AX and genetic instability in human bronchial epithelial cells (Thaiparambil et al., 2017); cobalt nanoparticles which, when instilled intratracheally in mice, resulted in increased proliferation, gamma-H2AX immunostaining, 8-OHdG and point mutations in bronchiolar epithelial cells and pneumocytes (Wan et al., 2017); and diesel exhaust particles, which induced genotoxic effects such as increased levels of single-strand breaks, increased frequencies of micronuclei, or deregulated expression of genes involved in DNA damage signaling pathways but failed to induce oxidative DNA damage or DNA double-strand breaks, as assessed by gamma-H2AX (Kowalska et al., 2017).

9.5 Conclusion

There are nowadays more or less field-friendly solutions available for all investigated markers. Minimum detectable doses and signal stability over time vary drastically between the different markers, with gamma-H2AX/53BP1 DNA damage foci being able to detect exposures down to a few milligray shortly after the exposure but the signal being lost within hours to days (even at doses of several gray). Gene expression changes (not covered in this chapter) also tend to be short-lived, whereas dicentrics and micronuclei persist for several months and are thus better suited for chronic exposure scenarios, with minimum dose limits of several dozens to hundreds of milligray. Reciprocal translocations can be passed through successive cell divisions and may therefore be used for retrospective dose assessments decades after the exposure, albeit with the caveat of increased uncertainties due to the accumulation of 'spontaneous', i.e. radiation-independent translocations with age.

In summary, there are therefore numerous DNA damage-based markers of radiation exposure available, which all perform reasonably well in a setting where radiation is the only or predominant source of DNA damage. However, it remains a major challenge for all biomarkers of radiation exposure to deliver radiation-specific exposure information in uncontrolled environments that expose organisms to all kinds of agents. Therefore, the only way forward is the comprehensive assessment of ALL components in an ecosystem that could contribute to the total hazard burden. One of these may be ionising radiation. But it is rarely the only one, and it may well interact with others. In the long term, it would be useful to develop a detailed interaction matrix that could provide a robust basis for interpreting environmental radiation exposure biomarker measurements. Similar interactions may, of course, also contribute to the overall health status of an ecosystem.

References

Ainsbury, E. A., Al-Hafidh, J., Bajinskis, A., Barnard, S., Barquinero, J. F., Beinke, C., de Gelder, V., Gregoire, E., Jaworska, A., Lindholm, C., Lloyd, D., Moquet, J., Nylund, R., Oestreicher, U., Roch-Lefévre, S., Rothkamm, K., Romm, H., Scherthan, H., Sommer, S., ... Wojcik, A. (2014a). Inter- and intra-laboratory comparison of a multibiodosimetric approach to triage in a simulated, large scale radiation emergency. *International Journal of Radiation Biology, 90*(2), 193–202.

Ainsbury, E. A., Moquet, J., Rothkamm, K., Darroudi, F., Vozilova, A., Degteva, M., Azizova, T. V., Lloyd, D. C., & Harrison, J. (2014b). What radiation dose does the FISH translocation assay measure in cases of incorporated radionuclides for the Southern Urals populations? *Radiation Protection Dosimetry, 159*(1–4), 26–33.

Baeyens, A., Swanson, R., Herd, O., Ainsbury, E., Mabhengu, T., Willem, P., Thierens, H., Slabbert, J. P., & Vral, A. (2011). A semi-automated micronucleus-centromere assay to assess low-dose radiation exposure in human lymphocytes. *International Journal of Radiation Biology, 87*(9), 923–931.

Barber, R. C., Hickenbotham, P., Hatch, T., Kelly, D., Topchiy, N., Almeida, G. M., Jones, G. D., Johnson, G. E., Parry, J. M., Rothkamm, K., & Dubrova, Y. E. (2006). Radiation-induced transgenerational alterations in genome stability and DNA damage. *Oncogene, 25*(56), 7336–7342.

Barnard, S., Ainsbury, E. A., Al-hafidh, J., Hadjidekova, V., Hristova, R., Lindholm, C., Gil, O. M., Moquet, J., Moreno, M., Rößler, U., Thierens, H., Vandevoorde, C., Vral, A., Wojewódzka, M., & Rothkamm, K. (2015). The first gamma-H2AX biodosimetry intercomparison exercise of the developing European biodosimetry network RENEB. *Radiation Protection Dosimetry, 164*(3), 265–270.

Bauerschmidt, C., Arrichiello, C., Burdak-Rothkamm, S., Woodcock, M., Hill, M. A., Stevens, D. L., & Rothkamm, K. (2010). Cohesin promotes the repair of ionizing radiation-induced DNA double-strand breaks in replicated chromatin. *Nucleic Acids Research, 38*(2), 477–487.

Bhatti, P., Preston, D. L., Doody, M. M., Hauptmann, M., Kampa, D., Alexander, B. H., Petibone, D., Simon, S. L., Weinstock, R. M., Bouville, A., Yong, L. C., Freedman, D. M., Mabuchi, K., Linet, M. S., Edwards, A. A., Tucker, J. D., & Sigurdson, A. J. (2007). Retrospective biodosimetry among United States radiologic technologists. *Radiation Research, 167*(6), 727–734.

Brooks, R. A., & Winton, D. J. (1996). Determination of spatial patterns of DNA damage and repair in intestinal crypts by multi-cell gel electrophoresis. *Journal of Cell Science, 109*(Pt 8), 2061–2068.

Burdak-Rothkamm, S., Rothkamm, K., Folkard, M., Patel, G., Hone, P., Lloyd, D., Ainsbury, L., & Prise, K. M. (2009). DNA and chromosomal damage in response to intermittent extremely low-frequency magnetic fields. *Mutation Research, 672*(2), 82–89.

Charbonnel, C., Allain, E., Gallego, M. E., & White, C. I. (2011). Kinetic analysis of DNA double-strand break repair pathways in Arabidopsis. *DNA Repair, 10*(6), 611–619.

Darroudi, F. (1998). Detection of total- and partial-body irradiation in a monkey model: A comparative study of chromosomal aberration, micronucleus and premature chromosome condensation assays. *International Journal of Radiation Biology, 74*(2), 207–215.

Degteva, M. O., Shagina, N. B., Shishkina, E. A., Vozilova, A. V., Volchkova, A. Y., Vorobiova, M. I., Wieser, A., Fattibene, P., Monaca, S. D., Ainsbury, E., Moquet, J., Anspaugh, L. R., & Napier, B. A. (2015). Analysis of EPR and FISH studies of radiation doses in persons who lived in the upper reaches of the Techa River. *Radiation and Environmental Biophysics, 54*(4), 433–444.

Fenech, M., & Morley, A. A. (1986). Cytokinesis-block micronucleus method in human lymphocytes: Effect of in vivo ageing and low dose X-irradiation. *Mutation Research, 161*(2), 193–198.

Flegal, F. N., Devantier, Y., McNamee, J. P., & Wilkins, R. C. (2010). Quickscan dicentric chromosome analysis for radiation biodosimetry. *Health Physics, 98*(2), 276–281.

Flegal, F. N., Devantier, Y., Marro, L., & Wilkins, R. C. (2012). Validation of QuickScan dicentric chromosome analysis for high throughput radiation biological dosimetry. *Health Physics, 102*(2), 143–153.

Georgakilas, A. (2011). Detection of clustered DNA lesions: Biological and clinical applications. *World Journal of Biological Chemistry, 2*(7), 173–176.

Glei, M., Schneider, T., & Schlormann, W. (2016). Comet assay: An essential tool in toxicological research. *Archives of Toxicology, 90*(10), 2315–2336.

Gotoh, E., Tanno, Y., & Takakura, K. (2005). Simple biodosimetry method for use in cases of high-dose radiation exposure that scores the chromosome number of Giemsa-stained drug-induced prematurely condensed chromosomes (PCC). *International Journal of Radiation Biology, 81*(1), 33–40.

Hayata, I., Kanda, R., Minamihisamatsu, M., Furukawa, A., & Sasaki, M. S. (2001). Cytogenetical dose estimation for 3 severely exposed patients in the JCO criticality accident in Tokai-mura. *Journal of Radiation Research, 42*(Suppl), S149–S155.

Horn, S., Barnard, S., & Rothkamm, K. (2011). Gamma-H2AX-based dose estimation for whole and partial body radiation exposure. *PLoS One, 6*(9), e25113.

IAEA. (2011). *Cytogenetic Dosimetry: Applications in preparedness for and response to radiation emergencies*. International Atomic Energy Agency.

Kowalska, M., Wegierek-Ciuk, A., Brzoska, K., Wojewodzka, M., Meczynska-Wielgosz, S., Gromadzka-Ostrowska, J., Mruk, R., Øvrevik, J., Kruszewski, M., & Lankoff, A. (2017). Genotoxic potential of diesel exhaust particles from the combustion of first- and second-generation biodiesel fuels-the FuelHealth project. *Environmental and Molecular Mutagenesis, 24*(31), 24223–24234.

Lobrich, M., Rief, N., Kuhne, M., Heckmann, M., Fleckenstein, J., Rube, C., & Uder, M. (2005). In vivo formation and repair of DNA double-strand breaks after computed tomography examinations. *Proceedings of the National Academy of Sciences of the United States of America, 102*(25), 8984–8989.

M'kacher, R., Maalouf, E. E. L., Ricoul, M., Heidingsfelder, L., Laplagne, E., Cuceu, C., Hempel, W. M., Colicchio, B., Dieterlen, A., & Sabatier, L. (2014). New tool for biological dosimetry: Reevaluation and automation of the gold standard method following telomere and centromere staining. *Mutation Research, 770*, 45–53.

Moquet, J., Barnard, S., Staynova, A., Lindholm, C., Monteiro Gil, O., Martins, V., Rößler, U., Vral, A., Vandevoorde, C., Wojewódzka, M., & Rothkamm, K. (2017). The second gamma-H2AX assay inter-comparison exercise carried out in the framework of the European biodosimetry network (RENEB). *International Journal of Radiation Biology, 93*(1), 58–64.

Nakamura, T. M., Du, L. L., Redon, C., & Russell, P. (2004). Histone H2A phosphorylation controls Crb2 recruitment at DNA breaks, maintains checkpoint arrest, and influences DNA repair in fission yeast. *Molecular and Cellular Biology, 24*(14), 6215–6230.

Pantelias, G. E., & Maillie, H. D. (1984). The use of peripheral blood mononuclear cell prematurely condensed chromosomes for biological Dosimetry. *Radiation Research, 99*(1), 140–150.

Romm, H., Ainsbury, E., Barnard, S., Barrios, L., Barquinero, J. F., Beinke, C., Deperas, M., Gregoire, E., Koivistoinen, A., Lindholm, C., Moquet, J., Oestreicher, U., Puig, R., Rothkamm, K., Sommer, S., Thierens, H., Vandersickel, V., Vral, A., & Wojcik, A. (2013). Automatic scoring of dicentric chromosomes as a tool in large scale radiation accidents. *Mutation Research, 756*(1–2), 174–183.

Romm, H., Ainsbury, E., Bajinskis, A., Barnard, S., Barquinero, J. F., Barrios, L., Beinke, C., Puig-Casanovas, R., Deperas-Kaminska, M., Gregoire, E., Oestreicher, U., Lindholm, C., Moquet, J., Rothkamm, K., Sommer, S., Thierens, H., Vral, A., Vandersickel, V., & Wojcik, A. (2014a). Web-based scoring of the dicentric assay, a collaborative biodosimetric scoring strategy for population triage in large scale radiation accidents. *Radiation and Environmental Biophysics, 53*(2), 241–254.

Romm, H., Ainsbury, E., Barnard, S., Barrios, L., Barquinero, J. F., Beinke, C., Deperas, M., Gregoire, E., Koivistoinen, A., Lindholm, C., Moquet, J., Oestreicher, U., Puig, R., Rothkamm, K., Sommer, S., Thierens, H., Vandersickel, V., Vral, A., & Wojcik, A. (2014b). Validation of semi-automatic scoring of dicentric chromosomes after simulation of three different irradiation scenarios. *Health Physics, 106*(6), 764–771.

Romm, H., Oestreicher, U., Ainsbury, E. A., Moquet, J., Barquinero, J. F., Barrios, L., Beinke, C., Cucu, A., Noditi, M., Popescu, I., Domene, M. M., Prieto, M. J., Filippi, S., Palitti, F., Monteiro Gil, O., Gregoire, E., Hadjidekova, V., Hatzi, V., Pantelias, G., ... Wojcik, A. (2017). Web based scoring is useful for validation and harmonisation of scoring criteria within RENEB. *International Journal of Radiation Biology, 93*(1), 110–117.

Rothkamm, K., & Horn, S. (2009). Gamma-H2AX as protein biomarker for radiation exposure. *Annali dell'Istituto superiore di sanita, 45*(3), 265–271.

Rothkamm, K., & Lloyd, D. (2014). *7.14 – Established and emerging methods of biological dosimetry A2 – Brahme, Anders. Comprehensive biomedical physics* (pp. 289–310). Elsevier.

Rothkamm, K., & Lobrich, M. (2002). Misrepair of radiation-induced DNA double-strand breaks and its relevance for tumorigenesis and cancer treatment (review). *International Journal of Oncology, 21*(2), 433–440.

Rothkamm, K., & Lobrich, M. (2003). Evidence for a lack of DNA double-strand break repair in human cells exposed to very low x-ray doses. *Proceedings of the National Academy of Sciences of the United States of America, 100*(9), 5057–5062.

Rothkamm, K., Kuhne, M., Jeggo, P. A., & Lobrich, M. (2001). Radiation-induced genomic rearrangements formed by nonhomologous end-joining of DNA double-strand breaks. *Cancer Research, 61*(10), 3886–3893.

Rothkamm, K., Kruger, I., Thompson, L. H., & Lobrich, M. (2003). Pathways of DNA double-strand break repair during the mammalian cell cycle. *Molecular and Cellular Biology, 23*(16), 5706–5715.

Rothkamm, K., Balroop, S., Shekhdar, J., Fernie, P., & Goh, V. (2007). Leukocyte DNA damage after multi-detector row CT: A quantitative biomarker of low-level radiation exposure. *Radiology, 242*(1), 244–251.

Rothkamm, K., Barnard, S., Ainsbury, E. A., Al-Hafidh, J., Barquinero, J.-F., Lindholm, C., Moquet, J., Perälä, M., Roch-Lefèvre, S., Scherthan, H., Thierens, H., Vral, A., & Vandersickel, V. (2013a). Manual versus automated gamma-H2AX foci analysis across five European laboratories: Can this assay be used for rapid biodosimetry in a large scale radiation accident? *Mutation Research, 756*(1–2), 170–173.

Rothkamm, K., Beinke, C., Romm, H., Badie, C., Balagurunathan, Y., Barnard, S., Bernard, N., Boulay-Greene, H., Brengues, M., De Amicis, A., De Sanctis, S., Greither, R., Herodin, F., Jones, A., Kabacik, S., Knie, T., Kulka, U., Lista, F., Martigne, P., ... Abend, M. (2013b).

Comparison of established and emerging biodosimetry assays. *Radiation Research, 180*(2), 111–119.

Rothkamm, K., Horn, S., Scherthan, H., Rössler, U., De Amicis, A., Barnard, S., Kulka, U., Lista, F., Meineke, V., Braselmann, H., Beinke, C., & Abend, M. (2013c). Laboratory intercomparison on the gamma-H2AX foci assay. *Radiation Research, 180*(2), 149–155.

Rothkamm, K., Barnard, S., Moquet, J., Ellender, M., Rana, Z., & Burdak-Rothkamm, S. (2015). DNA damage foci: Meaning and significance. *Environmental and Molecular Mutagenesis, 56*(6), 491–504.

Sak, A., Grehl, S., Erichsen, P., Engelhard, M., Grannass, A., Levegrün, S., Pöttgen, C., Groneberg, M., & Stuschke, M. (2007). Gamma-H2AX foci formation in peripheral blood lymphocytes of tumor patients after local radiotherapy to different sites of the body: Dependence on the dose-distribution, irradiated site and time from start of treatment. *International Journal of Radiation Biology, 83*(10), 639–652.

Siddiqui, M. S., Filomeni, E., François, M., Collins, S. R., Cooper, T., Glatz, R. V., Taylor, P. W., Fenech, M., & Leifert, W. R. (2013). Exposure of insect cells to ionising radiation in vivo induces persistent phosphorylation of a H2AX homologue (H2AvB). *Mutagenesis, 28*(5), 531–541.

Sigurdson, A. J., Bhatti, P., Preston, D. L., Doody, M. M., Kampa, D., Alexander, B. H., Petibone, D., Yong, L. C., Edwards, A. A., Ron, E., & Tucker, J. D. (2008a). Routine diagnostic X-ray examinations and increased frequency of chromosome translocations among U.S. radiologic technologists. *Cancer Research, 68*(21), 8825–8831.

Sigurdson, A. J., Ha, M., Hauptmann, M., Bhatti, P., Sram, R. J., Beskid, O., Tawn, E. J., Whitehouse, C. A., Lindholm, C., Nakano, M., Kodama, Y., Nakamura, N., Vorobtsova, I., Oestreicher, U., Stephan, G., Yong, L. C., Bauchinger, M., Schmid, E., Chung, H. W., ... Tucker, J. D. (2008b). International study of factors affecting human chromosome translocations. *Mutation Research, 652*(2), 112–121.

Sotnik, N. V., Azizova, T. V., Darroudi, F., Ainsbury, E. A., Moquet, J. E., Fomina, J., Lloyd, D. C., Hone, P. A., & Edwards, A. A. (2015). Verification by the FISH translocation assay of historic doses to Mayak workers from external gamma radiation. *Radiation and Environmental Biophysics, 54*(4), 445–451.

Terzoudi, G. I., Pantelias, G., Darroudi, F., Barszczewska, K., Buraczewska, I., Depuydt, J., Georgieva, D., Hadjidekova, V., Hatzi, V. I., Karachristou, I., Karakosta, M., Meschini, R., M'Kacher, R., Montoro, A., Palitti, F., Pantelias, A., Pepe, G., Ricoul, M., Sabatier, L., ... Wojcik, A. (2017). Dose assessment intercomparisons within the RENEB network using G0-lymphocyte prematurely condensed chromosomes (PCC assay). *International Journal of Radiation Biology, 93*(1), 48–57.

Thaiparambil, J., Mansour, O., & El-Zein, R. (2017). Effect of benzo[a]pyrene on spindle misorientation and fidelity of chromosome segregation in lung epithelial BEAS-2B cells. *Journal of Toxicology*.

Thierens, H., Vral, A., Barbe, M., Aousalah, B., & De Ridder, L. (1999). A cytogenetic study of nuclear power plant workers using the micronucleus-centromere assay. *Mutation Research, 445*(1), 105–111.

Thierens, H., Vral, A., Morthier, R., Aousalah, B., & De Ridder, L. (2000). Cytogenetic monitoring of hospital workers occupationally exposed to ionizing radiation using the micronucleus centromere assay. *Mutagenesis, 15*(3), 245–249.

Thierens, H., Vral, A., Vandevoorde, C., Vandersickel, V., de Gelder, V., Romm, H., Oestreicher, U., Rothkamm, K., Barnard, S., Ainsbury, E., Sommer, S., Beinke, C., & Wojcik, A. (2014). Is a semi-automated approach indicated in the application of the automated micronucleus assay for triage purposes? *Radiation Protection Dosimetry, 159*(1–4), 87–94.

Tucker, J. D., & Luckinbill, L. S. (2011). Estimating the lowest detectable dose of ionizing radiation by FISH whole-chromosome painting. *Radiation Research, 175*(5), 631–637.

Vozilova, A. V., Shagina, N. B., Degteva, M. O., Edwards, A. A., Ainsbury, E. A., Moquet, J. E., Hone, P., Lloyd, D. C., Fomina, J. N., & Darroudi, F. (2012). Preliminary FISH-based assessment of external dose for residents exposed on the Techa River. *Radiation Research, 177*(1), 84–91.

Vozilova, A. V., Shagina, N. B., Degteva, M. O., Moquet, J., Ainsbury, E. A., & Darroudi, F. (2014). FISH analysis of translocations induced by chronic exposure to Sr radioisotopes: Second set of analysis of the Techa River Cohort. *Radiation Protection Dosimetry, 159*(1–4), 34–37.

Vral, A., Thierens, H., & De Ridder, L. (1997). In vitro micronucleus-centromere assay to detect radiation-damage induced by low doses in human lymphocytes. *International Journal of Radiation Biology, 71*(1), 61–68.

Vral, A., Decorte, V., Depuydt, J., Wambersie, A., & Thierens, H. (2016). A semiautomated FISHbased micronucleuscentromere assay for biomonitoring of hospital workers exposed to low doses of ionizing radiation. *Molecular Medicine Reports, 14*(1), 103–110.

Wahab, M. A., Nickless, E. M., Najar-M'kacher, R., Parmentier, C., Podd, J. V., & Rowland, R. E. (2008). Elevated chromosome translocation frequencies in New Zealand nuclear test veterans. *Cytogenetic and Genome Research, 121*(2), 79–87.

Wan, R., Mo, Y., Zhang, Z., Jiang, M., Tang, S., & Zhang, Q. (2017). Cobalt nanoparticles induce lung injury. *DNA Damage and Mutations in Mice, 14*(1), 38.

Yong, L. C., Sigurdson, A. J., Ward, E. M., Waters, M. A., Whelan, E. A., Petersen, M. R., Bhatti, P., Ramsey, M. J., Ron, E., & Tucker, J. D. (2009). Increased frequency of chromosome translocations in airline pilots with long-term flying experience. *Occupational and Environmental Medicine, 66*(1), 56–62.

Chapter 10
The Development of Bio-assays Based on Non-targeted Effects of Radiation; a Potential Worm-Hole into Ecosystem Level Biomarkers

Carmel Mothersill, Andrej Rusin, and Colin Seymour

Abstract The role of non-targeted effects (NTE) in radiation biology and radiation protection is problematic and controversial. These effects include bystander signaling between irradiated and non-irradiated cells and also describe effects in progeny of irradiated progenitors which have recovered – a form of non-clonal genomic instability. They dominate the low dose response but saturate after very low doses yielding flat dose response curves meaning that low doses can have disproportionately large effects. In vivo many of the impacts of NTE are adaptive or protective. There are many well developed assays for NTE but because they are not based on measurement of primary DNA damage, they are not usually used to monitor radiation effects in humans or the environment. In this paper, we will review the use of NTE assays in human and environmental radiobiology and will propose that far from being irrelevant, these assays may provide the much needed system level biomarkers which could predict perturbations at the ecosystem level.

Keywords Non-targeted effects · Bystander effects · Radiation environment · Ecosystem · Biomarkers

---

C. Mothersill (✉) · A. Rusin · C. Seymour
Department of Biology, McMaster University, Hamilton, ON, Canada
e-mail: mothers@mcmaster.ca

M. D. Wood et al. (eds.), *Biomarkers of Radiation in the Environment*, NATO Science for Peace and Security Series A: Chemistry and Biology,
https://doi.org/10.1007/978-94-024-2101-9_10

10.1 Background and Potential Relevance of NTE

10.1.1 What Are NTE?

Non-targeted effects (NTE) mainly refer to biological effects of radiation which are seen in cells, tissues or organisms which did not receive any energy deposition due to radiation track traversal (Prise et al., 2002; Mothersill & Seymour, 2005; Mothersill et al., 2017a, b). They include bystander effects (effects in neighbours receiving signals from irradiated entities) (Nagasawa & Little, 1992; Mothersill & Seymour, 1997a) and genomic instability (delayed expression of effects in progeny of irradiated survivors) (Seymour et al., 1986; Pampfer & Streffer, 1989; Kadhim et al., 1992). Low dose hypersensitivity (HRS) and adaptive responses (AR) are sometimes included but they are more properly called "low dose effects" as they occur in targeted or non-targeted scenarios (Olivieri et al., 1984; Bosi & Olivieri, 1989; Marples et al., 1994; Skov et al., 1994; Marples & Joiner, 1995; Mothersill & Seymour, 2001, 2004, 2006a; Mothersill et al., 2002; Matsumoto et al., 2004; Kadhim et al., 2013). Figure 10.1 shows the types of NTE and how they alter our perceptions of radiation "targets". Most of the early work in the NTE field was done with cell cultures but it is now clear they occur in vivo both within and between organisms and thus have important implications for both radiotherapy and for radiation protection (Belyakov et al., 2000; Audette-Stuart et al., 2011; Barescut et al., 2011; Mothersill et al., 2017a, b; Rusin et al., 2019; Schofield et al., 2018). This discussion paper focuses on the potential importance of these effects in environmental radiation protection.

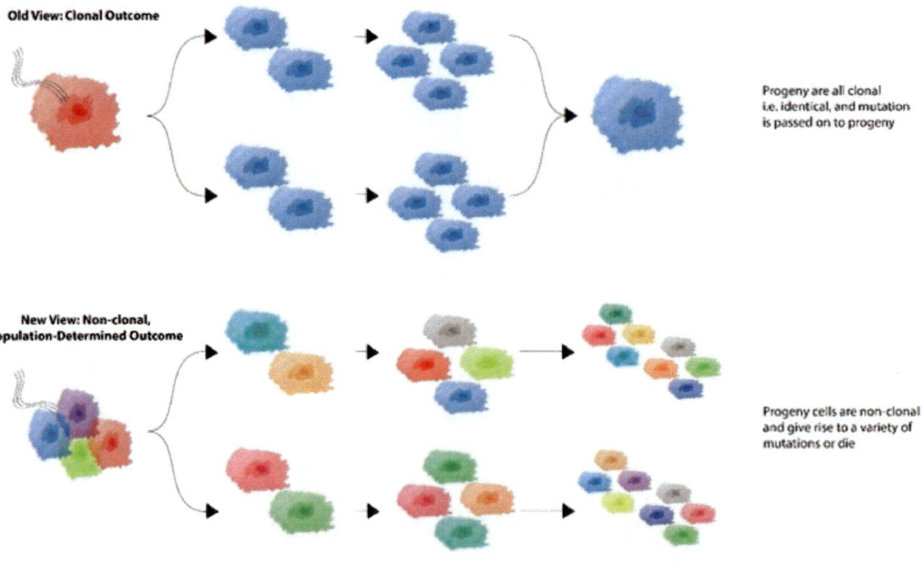

Fig. 10.1 Types of non-targeted effect and how they change our perception of the "target" for radiation damage

10.1.2 *Environmental Studies and Prevalence in Species*

Initial studies of NTE were mainly concerned with the potential importance of these effects for cancer – both carcinogenesis, where they were thought to increase the size of the target for radiation induced mutations (Iyer & Lehnert, 2000; Watson et al., 2000; Camphausen et al., 2003; Mothersill & Seymour 2003; Nagar et al., 2003; Azzam & Little, 2004) and as possible novel targets for cancer therapy where it was thought that if the bystander effect, which occurs preferentially in normal tissue, could be inhibited it would result in a more favourable therapeutic ratio (Azzam & Little, 2004; Mothersill et al., 2004; Zwicker et al., 2004; Mothersill & Seymour, 2006b; Hamada, 2008; Prise & O'sullivan, 2009; Wideł et al., 2009; Marín et al., 2015) However perhaps it is the area of environmental radiation protection rather than cancer that NTE could have most impact. Numerous studies have shown NTE in fish, amphibians, birds and mammals. They have also been documented in annelids, molluscs, crustaceans and in plants and yeast (Nagasawa & Little, 1992; Mothersill & Seymour, 1997a, b, 2013; Prise, 1998; Zhou et al., 2000; Mothersill et al., 2001a, b, 2005, 2006, 2007a, b; Surinov et al., 2004a, b; Dowling et al., 2005; Olwell et al., 2005; DeVeaux et al., 2006; Yang et al., 2007; Smith et al., 2013a, b; Múčka et al., 2015; Mavragani et al., 2016; Rusin et al., 2019). In fact, all species which have been tested for NTE have shown evidence of bystander effects or genomic instability. AR and HRS have a similar wide distribution in biota (Joiner et al., 1985, 1996; Shadley et al., 1987; Joiner & Johns, 1988; Marples & Joiner, 1993; Azzam et al., 1994; Hamilton et al., 1996; Marples et al., 1997, 2004; Mothersill & Seymour, 2006a; Marples & Collis, 2008; Ryan et al., 2008; Fernandez-Palomo et al., 2016; Mothersill et al., 2017a, b; Vo et al., 2017). While this means they have undoubtedly a role to play in determining the outcome after low and chronic exposures, their relevance may go much deeper.

The big challenges in environmental radiation protection are:

1. Impacts should be measured at the level of the population and ecosystem rather than at the level of the individual, but biomarkers only really predict at present for individual impact (Brechignac et al., 2008; Salbu, 2009; Caffrey et al., 2014; Schofield et al., 2018).
2. Wild organisms are exposed to multiple stressors, physical, chemical, biological and psychological (Salbu, 2009; Bréchignac, 2012; Vanhoudt et al., 2012; Schofield et al., 2018). Little is known about how these stressors interact.
3. Much of the available data for non-human species relate to laboratory experiments on a few model organisms and where comparisons have been made between field and laboratory data, field conditions lead to a much greater than expected dose effect than would be predicted from the available laboratory data (Mothersill & Seymour, 2018; Omar-Nazir et al., 2018; Schofield et al., 2018).

Consideration of NTE, rather than further complicate these issues, may in fact provide new ways of thinking and new solutions. These ideas are explored in the next few paragraphs.

10.1.2.1 Inter-organism Communication

A key outcome of bystander signalling is that it spreads messages from the impacted individual to others in the vicinity. Initially the signalling was thought to be confined to the cell population or the tissues within an organism but there are now multiple reports of inter-organism signalling. This has been shown in 3 different fish species and rodents by our group (Dowling et al., 2005; Mothersill et al., 2005, 2007a, b; Smith et al., 2007, 2013b; Saroya et al., 2010), in plants (Yang et al., 2007; Boyko & Kovalchuk, 2011), in frogs (Audette-Stuart et al., 2011) and in zebrafish embryos (Yum et al., 2009). We have also reported communication between bloodworm prey and the salmonid predator (Smith et al., 2013a, b). Outside of the radiation field, chemical ecology is a vibrant discipline dealing with chemical communication between different species, resulting in both beneficial and adverse impacts on survival (reviewed in Mothersill & Seymour, 2004, 2005). This widespread use of chemicals, to warn, evade, upregulate defences or attract made us consider whether bystander signalling induced by radiation exposure might be a means of coordinating a system level response to a changed environment (i.e. one now containing radiation). If so, then measurement of bystander signal production could be a system level biomarker of population "stress" or perception of change. Figure 10.2 is an attempt to depict this idea.

Similarly, what we call "genomic instability" may be actually an increased tolerance for mutation burden in a population. In static species in a controlled stable environment, mutations should be eliminated efficiently but in a changing environment, adaptation cannot occur in the absence of genetic plasticity. Therefore, measurement of genomic instability could be another indicator of "stress". This measure would have the great advantage of addressing the issue of multigenerational populations and ecosystems. Figure 10.3 is an attempt to convey this concept.

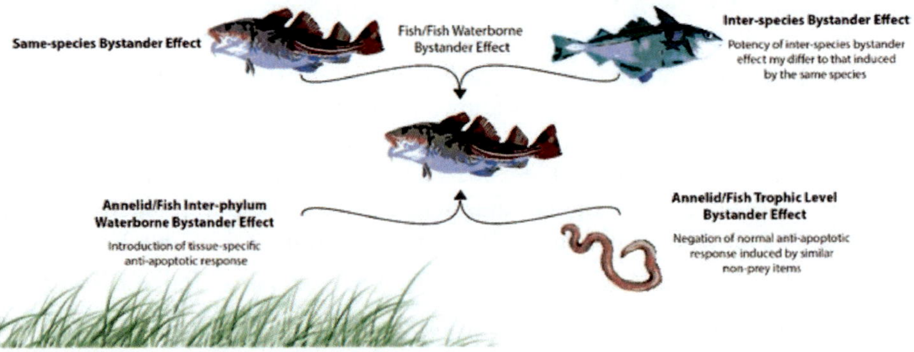

Fig. 10.2 Bystander signals as a measure of population "stress" or perception of environmental change

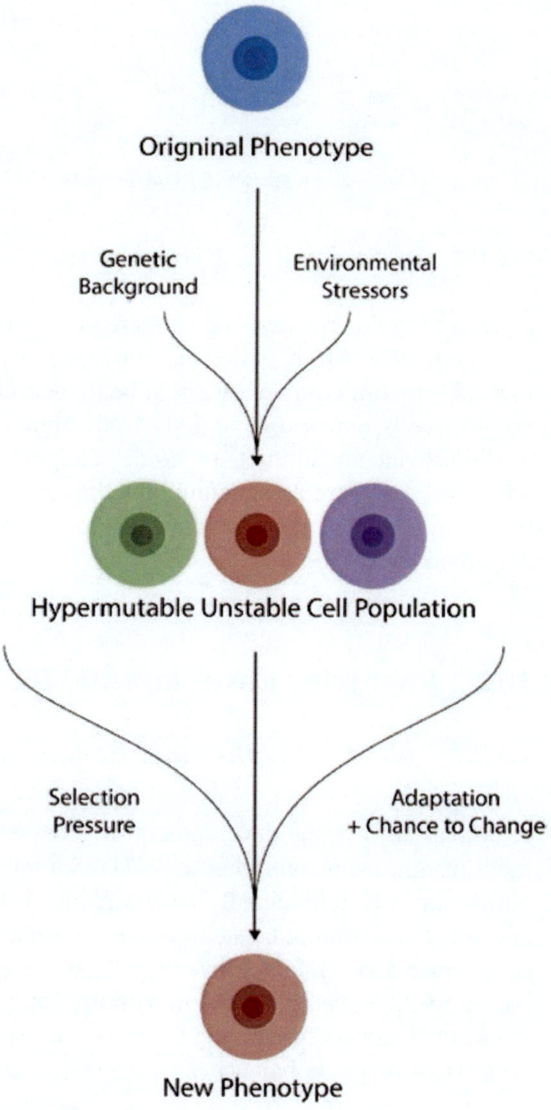

Fig. 10.3 Non-clonal delayed genomic instability results from multigenerational stress and facilitates adaptation

10.1.2.2 Relevance for Other Stressors

In relation to the second challenge, both bystander effects and genomic instability have been demonstrated after exposure to a variety of other stressors, including heavy metals (Coen et al., 2003; Dowling et al., 2005; Mothersill et al., 2007a, b; Salbu et al., 2008), organics such as prochloros and nonoxynol (Dowling & Mothersill, 1999; Dowling et al., 2005; Gowans et al., 2005; Mothersill & Seymour, 2009), and ultraviolet light (O'Reilly & Mothersill, 1997; Dahle et al., 2005a, b; Whiteside & McMillan, 2009; Widel et al., 2014; Le et al., 2015). The dose effect relationship for NTE after radiation exposure saturates after 0.5 Gy acute x-ray

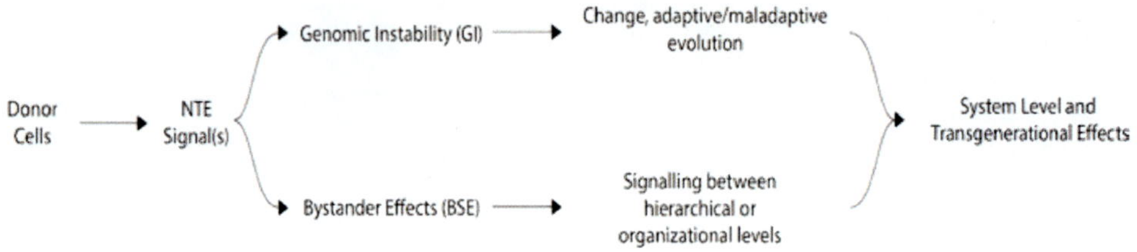

Fig. 10.4 Generic scheme for a NTE reporter assay

exposure but this relationship is altered by co-exposure to chemicals (Mothersill & Seymour, 1998; Mothersill et al., 2007a, b). A possible approach to simplifying the multiple stressor complexity might be to measure the level of NTE dose response in reporter cells exposed to signals from organisms collected from test sites. This would provide an "ultimate outcome" endpoint without the need for understanding what stressors were present and how they interacted. Figure 10.4 provides a possible assay scheme. Case studies where this has been tried as an approach are described later.

10.2 Improving Environmental Biomarkers

10.2.1 Need for Non-lethal Sampling

A critical need in the development of ecosystem level biomarkers is the development of non-lethal sampling methods. Sampling to inform about ecosystem level parameters will undoubtedly have a focus on keystone species and rare or critical species. These cannot be sampled in the usual way by harvesting them using traps or other devices which remove them from the ecosystem. Development of methods based on blood samples, sperm samples, hair, feathers or scales would minimise stress and instead of removing the organism from its environment would not disrupt delicate biodiversity balances.

10.2.2 Need for Population and Ecosystem Level Markers

While the above suggestions give some possible ways to investigate coordination of responses and adaptation in populations, they still use measurements on individuals to predict population and ecosystem level impacts. They could be a small step in the right direction but ultimately it is necessary to identify and develop predictors of ecosystem health. These will probably require modelling of ecosystem structure with identification of drifts in population ranges and in biodiversity indices in relation to a basket of environmental stressors of which radiation will be one. The NTEI or non-targeted effect index for the system, relative to a control could be a useful parameter to indicate stress burden in a multi-generational and multiple species context. Another conceptual approach we are developing is to adapt image analysis techniques used in medicine such as computed tomography (CT) and magnetic resonance imaging (MRI) to large scale environmental image analysis. This will require the development of algorithms suited to macro-scale analysis but if successful it could be an immensely powerful method for establishing the relative health of ecosystems of any scale.

10.3 Case Studies

10.3.1 How NTE Markers Relate to Ecosystem

The remaining part of this discussion paper relates to specific case studies by our group and others where NTE effects could be relevant biomarkers for environmental radiation protection. As stated earlier these studies are a "work-in-progress" and are not being presented as finished definitive new approaches.

10.3.2 Fish Case Study

Many studies from our laboratory used an explant culture system to look at effects of various in vivo irradiation scenarios on the production of NTE (Mothersill et al., 1988a, b, 1992, 1995a, b, 1999, 2001b, 2005, 2006; Sheridan et al., 1997; Belyakov et al., 2002, 2003, 2006; O'Dowd et al., 2006). The protocol is shown in Fig. 10.5. Basically, after irradiation in vivo under various conditions, tissues are taken from the fish, chopped up and plated as explants in tissue culture medium. The medium is harvested after 24–48 h by which time the explants have adhered and started to grow.

The harvested medium is filtered and placed on unirradiated reporter cells seeded in culture flasks for assay of reproductive survival, or in multiwall plates for assay of more immediate endpoints such as mitochondrial activity or calcium flux. The original explants can be cultured on to give information about direct effects of the radiation dose using immunocytochemistry or proteomics techniques. As a further extension of this technique, "bystander" fish were put swimming with the irradiated fish to examine inter-animal communication in vivo. A summary of some relevant data obtained from this type of experiment over several years is shown in Table 10.1. The technique can be used for non-lethal sampling as fin clip explants grow particularly well and so not require killing of the fish.

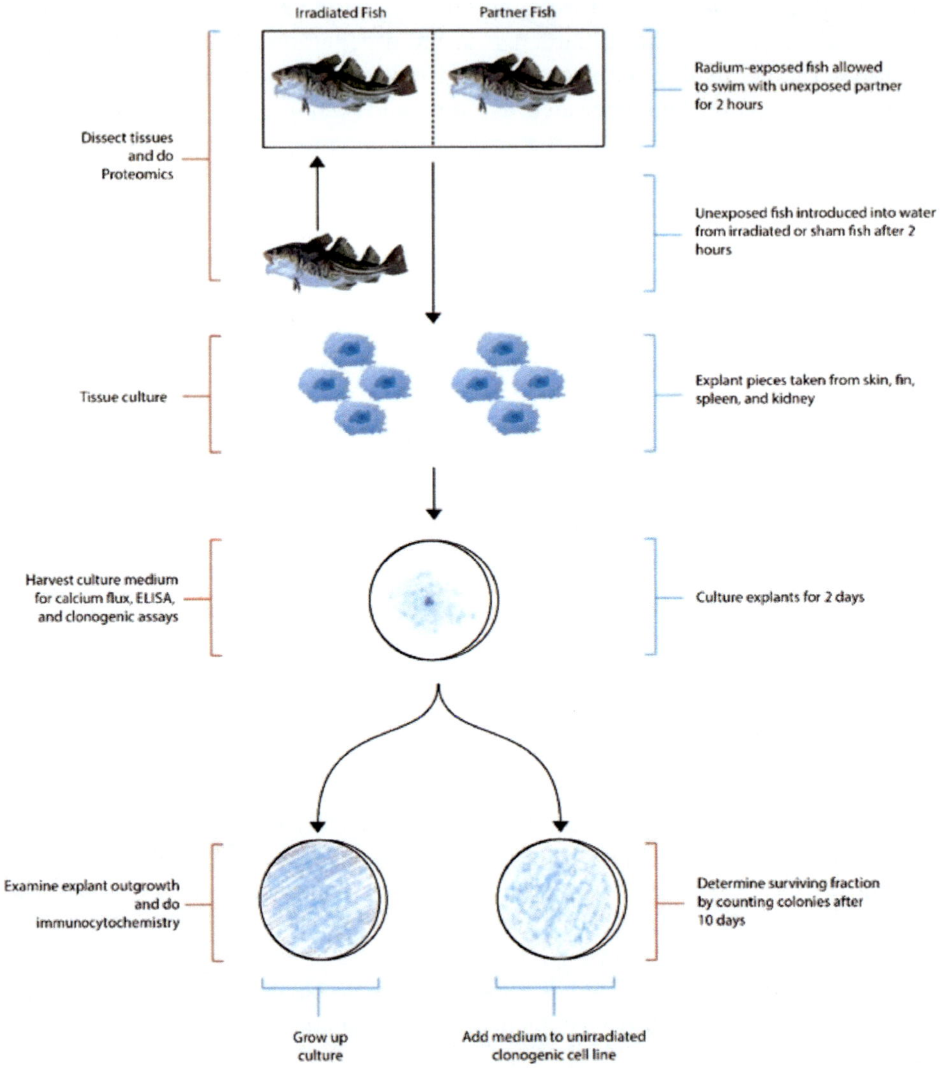

Fig. 10.5 Application of the reporter assay for NTE to fish populations

10.3.3 Worm Case Study

Work by Hertel-Aas et al. (2011a, b) and our group (Rusin et al., 2019), used two approaches to look at delayed and non-targeted effects of low doses in earthworms. The Norwegian group studied hatching success in worms over several generations and found that after a once off acute exposure, it took four generations before effects were seen in the cocoon hatching rates. This is what we see in cases of delayed genomic instability. In the study from our group in collaboration with Oughton's group, worms were collected from contaminated sites and control sites. All group were exposed to in vivo irradiation in Norway. Worms were dissected after 24 h and

Table 10.1 Highlights from fish bystander studies

Result	Conclusion
Bystander effect induced in 3 species of fish exposed to irradiated fish or their water	Evolutionary conserved mechanism
Attenuation of signal only seen after fish removed for 6 h from water	Stable water-soluble signal
Live fish continue to emit signal for over 12 h	
Chronically exposed Medaka confer an adaptive response on reported cells	Chronic radiation effect different to acute effect
Multiple stressors appear to have sub-additive effects	Suggests a saturable or antagonistic mechanism
Bystander proteome and direct irradiation proteome very different	Importance for understanding potential risk outcomes
Effect can be demonstrated in trout as early as the eyed egg stage and is still there in retested adults one year on	Persistent effect once induced
Serotonin involved *in vivo* and *in vitro* in fish and mammalian cells	Conserved mechanism

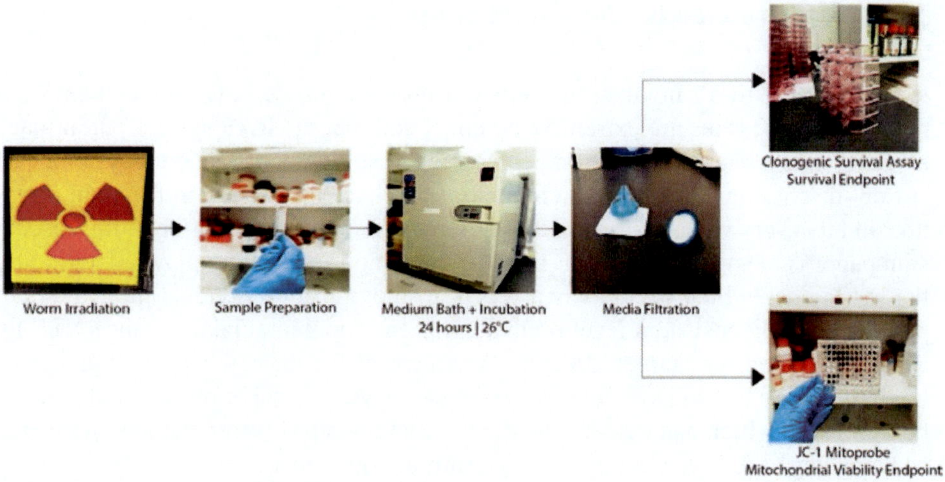

Fig. 10.6 Application of the reporter assay for NTE to worm populations

different body parts were incubated in tissue culture medium for a further 24 h. The medium was then pipetted off and filtered and tested for expression of bystander signals using the techniques described above for fish. In this pilot study, differences in radiation responses between species and body parts were seen, giving proof of principle that bystander responses can be measured in this way. The technique is shown in Fig. 10.6.

10.3.4 Frog Case Study

The final case study concerns amphibians taken from clean and tritium contaminated ponds at Canadian Nuclear Laboratories (then Atomic Energy Canada Ltd) by Audette-Stuart and Yankovich (Audette-Stuart et al., 2011; Barescut et al., 2011). In an early study this group showed that cells cultured from frogs taken from the tritium contaminated ponds were protected against an acute dose of 4 Gy gamma rays delivered in vivo or in vitro compared with frogs with no prior radiation exposure. In the 2011 study, they showed that if tadpoles from the contaminated site swam with tadpoles from clean sites, they conferred the adaptive response on the "clean" tadpoles. This adds a further layer of complexity to interpretation of NTE in the environment because it shows both the ability of the tadpoles to transmit signals inducing an adaptive response, but further, it suggests that chronically irradiated organisms may be more resistant to acute exposures that organisms from pristine environments.

10.4 Future Needs and Conclusion

It is very clear to all involved in radiation protection of ecosystems and environments, that the issues are extremely complex and that the RAP system put in place by ICRP was a backstop position pending the development of a better system. The current discussion about ecosystem approaches is clearly a step in the right direction and it is very valuable to have these discussions. The contribution of this discussion paper is to suggest that NTE may have an important role to play in facilitating the bottle neck which occurs when we try to move from individual biomarkers to population and ecosystem level predictors of environmental health. Since NTE by definition widen the "target" from the recipient of the energy deposited along the track to a more holistic system level response they also enable models and ideas to be developed which can help bridge the conceptual gap between the dose delivered as being key to the system response determining outcome.

Acknowledgements We acknowledge continued support from the National Sciences and Engineering Research Council 's Discovery and Cooperative Research and Development Programmes, the National CFIDS (Chronic Fatigue and Immune Deficiency Syndrome) Foundation Inc. and the Canada Research Council Chairs programme.

References

Audette-Stuart, M., Kim, S. B., McMullin, D., Festarini, A., Yankovich, T. L., Carr, J., & Mulpuru, S. (2011). Adaptive response in frogs chronically exposed to low doses of ionizing radiation in the environment. *Journal of Environmental Radioactivity, 102*, 566–573.

Azzam, E. I., & Little, J. B. (2004). The radiation-induced bystander effect: Evidence and significance. *Human & Experimental Toxicology, 23*, 61–65.

Azzam, E. I., Raaphorst, G. P., & Mitchel, R. E. J. (1994). Radiation-induced adaptive response for protection against micronucleus formation and neoplastic transformation in C3H 10T1/2 mouse embryo cells. *Radiation Research, 138*, S28–S31.

Barescut, J., Lariviere, D., Stocki, T., Audette-Stuart, M., & Yankovich, T. (2011). Bystander effects in bullfrog tadpoles. *Radioprotection, 46*, S497–S497.

Belyakov, O. V., Folkard, M., Mothersill, C., Prise, K. M., & Michael, B. D. (2000). Bystander effect and genomic instability. Challenging the classic paradigm of radiobiology. In *Modern problems of radiobiology, radioecology and evolution. Proceedings of the International conference dedicated to the centenary of the birth of NW Timofeeff-Ressovsky. Dubna, Russia* (pp. 6–9).

Belyakov, O. V., Folkard, M., Mothersill, C., Prise, K. M., & Michael, B. D. (2002). Bystander-induced apoptosis and premature differentiation in primary urothelial explants after charged particle microbeam irradiation. *Radiation Protection Dosimetry, 99*, 249–251.

Belyakov, O. V., Folkard, M., Mothersill, C., Prise, K. M., & Michael, B. D. (2003). A proliferation-dependent bystander effect in primary porcine and human urothelial explants in response to targeted irradiation. *British Journal of Cancer, 88*, 767–774. https://doi.org/10.1038/sj.bjc.6600804

Belyakov, O. V., Folkard, M., Mothersill, C., Prise, K. M., & Michael, B. D. (2006). Bystander-induced differentiation: A major response to targeted irradiation of a urothelial explant model. *Mutation Research – Fundamental and Molecular Mechanisms of Mutagenesis, 597*, 43–49. https://doi.org/10.1016/j.mrfmmm.2005.08.012

Bosi, A., & Olivieri, G. (1989). Variability of the adaptive response to ionizing radiations in humans. *Mutation Research – Fundamental and Molecular Mechanisms of Mutagenesis, 211*, 13–17.

Boyko, A., & Kovalchuk, I. (2011). Genetic and epigenetic effects of plant–pathogen interactions: An evolutionary perspective. *Molecular Plant, 4*, 1014–1023.

Bréchignac, F. (2012). Environment protection: The current challenge in radioecology. In *EPJ web of conferences* (Vol. 24, p. 1001). EDP Sciences.

Brechignac, F., Alexakhin, R., Godoy, J. M., Oughton, D., Sheppard, S., & Strand, P. (2008). Integrating environment protection, a new challenge: Strategy of the International Union of Radioecology. *Radioprotection, 43*, 339–356.

Caffrey, E. A., Leonard, M. E., Napier, J. B., Neville, D. R., & Higley, K. A. (2014). Radioecology: why bother? *Journal of Environmental Protection (Irvine,. Calif), 5*, 181.

Camphausen, K., Moses, M. A., Ménard, C., Sproull, M., Beecken, W.-D., Folkman, J., & O'Reilly, M. S. (2003). Radiation Abscopal antitumor effect is mediated through p53. *Cancer Research, 63*, 1990 LP-1993.

Coen, N., Kadhim, M. A., Wright, E. G., Case, C. P., & Mothersill, C. E. (2003). Particulate debris from a titanium metal prosthesis induces genomic instability in primary human fibroblast cells. *British Journal of Cancer, 88*, 548–552. https://doi.org/10.1038/sj.bjc.6600758

Dahle, J., Kaalhus, O., Stokke, T., & Kvam, E. (2005a). Bystander effects may modulate ultraviolet A and B radiation-induced delayed mutagenesis. *Radiation Research, 163*, 289–295.

Dahle, J., Kvam, E., & Stokke, T. (2005b). Bystander effects in UV-induced genomic instability: Antioxidants inhibit delayed mutagenesis induced by ultraviolet A and B radiation. *Journal of Carcinogenesis, 4*, 11.

DeVeaux, L. C., Durtschi, L. S., Case, J. G., & Wells, D. P. (2006). Bystander effects in unicellular organisms. *Mutation Research – Fundamental and Molecular Mechanisms of Mutagenesis, 597*, 78–86.

Dowling, K., & Mothersill, C. (1999). Use of rainbow trout primary epidermal cell cultures as an alternative to immortalized cell lines in toxicity assessment: A study with nonoxynol. *Environmental Toxicology and Chemistry, 18*, 2846–2850. https://doi.org/10.1897/1551-5028

Dowling, K., Seymour, C., & Mothersill, C. (2005). Delayed cell death and bystander effects in the progeny of Chinook Salmon Embryo cells exposed to radiation and a range of aquatic pollutants. *International Journal of Radiation Biology, 81*, 89–96. https://doi.org/10.1080/09553000400017606

Fernandez-Palomo, C., Seymour, C., & Mothersill, C. (2016). Inter-relationship between low-dose hyper-Radiosensitivity and radiation-induced bystander effects in the human T98G glioma and the epithelial HaCaT cell line. *Radiation Research, 185*, 124–133. https://doi.org/10.1667/RR14208.1

Gowans, I. D., Lorimore, S. A., McIlrath, J. M., & Wright, E. G. (2005). Genotype-dependent induction of transmissible chromosomal instability by γ-radiation and the benzene metabolite hydroquinone. *Cancer Research, 65*, 3527–3530.

Hamada, N. (2008). Recent insights into the biological action of heavy-ion radiation. *Journal of Radiation Research, 50*, 1–9.

Hamilton, C. S., Denham, J. W., O'Brien, M., Ostwald, P., Kron, T., Wright, S., & Dörr, W. (1996). Underprediction of human skin erythema at low doses per fraction by the linear quadratic model. *Radiotherapy and Oncology, 40*, 23–30.

Hertel-Aas, T., Brunborg, G., Jaworska, A., Salbu, B., & Oughton, D. H. (2011a). Effects of different gamma exposure regimes on reproduction in the earthworm Eisenia fetida (Oligochaeta). *Science of the Total Environment, 412*, 138–147.

Hertel-Aas, T., Oughton, D. H., Jaworska, A., & Brunborg, G. (2011b). Induction and repair of DNA strand breaks and oxidised bases in somatic and spermatogenic cells from the earthworm Eisenia fetida after exposure to ionising radiation. *Mutagenesis, 26*, 783–793.

Iyer, R., & Lehnert, B. E. (2000). Factors underlying the cell growth-related bystander responses to α particles. *Cancer Research, 60*, 1290–1298.

Joiner, M. C., & Johns, H. (1988). Renal damage in the mouse: The response to very small doses per fraction. *Radiation Research, 114*, 385–398.

Joiner, M. C., Denekamp, J., & Maughan, R. L. (1985). The use of "top-up"experiments to investigate the effect of very small doses per fraction in mouse skin. *International Journal of Radiation Biology and Related Studies in Physics, Chemistry, and Medicine, 49*, 565–580.

Joiner, M. C., Lambin, P., Malaise, E. P., Robson, T., Arrand, J. E., Skov, K. A., & Marples, B. (1996). Hypersensitivity to very-low single radiation doses: Its relationship to the adaptive response and induced radioresistance. *Mutation Research – Fundamental and Molecular Mechanisms of Mutagenesis, 358*, 171–183.

Kadhim, M. A., Macdonald, D. A., Goodhead, D. T., Lorimore, S. A., Marsden, S. J., & Wright, E. G. (1992). Transmission of chromosomal instability after plutonium [alpha]-particle irradiation. *Nature, 355*, 738–740.

Kadhim, M., Salomaa, S., Wright, E., Hildebrandt, G., Belyakov, O. V., Prise, K. M., & Little, M. P. (2013). Non-targeted effects of ionising radiation—Implications for low dose risk. *Mutation Research – Reviews in Mutation Research, 752*, 84–98. https://doi.org/10.1016/j.mrrev.2012.12.001

Le, M., McNeill, F. E., Seymour, C., Rainbow, A. J., & Mothersill, C. E. (2015). An observed effect of ultraviolet radiation emitted from beta-irradiated HaCaT cells upon non-beta-irradiated bystander cells. *Radiation Research, 183*, 279–290. https://doi.org/10.1667/RR13827.1

Marín, A., Martín, M., Liñán, O., Alvarenga, F., López, M., Fernández, L., Büchser, D., & Cerezo, L. (2015). Bystander effects and radiotherapy. *Reports of Practical Oncology and Radiotherapy, 20*, 12–21.

Marples, B., & Collis, S. J. (2008). Low-dose hyper-radiosensitivity: Past, present, and future. *International Journal of Radiation Oncology, Biology, Physics, 70*, 1310–1318.

Marples, B., & Joiner, M. C. (1993). The response of Chinese hamster V79 cells to low radiation doses: Evidence of enhanced sensitivity of the whole cell population. *Radiation Research, 133*, 41–51.

Marples, B., & Joiner, M. C. (1995). The elimination of low-dose hypersensitivity in Chinese hamster V79-379A cells by pretreatment with X rays or hydrogen peroxide. *Radiation Research, 141*, 160–169.

Marples, B., Joiner, M. C., & Skov, K. A. (1994). The effect of oxygen on low-dose hypersensitivity and increased radioresistance in Chinese hamster V79-379A cells. *Radiation Research, 138*, S17–S20.

Marples, B., Lambin, P., Skov, K. A., & Joiner, M. C. (1997). Low dose hyper-radiosensitivity and increased radioresistance in mammalian cells. *International Journal of Radiation Biology, 71*, 721–735.

Marples, B., Wouters, B. G., Collis, S. J., Chalmers, A. J., & Joiner, M. C. (2004). Low-dose hyper-radiosensitivity: A consequence of ineffective cell cycle arrest of radiation-damaged G2-phase cells. *Radiation Research, 161*, 247–255.

Matsumoto, H., Takahashi, A., & Ohnishi, T. (2004). Radiation-induced adaptive responses and bystander effects. *Biological Sciences in Space, 18*, 247–254.

Mavragani, I. V., Laskaratou, D. A., Frey, B., Candéias, S. M., Gaipl, U. S., Lumniczky, K., & Georgakilas, A. G. (2016). Key mechanisms involved in ionizing radiation-induced systemic effects. A current review. *Toxicology Research (Camb), 5*, 12–33.

Mothersill, C., & Seymour, C. (1997a). Medium from irradiated human epithelial cells but not human fibroblasts reduces the clonogenic survival of unirradiated cells. *International Journal of Radiation Biology, 71*, 421–427.

Mothersill, C., & Seymour, C. (1997b). Lethal mutations and genomic instability. *International Journal of Radiation Biology, 71*, 751–758.

Mothersill, C., & Seymour, C. B. (1998). Cell-cell contact during gamma irradiation is not required to induce a bystander effect in normal human keratinocytes: Evidence for release during irradiation of a signal controlling survival into the medium. *Radiation Research, 149*, 256–262. https://doi.org/10.2307/3579958

Mothersill, C., & Seymour, C. (2001). Radiation-induced bystander effects: Past history and future directions. *Radiation Research, 155*, 759–767. https://doi.org/10.1667/0033-7587

Mothersill, C., & Seymour, C. (2003). Radiation-induced bystander effects, carcinogenesis and models. *Oncogene, 22*, 7028–7033. https://doi.org/10.1038/sj.onc.1206882

Mothersill, C., & Seymour, C. (2004). Radiation-induced bystander effects and adaptive responses – The Yin and Yang of low dose radiobiology? *Mutation Research – Fundamental and Molecular Mechanisms of Mutagenesis, 568*, 121–128. https://doi.org/10.1016/j.mrfmmm.2004.06.050

Mothersill, C., & Seymour, C. (2005). Radiation-induced bystander effects: Are they good, bad or both? *Medicine, Conflict, and Survival, 21*, 101–110.

Mothersill, C., & Seymour, C. (2006a). Radiation-induced bystander effects: Evidence for an adaptive response to low dose exposures? *Dose-Response, 4*, 283–290.

Mothersill, C., & Seymour, C. (2006b). Radiation-induced bystander and other non-targeted effects: Novel intervention points in cancer therapy? *Current Cancer Drug Targets, 6*, 447–454. https://doi.org/10.2174/156800906777723976

Mothersill, C., & Seymour, C. (2009). Communication of ionising radiation signals – A tale of two fish. *International Journal of Radiation Biology, 85*, 909–919. https://doi.org/10.3109/09553000903105452

Mothersill, C., & Seymour, C. (2013). Uncomfortable issues in radiation protection posed by low-dose radiobiology. *Radiation and Environmental Biophysics, 52*, 293–298. https://doi.org/10.1007/s00411-013-0472-y

Mothersill, C., & Seymour, C. (2018). Old data—New concepts: Integrating "indirect effects" into radiation protection. *Health Physics, 115*, 170–178.

Mothersill, C., Cusack, A., MacDonnell, M., Hennessy, T. P., & Seymour, C. B. (1988a). Differential response of normal and tumour oesophageal explant cultures to radiation. *Acta Oncologica (Madr), 27*, 275–280.

Mothersill, C., Cusack, A., & Seymour, C. B. (1988b). Radiation-induced outgrowth inhibition in explant cultures from surgical specimens of five human organs. *The British Journal of Radiology, 61*, 226–230.

Mothersill, C., Seymour, C. B., Mulvin, D., & Hennessy, T. P. (1992). Endothelial cell proliferation is induced by radiation in cultured explants of human urothelium and oesophageal mucosa. In *Angiogenesis* (pp. 407–410). Springer.

Mothersill, C., Harney, J., Lyng, F., Cottell, D., Parsons, K., Murphy, D. M., & Seymour, C. B. (1995a). Primary explants of human uroepithelium show an unusual response to low-dose irradiation with cobalt-60 gamma rays. *Radiation Research, 142*, 181–187.

Mothersill, C., Lyng, F., Lyons, M., & Cottell, D. (1995b). Growth and differentiation of epidermal-cells from the rainbow-trout established as explants and maintained in various media. *Journal of Fish Biology, 46*, 1011–1025. https://doi.org/10.1111/j.1095-8649.1995.tb01406.x

Mothersill, C. E., O'Malley, K. J., Murphy, D. M., Seymour, C. B., Lorimore, S. A., & Wright, E. G. (1999). Identification and characterization of three subtypes of radiation response in normal human urothelial cultures exposed to ionizing radiation. *Carcinogenesis, 20*, 2273–2278. https://doi.org/10.1093/carcin/20.12.2273

Mothersill, C., Lyng, F., Mulford, A., Seymour, C., Cottell, D., Lyons, M., & Austin, B. (2001a). Effect of low doses of ionizing radiation on cells cultured from the hematopoietic tissue of the Dublin Bay prawn, Nephrops norvegicus. *Radiation Research, 156*, 241–250. https://doi.org/10.1667/0033-7587

Mothersill, C., Rea, D., Wright, E. G., Lorimore, S. A., Murphy, D., Seymour, C. B., & O'malley, K. (2001b). Individual variation in the production of abystander signal'following irradiation of primary cultures of normal human urothelium. *Carcinogenesis, 22*, 1465–1471.

Mothersill, C., Seymour, C. B., Joiner, M. C., May, N., Mothersill, C., Seymoura, I. C. B., & Joinerc, M. C. (2002). Relationship between radiation-induced low-dose hypersensitivity and the bystander effect. *Radiation Research, 157*, 526–532.

Mothersill, C. E., Moriarty, M. J., & Seymour, C. B. (2004). Radiotherapy and the potential exploitation of bystander effects. *International Journal of Radiation Oncology, Biology, Physics, 58*, 575–579. https://doi.org/10.1016/j.ijrobp.2003.09.038

Mothersill, C., Lyng, F., Seymour, C., Maguire, P., Lorimore, S., & Wright, E. (2005). Genetic factors influencing bystander signaling in murine bladder epithelium after low-dose irradiation in vivo. *Radiation Research, 163*, 391–399. https://doi.org/10.1667/RR3320

Mothersill, C., Bucking, C., Smith, R. W., Agnihotri, N., O'Neill, A., Kilemade, M., & Seymour, C. B. (2006). Communication of radiation-induced stress or bystander signals between fish in vivo. *Environmental Science & Technology, 40*, 6859–6864. https://doi.org/10.1021/es061099v

Mothersill, C., Salbu, B., Heier, L. S., Teien, H. C., Denbeigh, J., Oughton, D., Rosseland, B. O., & Seymour, C. B. (2007a). Multiple stressor effects of radiation and metals in salmon (Salmo salar). *Journal of Environmental Radioactivity, 96*, 20–31. https://doi.org/10.1016/j.jenvrad.2007.01.025

Mothersill, C., Smith, R. W., Agnihotri, N., & Seymour, C. B. (2007b). Characterization of a radiation-induced stress response communicated in vivo between zebrafish. *Environmental Science & Technology, 41*, 3382–3387. https://doi.org/10.1021/es062978n

Mothersill, C. E., Rusin, A., Fernandez-Palomo, C., & Seymour, C. B. (2017a). History of bystander effects research 1905-present; What's in a name? *International Journal of Radiation Biology*, 1–44.

Mothersill, C., Rusin, A., & Seymour, C. (2017b). Low doses and non-targeted effects in environmental radiation protection; where are we now and where should we go? *Environmental Research, 159*, 484–490.

Múčka, V., Červenák, J., Čuba, V., & Bláha, P. (2015). Determination of the survival of yeast and bacteria under the influence of gamma or UV radiation in the presence of some scavengers of OH radicals. *Journal of Radioanalytical and Nuclear Chemistry, 304*, 237–244.

Nagar, S., Smith, L. E., & Morgan, W. F. (2003). Characterization of a novel epigenetic effect of ionizing radiation: The death-inducing effect. *Cancer Research, 63*, 324–328.

Nagasawa, H., & Little, J. B. (1992). Induction of sister chromatid exchanges by extremely low doses of α-particles. *Cancer Research, 52*, 6394 LP-6396.

O'Dowd, C., Mothersill, C. E., Cairns, M. T., Austin, B., McClean, B., Lyng, F. M., & JEJ, M. (2006). The release of bystander factor(s) from tissue explant cultures of rainbow trout (Onchorhynchus mykiss) after exposure to gamma radiation. *Radiation Research, 166*, 611–617. https://doi.org/10.1667/RR0606.1

O'Reilly, P., & Mothersill, C. (1997). Comparative effects of UV a and UV B on clonogenic survival and delayed cell death in skin cell lines from humans and fish. *International Journal of Radiation Biology, 72*, 111–119.

Olivieri, G., Bodycote, J., & Wolff, S. (1984). Adaptive response of human lymphocytes to low concentrations of radioactive thymidine. *Science (80-.), 223*, 594–597.

Olwell, P. M., Cottell, D. C., Ni Shuilleabhain, S., Maderna, P., Seymour, C., Mothersill, C., & Lyng, F. M. (2005). Cytoskeletal reorganization and altered phagocytotic ability in primary cultures of rainbow trout hemopoietic tissue exposed to low-level ionizing radiation. *Radiation Research, 164*, 45–52. https://doi.org/10.1667/RR3386

Omar-Nazir, L., Shi, X., Moller, A., Mousseau, T., Byun, S., Hancock, S., Seymour, C., & Mothersill, C. (2018). Long-term effects of ionizing radiation after the Chernobyl accident: Possible contribution of historic dose. *Environmental Research, 165*, 55–62.

Pampfer, S., & Streffer, C. (1989). Increased chromosome aberration levels in cells from mouse fetuses after zygote X-irradiation. *International Journal of Radiation Biology, 55*, 85–92. https://doi.org/10.1080/09553008914550091

Prise, K. M. (1998). Studies of bystander effects in human fibroblasts using a charged particle microbeam. *International Journal of Radiation Biology, 74*, 793–798.

Prise, K. M., & O'sullivan, J. M. (2009). Radiation-induced bystander signalling in cancer therapy. *Nature Reviews. Cancer, 9*, 351–360.

Prise, K. M., Belyakov, O. V., Newman, H. C., Patel, S., Schettino, G., Folkard, M., & Michael, B. D. (2002). Non-targeted effects of radiation: Bystander responses in cell and tissue models. *Radiation Protection Dosimetry, 99*, 223–226.

Rusin, A., Lapied, E., Le, M., Seymour, C., Oughton, D., Haanes, H., & Mothersill, C. (2019). Effect of gamma radiation on the production of bystander signals from three earthworm species irradiated in vivo. *Environmental Research, 168*, 211–221.

Ryan, L. A., Seymour, C. B., O'Neill-Mehlenbacher, A., & Mothersill, C. E. (2008). Radiation-induced adaptive response in fish cell lines. *Journal of Environmental Radioactivity, 99*, 739–747. https://doi.org/10.1016/j.jenvrad.2007.10.001

Salbu, B. (2009). Challenges in radioecology. *Journal of Environmental Radioactivity, 100*, 1086–1091.

Salbu, B., Denbeigh, J., Smith, R. W., Heier, L. S., Teien, H. C., Rosseland, B. O., Oughton, D., Seymour, C. B., & Mothersill, C. (2008). Environmentally relevant mixed exposures to radiation and heavy metals induce measurable stress responses in Atlantic salmon. *Environmental Science & Technology, 42*, 3441–3446. https://doi.org/10.1021/es7O27394

Saroya, R., Smith, R., Seymour, C., & Mothersill, C. (2010). Injection of resperpine into zebrafish, prevents fish to fish communication of radiation-induced bystander signals: Confirmation in vivo of a role for serotonin in the mechanism. *Dose-Response, 8*, 317–330. https://doi.org/10.2203/dose-response.09-043.Saroya

Schofield, P. N., Mothersill, C., Abend, M., Brechignac, F., Iliakis, G., Impens, N., Kadhim, M., Moller, A., Oughton, D., & Powathil, G. (2018). When a duck is not a duck; a new interdisciplinary synthesis for environmental radiation protection. *Environmental Research, 162*, 318–324.

Seymour, C. B., Mothersill, C., Michael, B. D., & Alper, T. (1986). Further-studies on lethal mutations expressed by distant descendants of somatic-cells that have survived radiation. *International Journal of Radiation Biology, 50*, 922–923.

Shadley, J. D., Afzal, V., & Wolff, S. (1987). Characterization of the adaptive response to ionizing radiation induced by low doses of X rays to human lymphocytes. *Radiation Research, 111*, 511–517.

Sheridan, M. T., O'Dwyer, T., Seymour, C. B., & Mothersill, C. E. (1997). Potential indicators of radiosensitivity in squamous cell carcinoma of the head and neck. *Radiation Oncology Investigations: Clinical and Basic Research, 5*, 180–186.

Skov, K., Marples, B., Matthews, J. B., Joiner, M. C., & Zhou, H. (1994). A preliminary investigation into the extent of increased radioresistance or hyper-radiosensitivity in cells of hamster cell lines known to be deficient in DNA repair. *Radiation Research, 138*, S126–S129.

Smith, R. W., Wang, J., Bucking, C. P., Mothersill, C. E., & Seymour, C. B. (2007). Evidence for a protective response by the gill proteome of rainbow trout exposed to X-ray induced bystander signals. *Proteomics, 7*, 4171–4180. https://doi.org/10.1002/pmic.200700573

Smith, R. W., Seymour, C. B., Moccia, R. D., Hinton, T. G., & Mothersill, C. E. (2013a). The induction of a radiation-induced bystander effect in fish transcends taxonomic group and trophic level. *International Journal of Radiation Biology, 89*, 225–233. https://doi.org/10.3109/09553002.2013.754558

Smith, R. W., Seymour, C. B., & Mothersill, C. E. (2013b). Short and long term bystander effect induction by fathead minnows (Pimephales promelas, Rafinesque, 1820) injected with environmentally relevant whole body doses of 226Ra. *Journal of Environmental Radioactivity, 126*, 133–136.

Surinov, B. P., Isaeva, V. G., & Dukhova, N. N. (2004a). Postirradiation volatile secretions of mice: Syngeneic and allogeneic immune and behavioral effects. *Bulletin of Experimental Biology and Medicine, 138*, 384–386.

Surinov, B. P., Isaeva, V. G., & Dukhova, N. N. (2004b). [Communicative multiplication of secondary disorders in blood formula and immunity in groups of intact mice caused by volatile compounds excreted by irradiated animals]. *Radiatsionnaia Biologiia, Radioecologiia, 44*, 387–391.

Vanhoudt, N., Vandenhove, H., Real, A., Bradshaw, C., & Stark, K. (2012). A review of multiple stressor studies that include ionising radiation. *Environmental Pollution, 168*, 177–192.

Vo, N. T. K., Seymour, C. B., & Mothersill, C. E. (2017). Dose rate effects of low-LET ionizing radiation on fish cells. *Radiation and Environmental Biophysics, 56*, 433–441.

Watson, G. E., Lorimore, S. A., Macdonald, D. A., & Wright, E. G. (2000). Chromosomal instability in unirradiated cells induced in vivo by a bystander effect of ionizing radiation. *Cancer Research, 60*, 5608–5611.

Whiteside, J. R., & McMillan, T. J. (2009). A bystander effect is induced in human cells treated with UVA radiation but not UVB radiation. *Radiation Research, 171*, 204–211.

Wideł, M., Przybyszewski, W., & Rzeszowska-Wolny, J. (2009). Radiation-induced bystander effect: The important part of ionizing radiation response. Potential clinical implications. *Postępy Higieny i Medycyny Doświadczalnej (Online), 63*, 377–388.

Widel, M., Krzywon, A., Gajda, K., Skonieczna, M., & Rzeszowska-Wolny, J. (2014). Induction of bystander effects by UVA, UVB, and UVC radiation in human fibroblasts and the implication of reactive oxygen species. *Free Radical Biology & Medicine, 68*, 278–287.

Yang, G., Wu, L., Chen, L., Pei, B., Wang, Y., Zhan, F., Wu, Y., & Yu, Z. (2007). Targeted irradiation of shoot apical meristem of Arabidopsis embryos induces long-distance bystander/abscopal effects. *Radiation Research, 167*, 298–305.

Yum, E. H. W., Choi, V. W. Y., Nikezic, D., Li, V. W. T., Cheng, S. H., & Yu, K. N. (2009). Alpha-particle-induced bystander effects between zebrafish embryos in vivo. *Radiation Measurements, 44*, 1077–1080.

Zhou, H., Randers-Pehrson, G., Waldren, C. A., Vannais, D., Hall, E. J., & Hei, T. K. (2000). Induction of a bystander mutagenic effect of alpha particles in mammalian cells. *Proceedings of the National Academy of Sciences, 97*, 2099–2104.

Zwicker, R. D., Meigooni, A., & Mohiuddin, M. (2004). Therapeutic advantage of GRID irradiation for large single fractions. *International Journal of Radiation Oncology, Biology, Physics, 58*, 1309–1315.

Part IV
Biomarkers in Wildlife

Chapter 11
Birds as Bioindicators of Radioactive Contamination and Its Effects

Andrea Bonisoli-Alquati, Anders P. Møller, Geir Rudolfsen, and Timothy A. Mousseau

Abstract Birds are regarded as excellent bioindicators of anthropogenic environmental change, including changes due to the release of toxicants into ecosystems. Consistent with this, birds are among the best-studied groups of organisms under conditions of radioactive contamination. This includes the study of radionuclide transfer to their bodies as well as the estimation of the associated radiation doses. This paper reviews the literature on the use of birds in studies of radiation exposure, and the effects of low-dose, low-dose rate ionizing radiation on birds. A variety of effects have been documented on birds exposed to ionizing radiation under ecological conditions. These studies, mostly in the context of the nuclear accidents of Chernobyl and Fukushima, are organized based on the level of biological organization analyzed, from molecular and cytological studies, to physiological, behavioral and life-history effects. We also discuss the translation of these effects into consequences at the level of populations and communities, and examine how avian species vary in their exposure and susceptibility. We conclude by suggesting future avenues of research, including the development of molecular and –omics biomarkers that will increase the sensitivity of retrospective dosimetry, and aid the detection of low-dose effects.

A. Bonisoli-Alquati (✉)
Department of Biological Sciences, California State Polytechnic University - Pomona, Pomona, CA, USA
e-mail: aalquati@cpp.edu

A. P. Møller
Ecologie Systématique Evolution, Université Paris-Sud, CNRS, AgroParisTech, Université Paris-Saclay, Orsay, France

G. Rudolfsen
Arctic University Museum of Norway, Arctic University of Norway, Tromsø, Norway

T. A. Mousseau
Department of Biological Sciences, University of South Carolina, Columbia, SC, USA

M. D. Wood et al. (eds.), *Biomarkers of Radiation in the Environment*, NATO Science for Peace and Security Series A: Chemistry and Biology,
https://doi.org/10.1007/978-94-024-2101-9_11

11.1 Birds as Bioindicators

In modern times, the concept that organisms could serve as biological indicators of ecological changes has largely gained impulse from regulatory and stewardship needs, as an attempt to make the overload of ecological data useful at the management and political levels (Niemi & McDonald, 2004). Broadly defined, a biological indicator (or bioindicator) is a quantitative characteristic of any biological system, whether structural or functional, at any level of biological complexity; the characteristic has to conveys information about the underlying processes in the ecosystem where the biological system functions and lives. In ecological systems, structural characteristics tracked by bioindicators include genetic, population, species, and community-level composition. The functional aspects can also be measured at levels from genes to communities and ecosystems, by focusing on agents that are responsible for carrying out those processes, or are affected by them.

Birds are generally regarded as excellent indicators of the abundance of numerous taxa of animals and plants and of ecological and environmental processes (Furness, 1993; Armon & Hänninen, 2015). Bird species also track anthropogenic disturbance of those same processes, including due to a variety of pollutants (Becker, 2003; Burger & Gochfeld, 2004; Burger et al., 2013). The ubiquity of birds across biomes and ecosystems makes them efficient sentinels of the fate of those pollutants. The fact that they are relatively long-lived and mobile means that they can integrates signals of ecological change and disturbance over time and space (Burger & Gochfeld, 2004). Arguably, however, it also means that such signal might become hard to pinpoint. At the same time, the diversity of bird species and their relative approachability (especially when compared with mammals) grants opportunities for environmental monitoring and risk assessment, by guaranteeing that at least some species will be sensitive to the effects of the contaminants. This is the familiar concept of birds as 'canaries' in the proverbial coal mine. At all scales, birds alert us of the consequences of our physical and chemical modification of the environment.

To ask whether birds are reliable indicators in radioecological studies means different things depending on the context and aim of the study. With the goal of modeling the fate of radionuclides in an ecosystem, a reliable bioindicator is an organism in which radionuclides can be detected, at levels that mirror or magnify those found in some ecosystem compartment. In other words, the organism has to track the fate of contaminants, and anticipate their potential bioconcentration or biomagnification in other species or compartments. If the goal is instead to quantify the effects of radionuclides in the environment, a good bioindicator is a sensitive species where exposure routes and physiological sensitivity conspire to make that organism susceptible to the effects of ionizing radiation. On the other side, species that are resistant to ionizing radiation are also worthy of scientific interest. The interest in studying them relies in learning the mechanistic basis for their ability to withstand the insult of contamination, whether as a result of adaptation or exaptation (sensu Gould & Lloyd, 1999). These two uses of birds as bioindicators of radioactive contamination – their role as living monitors of radiation levels and of sentinels of radiation effects are discussed below, followed by a discussion of desirable future developments.

11.2 Birds as Indicators of the Fate of Radioactive Contaminants

The first question is whether birds are reliable at tracking levels of radioactive contamination in their environment. This is an articulated question, which includes two sub-questions. First, we can ask whether birds are at risk of being exposed to ionizing radiation in their environment, from external sources as well as due to uptake of significant amount of radioactivity. Second, we should also ask ourselves whether as radioecologists we can pick up the signal of past exposure by focusing our measurements on birds, particularly in the field. To answer the first question, scientists have long used birds to measure radionuclide concentrations in the environment (Brisbin, 1993). While the majority of such studies were conducted under controlled laboratory conditions on domesticated species (Tyler & Stearner, 1966), some notable early studies attempted to use more realistic settings and wild populations from several bird orders (Zach & Mayoh, 1986a, b). Radionuclides are not susceptible to biomagnification the way organic pollutants do. Thus, the relatively high trophic position that birds, especially marine birds and raptors, occupy in a food web is not necessarily useful in the case of radionuclides, as opposed to the tracking of organochlorine substances and other organic pollutants (Gómez-Ramírez et al., 2014).

Only a few of the methodologies for retrospective radiation dosimetry have been applied to birds, especially under conditions of environmental radiation exposure. A review of the existing techniques is beyond the scope of the present review. In addition, a review of dosimetry technologies for measuring external dose to wildlife was recently published (Aramrun et al., 2018), where available technologies and desirable developments are extensively discussed. Some of the most promising tools identified by Aramrun and collaborators (2018) are not easily applicable to birds, as is the case for direct ion storage dosimeters, which are suitable only to medium and large mammals. Yet, in recent years we have conducted extensive screening of radiation exposure in wild populations of birds in both Chernobyl and Fukushima, and it is useful to discuss our experience in retrospective dosimetry of birds. The advancements in dosimetry that we have promoted mirror recent changes in the approach to radiological protection of the environment. Other progresses await, especially with the coupling between dosimeters and GPS trackers, possibly one of the most promising developments in wildlife dosimetry, as argued by others authors too (Stark et al., 2017).

In the field in Chernobyl we tested whether estimates of environmental radiation obtained with a hand-held dosimeter accurately predicted past radiation exposure in understory birds captured in and around the Chernobyl Exclusion Zone in 2012. We mist-netted birds in abandoned collective farms and woodland sites that differed more than three orders of magnitude in their environmental radiation levels. We estimated whole-body burdens of radiation by conducting gamma spectrometry of individual birds in the field using a portable SAM 940 Radioisotope Identifier (Berkeley Nucleonics, San Rafael, CA) equipped with a sodium iodide (NaI) detector. In that same occasion, we also attached GR-200A LiF:Mg,Cu,P thermoluminescent dosimeters (TLDs) to standard aluminum bird bands in each captured birds. In a sub-sample of birds that we were able to recapture after approximately one month from the initial capture, we were able to estimate external radiation exposure using the TLDs. Environmental radiation levels measured in the spot of capture of each bird accurately predicted the internal dose received by the birds, as indexed by activity concentration in their body (Fig. 11.1). Environmental radiation levels also predicted the dose received by the TLDs during a month-long period between subsequent captures for a subsample of birds (Fig. 11.2). For both relationships the strength of the association varied among closely-related passerine species. The determinants of such interspecific variation in exposure remain to be investigated. Such systematic analysis of the differences among closely-related species in their exposure and uptake of radionuclides is one of the most promising developments in wildlife dosimetry, likely to contribute to a more ecologically realistic assessment of radiological risk (Bréchignac et al., 2016).

The existence of such variation across bird species also highlights the limitation of radiological protection approaches based on reference organisms. As previously observed (Bréchignac et al., 2016; Mothersill et al., 2020), such approaches based on reference organisms are crucial first steps in the understanding of radiological risk, but they can underestimate risk due to intrinsic limitations. Such limitations are at least two-fold. First, reference organisms might be inadvertently picked from the left tail of the distribution of concentration ratios (CRs), a tool commonly used to

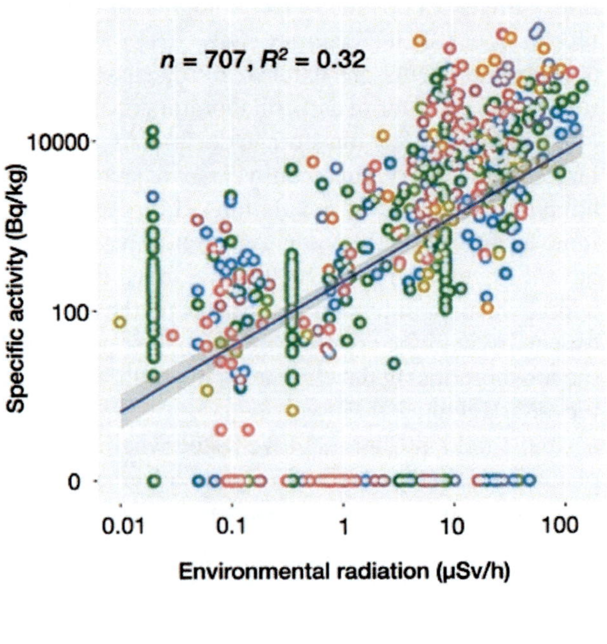

Fig. 11.1 Whole-body burdens of radiation in birds from the Chernobyl Exclusion Zone as a function of environmental radiation levels in their site of capture. Different colors indicate different species. The line represent a linear regression, while the shaded area represents the 95% confidence bands of the linear model prediction estimated by *ggplot* (R Core Team 2019)

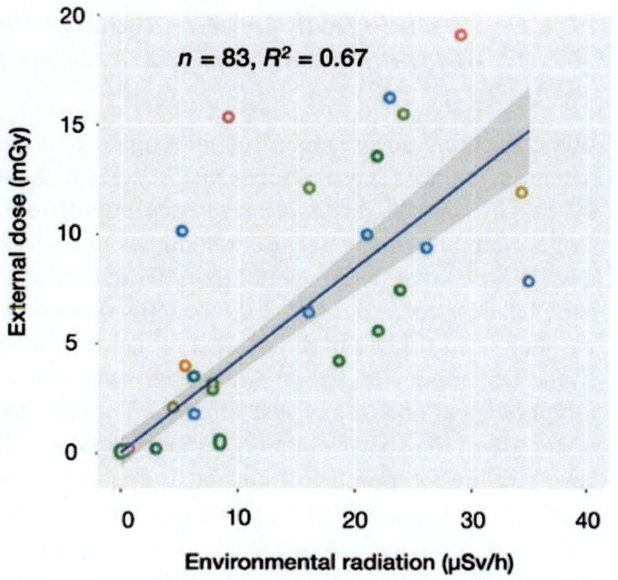

Fig. 11.2 External radiation exposure as measured by thermoluminescent dosimeters (TLDs) as a function of environmental radiation levels measured in the site of capture of each birds. TLDs were attached to the bands of birds, and retrieved during a later capture, approximately one month after they were attached. Different colors indicate different species. The line represent a linear regression, while the shaded area represents the 95% confidence bands of the linear model prediction estimated by ggplot (R Core Team 2019)

assess transfer of radionuclides from environmental matrices to wildlife (Strand et al., 2009). A recent phylogenetic analysis of CRs in marine biota disclosed extensive variation in transfer of radiocesium, in spite of close taxonomic relationship between some organisms (Brown et al., 2019). Similar limitations also apply to variation in sensitivity among species (discussed below). A second crucial limitation of reference organisms consists in their being studied under laboratory conditions, where radiological risk is likely to be underestimated (Garnier-Laplace et al., 2013).

11.3 Birds as Sentinels of Low-Dose Radiation Effects

The idea that birds are good sentinels of the effects of ionizing radiation has been around since the early studies where irradiation was typically conducted under laboratory conditions (Zach & Mayoh, 1986a, b). Early summaries of research conducted on the effects of radioactive contamination on birds would conclude that they are generally suitable, although developmental and physiological effects were typically more apparent than mortality effects and their translation on population trends (Brisbin, 1993). The discussion of whether a group of organisms is suitable as indicator of the effects of a contaminant might seem counterintuitive, as populations or species are typically used as bioindicators. The diversity of existing bird species in virtually any ecosystems is what ensures that some birds will be found that are affected by radiation exposure.

11.3.1 Physiological, Genetic and Morphological Effects of Radiation on Birds: A Role for Oxidative Stress

When selecting an organism for investigating the effects of exposure to ionizing radiation, it makes sense to focus on a species with known life history, which is easily caught, and has available reference populations outside of the contaminated range. Naturally, many of the early studies in Chernobyl focused on the barn swallow (*Hirundo rustica*), an aerial insectivore that nests in man-made structures. The barn swallow embodies all of the benefits of an ecological model species, since it has long been studied in the context of natural and sexual selection (Møller 1994). In the beginning, studies on Chernobyl swallows were especially focused on the morphological and behavioral effects of radiation exposure, mirroring the traditional use of the species in an ecological context. Those studies demonstrated that barn swallows exposed to radioactive contamination had increased frequency of albinistic feathers, which translated in their reduced success in acquiring mates (Møller & Mousseau, 2001, 2003). They also had lower reproductive success and survival (Møller et al., 2005a), which was found to drive a numerical decline in their colonies in radioactively contaminated sites, only partially compensated by immigration from the outside (Møller et al., 2006). The responsibility for this array of morphological and behavioral effects was initially attributed to deleterious mutations, consistent with the demonstration of a two to ten-fold increase in mutation rate in birds from Chernobyl, where mutation of germline origin were associated with abnormal phenotypes (Ellegren et al., 1997). After all, the evidence of increased mutation rate in barn swallows was consistent with similar tests in other species (Møller & Mousseau, 2006, 2015). Mutations were mostly interpreted as a result of direct damage to DNA by ionizing radiation.

In more recent years, oxidative stress has instead been more commonly highlighted as the agent driving many of the demonstrated differences in the physiology and reproductive success of birds from radioactively contaminated areas. Oxidative stress, the imbalance between the production of reactive oxygen species and their neutralization by the antioxidant system, is a well-known agent of the decline of physiological performance with ageing (Sohal et al., 2002). Extensive evidence also supports a role of ionizing radiation in generating oxidative damage to macromolecules (Einor et al., 2016). Another fundamental reason for this focus was the suggestion that oxidative stress underlies life history trade-offs, and their evolution across species (Dowling & Simmons, 2009).

We showed that barn swallow from within the Chernobyl Exclusion Zone (CEZ) has higher levels of oxidative damage, as indexed by the circulating levels of reactive oxygen metabolites (ROMs) (Bonisoli Alquati et al., 2010a). Unchecked by the antioxidant system, those ROMs were also leading swallows to higher levels of oxidative stress (Bonisoli Alquati et al., 2010a). We also demonstrated that swallows in radioactively contaminated areas in Ukraine and Belarus had higher levels of DNA damage, as indexed by the so-called 'comet assay' (Bonisoli Alquati et al., 2010b). Also known as the single-cell gel electrophoresis test, the 'comet assay' is an inexpensive cytogenetic test at the level of individual cells that is applicable to virtually any cell line, and is frequently employed as a sensitive biomarker of a variety of categories of damage to the DNA molecule (Tice et al., 2000). The oxidative status of individual barn swallows was also related to the quality of their sperm, suggesting the potential for effects of radiation to be transmitted to the next generation (Bonisoli Alquati et al., 2011). Given its special place in indicating radiation effects across generations, studies on sperm motility in bird species are outlined separately.

11.3.2 *Sperm Motility as a Sensitive Endpoint in Birds Exposed to Ionizing Radiation*

Both germ line and somatic indices may not always be very sensitive endpoints for assessing the effects of low dose radiation on reproduction in free-living populations. For instance, different selection pressure and variation in factors such as DNA repair mechanisms, genome size and life history strategies will have an impact on how a species will respond to elevated levels of ionizing radiation.

Sperm cells are generally highly vulnerable to oxidative stress, which in turn results in impaired ejaculate quality and male fertility (Aitken, 1999; Aitken et al., 2010). As sperm is vulnerable to oxidative stress, a number of antioxidants and antioxidant enzymes are found in the seminal fluid, where they act as a defence against ROS (Surai et al., 1998; Sikka, 2001; Murphey et al., 2013). Neutrophils and macrophages produce ROS in the male genital tract (Henkel et al., 1997; Whittington & Ford, 1999), and act as a defence mechanism against infection (Hang et al., 1999; Haraoka et al., 1999). However, such oxidative burst might damage spermatozoa and impair sperm motility (Armstrong et al., 1999). Further, exposure to low dose radiation increase oxidative stress and results in impaired ejaculate quality and sperm motility (Cheburakov & Cheburakova, 1993; Møller et al., 2008b; Sakharov et al., 2009; Bonisoli Alquati et al., 2011; Møller et al., 2014). This suggests that in addition to a damaging effect on DNA integrity due to radiation, excessive ROS production in exposure to radiation may overwhelm the antioxidant defences, and cause infertility or impaired sperm quality (Bonisoli Alquati et al., 2011).

Based on the expectations above, and that sperm quality is a major factor in the loss of fertility, males should invest more into the antioxidant protection of their sperm from oxidative stressors such as radiation (Bonisoli Alquati et al., 2011). However, among free-living organisms, environmental conditions are seldom optimal, and animals typically face multiple concomitant stressors, such as predators, parasites and food shortages. This in turn will lead to a non-optimal level of antioxidants and antioxidant enzymes, and make sperm quality a sensitive endpoint for reproduction. The limited number of studies of sperm quality in free-living birds under conditions of radioactive contamination has identified sperm responses to changes in background radiation, suggesting that males are not always able to compensate by investing in antioxidant protection (Møller et al., 2008b, 2014; Bonisoli

Alquati et al., 2011). However, the lack of studies that experimentally manipulated exposure to radiation either in the field or in the laboratory leaves this field with only correlative studies so far. Passerine birds are a suitable model system, as birds have high sperm production that is associated with a rapid spermatogenesis, variation in life history traits and relatively easy sampling.

Sperm motility measurement by sperm velocity tracking is a fast and non-invasive census method for reproductive potential in passerine birds (Kleven et al., 2009). By a gentle massage of the cloacal protuberance of male passerine birds, the ejaculate can be collected directly in a capillary tube. The ejaculate is then diluted in Dulbecco's Modified Eagle Medium before the diluted sperm sample are placed on a count slide. The sample should be stored at a temperature of 40 °C, and sperm swimming behaviour should be immediately video recorded or analysed by a Computer-assisted sperm analyser (CASA). The typical setup allows more than one video frame to estimate the overall sperm behaviour, having up to 9 replicates for each sample on one slides. The whole procedure takes less than 2 min per bird, and can be conducted on site in the field, given sufficient power supply and shelter.

Despite the fact that CASA cannot accurately predict male fertility from a semen sample, current CASA systems provide important information of ejaculates quality, and measure sperm responses to changes in the environment (Møller et al., 2008b, 2014; Bonisoli Alquati et al., 2011). Consequently, sperm motility is a valuable endpoint when analyzing the effects of low-dose radiation exposure on a male's reproductive potential.

11.3.3 Interspecific Variation in Susceptibility to Radioactive Contamination

Many studies on birds in Chernobyl and Fukushima encompassed a variety of species, often with the explicit aim of comparing their respective sensitivities to radioactive contamination. Multi-year censuses that expanded on the findings on the barn swallow had detected significant declines in the abundance and diversity of bird species in highly radioactively contaminated areas (Møller & Mousseau, 2007a). Perhaps unsurprisingly, not all bird species were equally affected by exposure to the same environmental radiation levels, which begged the question of what determined interspecific variation. The negative relationships between abundance and radiation linked to migration distance, dispersal distance, clutch size and carotenoid-based coloration (Møller & Mousseau, 2007b). This list of predictors was subsequently extended by phaeolemanin which links directly to glutathione as an intracellular antioxidant (Galván et al., 2011). Subsequent analyses suggested that body size, trophic level and plumage pigments accounted for additional variation. Finally, breeding bird surveys in Fukushima during 2011–2014 at Fukushima indicated that the negative effects of radiation on abundance and species richness accumulated over time because the relationship between abundance and radiation became more negative over time (Møller et al., 2015). These findings suggest that antioxidants and antioxidant use accounts of interspecific variation in effects of radiation on abundance of different species. Rather than being random effects, these differences in abundance were also linked directly to an active choice of breeding sites (Møller & Mousseau, 2007c). Bird species in Chernobyl had variable slopes of the relationship between abundance and radiation (mean (SE) = −0.021 (0.004), N = 80 species), implying that on average population size decreased with increasing level of background radiation. This relationship differed significantly from zero ($t = -4.70$, $df = 79$, $P < 0.0001$). A similar relationship has been found in Fukushima (Møller et al., 2012; Bonisoli-Alquati et al., 2015; Møller et al., 2015), even when special care was taken to estimate internal dose while considering size, shape and diet of bird species (Garnier-Laplace et al., 2015).

We surveyed breeding birds across 731 census points in Ukraine and Belarus to assess the relationship between abundance of different species and radiation (Møller & Mousseau, 2011a). Most bird species avoided contaminated sites and individuals were concentrated in relatively uncontaminated sites. While common species were recorded at a wide range of radiation levels, rare species were restricted to the least contaminated sites and to sites with a high biodiversity. Therefore, rare species were disproportionately impacted by the accident. Mutation rates of plants and animals have increased by up to a factor 20 due to release of radionuclides from Chernobyl. Given that each slightly deleterious mutation is expected to result in a selective genetic death, and that an average fruit fly under normal conditions may carry as many as 80 mutations, the number of mutations in animals and plants around Chernobyl and hence the number of selective deaths is bound to be much higher. There is empirical evidence for highly elevated mortality rates and dramatically increased rates of reproductive failure in contaminated areas, consistent with the expected high frequency of selective deaths due to mutations. The average slightly deleterious mutation is present for 33–167 generations in *Drosophila*, so if these estimates are qualitatively similar in other organisms, we can expect that mutants will disperse outside contaminated areas resulting in the spread of mutations well beyond the reach of contamination with radionuclides (Møller & Mousseau, 2011b).

In addition to being mediated by the effects of radiation on mutation rate, several of the documented effects were either confirmed or consistent with the disruption of the oxidative balance of the affected birds. Across studies in a variety of taxa, including humans, exposure to low-dose, low-dose rate radiation is generally responsible for (small) decreases in antioxidant protection and increases in oxidative damage and stress (Einor et al., 2016). The role of oxidative damage and stress in mediating the deleterious effects of ionizing radiation is consistent with radiation-induced generation of reactive oxygen species (ROS). As previously stated, ROS have been suggested as a currency underlying life history trade-offs (Costantini, 2008). A number of life history functions have been documented to be affected in birds exposed to radioactive contamination. These included various indices of reproductive success, including during sexual selection (Møller & Mousseau, 2001), as well as related to parental care, both pre and post-hatching (Møller et al., 2005b, 2008a).

For this reason, radiation has been endorsed as a tool for gaining insight into the mechanistic working of those trade-offs (Koch & Hill, 2018).

Uric acid is a marker of nonenzymatic antioxidant defence (Costantini, 2011). European bird species that were more common, on average, had higher concentrations of circulating uric acid ($F = 5.27$, $df = 1, 50$, $P = 0.026$, estimate (SE) = −16.12 (7.02)).

Bird species that migrated long distances from the breeding grounds to the winter quarters had higher circulating levels of malondialdehyde (MDA), which is an indicator of oxidative damage (Barja 2004; Halliwell & Gutteridge, 2007) that increase the level of damage to DNA and other molecules ($F = 10.71$, $df = 1, 70$, $r^2 = 0.12$, $P = 0.0017$, slope (SE) = 2.21 (0.68)). Likewise, bird species that dispersed long distances had high MDA levels ($F = 10.71$, $df = 1, 36$, $r^2 = 0.21$ $P = 0.00123$, slope (SE) = 2.57 (0.78)). This implies that individuals that moved long distances were preadapted in terms of physiology to such movement. Multifaceted negative effects of MDA can ultimately jeopardize mitochondrial, cellular and organismal functioning and accelerate ageing (Pamplona, 2008).

Cuckoos are prime indicators of species richness as a surrogate of biodiversity (Morelli et al., 2015). Background radiation in Chernobyl and Fukushima was negatively correlated with bird species richness and bird abundance in both sites, while the numbers of top predators and cuckoos were both positively correlated with bird species richness and abundance. However, models with number of cuckoos as a predictor was more performant than model with number of avian top predators. These differences in performance supports the hypothesis that cuckoos are a largely superior bio-indicators than top predators (Morelli et al., 2017).

A final layer in the question of whether birds are reliable bioindicators of the effects of ionizing radiation is whether birds can also track the evolutionary response of biological systems to radiation exposure. Certain bird species living in more radioactively contaminated areas have been shown to have higher levels of the intracellular antioxidant glutathione, and lower levels of DNA damage, as measured by the comet assay (Galván et al., 2014). In principle, this could be due to two mechanisms – phenotypic plasticity or evolutionary adaptation (Hoffmann and Willi 2008). An answer to this question is premature, as conclusive tests of whether microevolutionary adaptation occurred will require the demonstration of selective sweeps of genes that are associated with differential reproductive success under conditions of chronic exposure to ionizing radiation.

References

Aitken, R. J. (1999). The amoroso lecture the human spermatozoon – A cell in crisis? *Journal of Reproduction and Fertility, 115*, 1–7. https://doi.org/10.1530/jrf.0.1150001

Aitken, R. J., De Iuliis, G. N., Finnie, J. M., Hedges, A., & McLachlan, R. I. (2010). Analysis of the relationships between oxidative stress, DNA damage and sperm vitality in a patient population: Development of diagnostic criteria. *Human Reproduction, 25*, 2415–2426. https://doi.org/10.1093/humrep/deq214

Aramrun, P., Beresford, N. A., & Wood, M. D. (2018). Selecting passive dosimetry technologies for measuring the external dose of terrestrial wildlife. *Journal of Environmental Radioactivity, 182*, 128–137. https://doi.org/10.1016/j.jenvrad.2017.12.001

Armon, R. H., & Hänninen, O. (Eds.). (2015). *Environmental indicators*. Springer Netherlands.

Armstrong, J. S., Rajasekaran, M., Chamulitrat, W., Gatti, P., Hellstrom, W. J., & Sikka, S. C. (1999). Characterization of reactive oxygen species induced effects on human spermatozoa movement and energy metabolism. *Free Radical Biology & Medicine, 26*, 869–880. https://doi.org/10.1016/S0891-5849(98)00275-5

Barja, G. (2004). Aging in vertebrates, and the effect of caloric restriction: A mitochondrial free radical production-DNA damage mechanism? *Biological Reviews, 79*, 235–251. https://doi.org/10.1017/S1464793103006213

Becker, P. H. (2003). Biomonitoring with birds. In *Trace metals and other contaminants in the environment* (Vol. 6, pp. 677–736).

Bonisoli-Alquati, A., Mousseau, T. A., Møller, A. P., Caprioli, M., & Saino, N. (2010a). Increased oxidative stress in barn swallows from the Chernobyl region. *Comparative Biochemistry and Physiology. Part A, 155*, 205–210. https://doi.org/10.1016/j.cbpa.2009.10.041

Bonisoli-Alquati, A., Voris, A., Mousseau, T. A., Møller, A. P., Saino, N., & Wyatt, M. D. (2010b). DNA damage in barn swallows (*Hirundo rustica*) from the Chernobyl region detected by use of the comet assay. *Comparative Biochemistry and Physiology – C Toxicology and Pharmacology, 151*, 271–277. https://doi.org/10.1016/j.cbpc.2009.11.006

Bonisoli-Alquati, A., Møller, A. P., Rudolfsen, G., Saino, N., Caprioli, M., Ostermiller, S., & Mousseau, T. A. (2011). The effects of radiation on sperm swimming behavior depend on plasma oxidative status in the barn swallow (*Hirundo rustica*). *Comparative Biochemistry and Physiology. Part A, 159*, 105–112. https://doi.org/10.1016/j.cbpa.2011.01.018

Bonisoli-Alquati, A., Koyama, K., Tedeschi, D. et al. (2015). Abundance and genetic damage of barn swallows from Fukushima. *Sci Rep, 5*, 9432. https://doi.org/10.1038/srep09432

Bréchignac, F., Oughton, D., Mays, C., Barnthouse, L., Beasley, J. C., Bonisoli-Alquati, A., Bradshaw, C., Brown, J., Dray, S., Geras'kin, S., Glenn, T., Higley, K., Ishida, K., Kapustka, L., Kautsky, U., Kuhne, W., Lynch, M., Mappes, T., Mihok, S., … Tsukada, H. (2016). Addressing ecological effects of radiation on populations and ecosystems to improve protection of the environment against radiation: Agreed statements from a consensus symposium. *Journal of Environmental Radioactivity, 158*, 21–29.

Brisbin, I. L., Jr. (1993). Birds as monitors of radionuclide contamination. In *Birds as monitors of environmental change* (pp. 144–178). Springer.

Brown, J. E., Beresford, N. A., & Hevrøy, T. H. (2019). Exploring taxonomic and phylogenetic relationships to predict radiocaesium transfer to marine biota. *Science of the Total Environment, 649*, 916–928. https://doi.org/10.1016/j.scitotenv.2018.08.343

Burger, J., & Gochfeld, M. (2004). Marine birds as sentinels of environmental pollution. *EcoHealth, 1*, 263–274. https://doi.org/10.1007/s10393-004-0096-4

Burger, J., Gochfeld, M., Powers, C. W., Clarke, J. H., Brown, K., Kosson, D., Niles, L., Dey, A., Jeitner, C., & Pittfield, T. (2013). Determining environmental impacts for sensitive species: Using iconic species as bioindicators for management and policy. *Journal of Environmental Protection, 04*, 87. https://doi.org/10.4236/jep.2013.48A2011

Cheburakov, I. I. & Cheburakova, O. P. (1993) [Disorders of spermatogenesis in people working at the clean-up of the Chernobyl nuclear power plant accident]. *Radiatsionnaia biologiia, radioecologiia/Rossiĭskaia akademiia nauk, 33*, 771–774.

Costantini, D. (2008). Oxidative stress in ecology and evolution: lessons from avian studies. *Ecology Letters, 11*, 1238–1251. https://doi.org/10.1111/j.1461-0248.2008.01246.x

Costantini, D. (2011). On the measurement of circulating antioxidant capacity and the nightmare of uric acid. *Methods in Ecology and Evolution, 2*, 321–325. https://doi.org/10.1111/j.2041-210X.2010.00080.x

Dowling, D. K., & Simmons, L. W. (2009). Reactive oxygen species as universal constraints in life-history evolution. *Proceedings of the Royal Society of London – Series B: Biological Sciences, 276*, 1737–1745. https://doi.org/10.1098/rspb.2008.1791

Einor, D., Bonisoli-Alquati, A., Costantini, D., Mousseau, T. A., & Møller, A. P. (2016). Ionizing radiation, antioxidant response and oxidative damage: A meta-analysis. *Science of the Total Environment, 548–549*, 463–471. https://doi.org/10.1016/j.scitotenv.2016.01.027

Ellegren, H., Lindgren, G., Primmer, C. R., & Møller, A. P. (1997). Fitness loss and germline mutations in barn swallows breeding in Chernobyl. *Nature, 389*, 593–596.

Furness, R. W. (1993). *Birds as monitors of environmental change*. Chapman & Hall.

Galván, I., Mousseau, T. A., & Møller, A. P. (2011). Bird population declines due to radiation exposure at Chernobyl are stronger in species with pheomelanin-based coloration. *Oecologia, 165*, 827–835. https://doi.org/10.1007/s00442-010-1860-5

Galván, I., Bonisoli-Alquati, A., Jenkinson, S., Ghanem, G., Wakamatsu, K., Mousseau, T. A., & Møller, A. P. (2014). Chronic exposure to low-dose radiation at Chernobyl favours adaptation to oxidative stress in birds. *Functional Ecology, 28*, 1387–1403. https://doi.org/10.1111/1365-2435.12283

Garnier-Laplace, J., Geras'kin, S., Della-Vedova, C., Beaugelin-Seiller, K., Hinton, T. G., Real, A., & Oudalova, A. (2013). Are radiosensitivity data derived from natural field conditions consistent with data from controlled exposures? A case study of Chernobyl wildlife chronically exposed to low dose rates. *Journal of Environmental Radioactivity, 121*, 12–21. https://doi.org/10.1016/j.jenvrad.2012.01.013

Garnier-Laplace, J., Beaugelin-Seiller, K., Della-Vedova, C., Métivier, J.-M., Ritz, C., Mousseau, T. A., & Møller, A. P. (2015). Radiological dose reconstruction for birds reconciles outcomes of Fukushima with knowledge of dose-effect relationships. *Scientific Reports*, 1–13. https://doi.org/10.1038/srep16594

Gómez-Ramírez, P., Shore, R. F., van den Brink, N. W., van Hattum, B., Bustnes, J. O., Duke, G., Fritsch, C., García-Fernández, A. J., Helander, B. O., Jaspers, V., Krone, O., Martínez-López, E., Mateo, R., Movalli, P., & Sonne, C. (2014). An overview of existing raptor contaminant monitoring activities in Europe. *Environment International, 67*, 12–21. https://doi.org/10.1016/j.envint.2014.02.004

Gould, S. J., & Lloyd, E. A. (1999). Individuality and adaptation across levels of selection: How shall we name and generalize the unit of Darwinism? *Proceedings of the National Academy of Sciences USA, 96*, 11904–11909. https://doi.org/10.1073/pnas.96.21.11904

Halliwell, B., & Gutteridge, J. (2007). *Free radicals in biology and medicine* (4th ed.). Oxford University Press.

Hang, L., Haraoka, M., Agace, W. W., Leffler, H., Burdick, M., Strieter, R., & Svanborg, C. (1999). Macrophage inflammatory protein-2 is required for neutrophil passage across the epithelial barrier of the infected urinary tract. *The Journal of Immunology, 162*, 3037–3044.

Haraoka, M., Hang, L., Frendéus, B., Godaly, G., Burdick, M., Strieter, R., & Svanborg, C. (1999). Neutrophil recruitment and resistance to urinary tract infection. *The Journal of Infectious Diseases, 180*, 1220–1229. https://doi.org/10.1086/315006

Henkel, R., Ichikawa, T., Sánchez, R., Miska, W., Ohmori, H., & Schill, W. B. (1997). Differentiation of ejaculates showing reactive oxygen species production by spermatozoa or leukocytes. *Andrologia, 29*, 295–301. https://doi.org/10.1093/HUMREP/10.6.1444

Hoffmann, A. A., & Willi, Y. (2008). Detecting genetic responses to environmental change. *Nat Rev Genet, 9*, 421–432.

Kleven, O., Fossøy, F., Laskemoen, T., Robertson, R. J., Rudolfsen, G., & Lifjeld, J. T. (2009). Comparative evidence for the evolution of sperm swimming speed by sperm competition and female sperm storage duration in passerine birds. *Evolution, 63*, 2466–2473. https://doi.org/10.1111/j.1558-5646.2009.00725.x

Koch, R. E., & Hill, G. E. (2018). An assessment of techniques to manipulate oxidative stress in animals. *Functional Ecology*. https://doi.org/10.1111/1365-2435.12664

Møller, A. P. (1994). *Sexual selection and the barn swallow*. Oxford University Press.

Møller, A. P., & Mousseau, T. A. (2001). Albinism and phenotype of barn swallows (Hirundo rustica) from Chernobyl. *Evolution, 55*, 2097–2104.

Møller, A. P., & Mousseau, T. A. (2003). Mutation and sexual selection: A test using barn swallows from Chernobyl. *Evolution, 57*, 2139–2146.

Møller, A. P., & Mousseau, T. A. (2006). Biological consequences of Chernobyl: 20 years on. *Trends in Ecology & Evolution, 21*, 200–207. https://doi.org/10.1016/j.tree.2006.01.008

Møller, A. P., & Mousseau, T. A. (2007a). Species richness and abundance of forest birds in relation to radiation at Chernobyl. *Biology Letters, 3*, 483–486. https://doi.org/10.1098/rsbl.2007.0226

Møller, A. P., & Mousseau, T. A. (2007b). Determinants of interspecific variation in population declines of birds after exposure to radiation at Chernobyl. *Journal of Applied Ecology, 44*, 909–919. https://doi.org/10.1111/j.1365-2664.2007.01353.x

Møller, A. P., & Mousseau, T. A. (2007c). Birds prefer to breed in sites with low radioactivity in Chernobyl. *Proceedings of the Royal Society of London – Series B: Biological Sciences, 274*, 1443–1448. https://doi.org/10.1098/rspb.2007.0005

Møller, A. P., & Mousseau, T. A. (2011a). Efficiency of bio-indicators for low-level radiation under field conditions. *Ecological Indicators, 11*, 424–430. https://doi.org/10.1016/j.ecolind.2010.06.013

Møller, A. P., & Mousseau, T. A. (2011b). Conservation consequences of Chernobyl and other nuclear accidents. *Biological Conservation, 144*, 2787–2798. https://doi.org/10.1016/j.biocon.2011.08.009

Møller, A. P., & Mousseau, T. A. (2015). Strong effects of ionizing radiation from Chernobyl on mutation rates. *Scientific Reports, 5*, 8363. https://doi.org/10.1038/srep08363

Møller, A. P., Mousseau, T. A., Milinevsky, G., Peklo, A., Pysanets, E., & Szép, T. (2005a). Condition, reproduction and survival of barn swallows from Chernobyl. *The Journal of Animal Ecology, 74*, 1102–1111.

Møller, A. P., Surai, P. F., & Mousseau, T. A. (2005b). Antioxidants, radiation and mutation as revealed by sperm abnormality in barn swallows from Chernobyl. *Proceedings of the Royal Society of London – Series B: Biological Sciences, 272*, 247–253. https://doi.org/10.1098/rspb.2004.2914

Møller, A. P., Hobson, K., Mousseau, T. A., & Peklo, A. (2006). Chernobyl as a population sink for barn swallows: Tracking dispersal using stable-isotope profiles. *Ecological Applications, 16*, 1696–1705.

Møller, A. P., Karadas, F., & Mousseau, T. A. (2008a). Antioxidants in eggs of great tits *Parus major* from Chernobyl and hatching success. *Journal of Comparative Physiology B, 178*, 735–743. https://doi.org/10.1007/s00360-008-0262-z

Møller, A. P., Mousseau, T. A., Lynnn, C., Ostermiller, S., & Rudolfsen, G. (2008b). Impaired swimming behaviour and morphology of sperm from barn swallows *Hirundo rustica* in Chernobyl. *Mutation Research/Genetic Toxicology and Environmental Mutagenesis, 650*, 210–216. https://doi.org/10.1016/j.mrgentox.2007.12.006

Møller, A. P., Hagiwara, A., Matsui, S., Kasahara, S., Kawatsu, K., Nishiumi, I., Suzuki, H., Ueda, K., & Mousseau, T. A. (2012). Abundance of birds in Fukushima as judged from Chernobyl. *Environmental Pollution, 164*, 36–39. https://doi.org/10.1016/j.envpol.2012.01.008

Møller, A. P., Bonisoli Alquati, A., Mousseau, T. A., & Rudolfsen, G. (2014). Aspermy, sperm quality and radiation in Chernobyl birds. *PLoS One, 9*, e100296. https://doi.org/10.1371/journal.pone.0100296

Møller, A. P., Nishiumi, I., & Mousseau, T. A. (2015). Cumulative effects of radioactivity from Fukushima on the abundance and biodiversity of birds. *Journal für Ornithologie, 156*, 297–305. https://doi.org/10.1007/s10336-015-1197-2

Morelli, F., Jiguet, F., Reif, J., Plexida, S., Valli, A. S., Indykiewicz, P., Šímová, P., Tichit, M., Moretti, M., & Tryjanowski, P. (2015). Cuckoo and biodiversity: Testing the correlation between species occurrence and bird species richness in Europe. *Biological Conservation, 190*, 123–132. https://doi.org/10.1016/j.biocon.2015.06.003

Morelli, F., Mousseau, T. A., & Møller, A. P. (2017). Journal of environmental radioactivity. *Journal of Environmental Radioactivity, 177*, 158–164. https://doi.org/10.1016/j.jenvrad.2017.06.029

Mothersill, C. E. et al. (2020). From tangled banks to toxic bunnies; A reflection on the issues involved in developing an ecosystem approach for environmental radiation protection. *Int J Radiat Biol*, 1–57. https://doi.org/10.1080/09553002.2020.1793022.

Murphey, P., McLean, D. J., McMahan, C. A., Walter, C. A., & McCarrey, J. R. (2013). Enhanced genetic integrity in mouse germ cells. *Biology of Reproduction, 88*, 6. https://doi.org/10.1095/biolreprod.112.103481

Niemi, G. J., & McDonald, M. E. (2004). Application of ecological indicators. *Annual Review of Ecology, Evolution, and Systematics*, 89–111.

Pamplona, R. (2008). Membrane phospholipids, lipoxidative damage and molecular integrity: A causal role in aging and longevity. *Biochimica et Biophysica Acta (BBA) – Bioenergetics, 1777*, 1249–1262. https://doi.org/10.1016/j.bbabio.2008.07.003

R Core Team. (2019). R: A Language and Environment for Statistical Computing. *R Foundation for Statistical Computing*. Vienna, Austria.

Sakharov, I. Y., Evdokimov, V. V., & Atochina, E. A. (2009). Sperm angiotensin-converting enzyme activity in Chernobyl victims and patients with chronic prostatitis. *Andrologia, 28*, 259–260. https://doi.org/10.1111/j.1439-0272.1996.tb02793.x

Sikka, S. C. (2001). Relative impact of oxidative stress on male reproductive function. *Current Medicinal Chemistry, 8*, 851–862.

Sohal, R., Mockett, R., & Orr, W. (2002). Mechanisms of aging: An appraisal of the oxidative stress hypothesis. *Free Radical Biology & Medicine, 33*, 575–586.

Stark, K., Goméz-Ros, J. M., Batlle, J. V. I., Hansen, E. L., Beaugelin-Seiller, K., Kapustka, L. A., Wood, M. D., Bradshaw, C., Real, A., McGuire, C., & Hinton, T. G. (2017). Dose assessment in environmental radiological protection: State of the art and perspectives. *Journal of Environmental Radioactivity, 175–176*, 105–114. https://doi.org/10.1016/j.jenvrad.2017.05.001

Strand, P., Beresford, N., Copplestone, D., Godoy, J., Jianguo, L., Saxén, R., Yankovich, T., Brown, J., & International Commission on Radiological Protection. (2009). ICRP Publication 114. Environmental protection: Transfer parameters for reference animals and plants. *Annals of the ICRP, 39*, 1–111.

Surai, P. F., Blesbois, E., Grasseau, I., Chalah, T., Brillard, J. P., Wishart, G. J., Cerolini, S., & Sparks, N. H. (1998). Fatty acid composition, glutathione peroxidase and superoxide dismutase activity and total antioxidant activity of avian semen. *Comparative Biochemistry and Physiology. Part B, Biochemistry & Molecular Biology, 120*, 527–533. https://doi.org/10.1016/S0305-0491(98)10039-1

Tice, R. R., Agurell, E., Anderson, D., Burlinson, B., Hartmana, A., Kobayashi, H., Miyamae, Y., Rojas, E., Ryu, J. C., & Sasaki, Y. F. (2000). Single cell gel/comet assay: Guidelines for in vitro and in vivo genetic toxicology testing. *Environmental and Molecular Mutagenesis, 35*, 206–221.

Tyler, S. A., & Stearner, S. P. (1966). Growth in the chicken after high-rate and low-rate Cobalt-60 gamma irradiation. *Radiation Research, 29*, 257. https://doi.org/10.2307/3572010

Whittington, K., & Ford, W. C. L. (1999). Relative contribution of leukocytes and of spermatozoa to reactive oxygen species production in human sperm suspensions. *International Journal of Andrology, 22*, 229–235. https://doi.org/10.1046/j.1365-2605.1999.00173.x

Zach, R., & Mayoh, K. R. (1986a). Gamma-radiation effects on nestling house wrens: A field study. *Radiation Research, 105*, 49. https://doi.org/10.2307/3576724

Zach, R., & Mayoh, K. R. (1986b). Gamma irradiation of tree swallow embryos and subsequent growth and survival. *The Condor, 88*, 1. https://doi.org/10.2307/1367747

Chapter 12
Amphibians in Field Radioecology: A Review and Perspective

Germán Orizaola

Abstract Understanding the effects of the chronic exposure to ionizing radiation in wild organisms is essential in order to evaluate the environmental impact of nuclear accidents. Amphibians are an ideal group for the study of the effects that ionizing radiation has on wildlife, due to their relatively long lifespan, low dispersal capacities and use of both the aquatic and terrestrial environments. Here, I first summarize the current knowledge on the accumulated dose rates and biological effects of ionizing radiation on amphibians living in the wild. A total of 13 studies have provided data on radioactive concentration in amphibians collected in the field, covering 16 species, with maximum values of 188 and 205 kBq/kg for ^{134}Cs and ^{137}Cs in Japanese brown frog (*Rana japonica*) collected at Fukushima, and 3090 kBq/kg ^{90}Sr in common spadefoot toads (*Pelobates fuscus*) inside Chernobyl Exclusion Zone. A total of 21 studies have reported biological effects on amphibians exposed to radiation in the wild, ranging from alteration in chromosome structure and the presence of micronuclei, to changes in pigments levels and alteration of the immune system. Finally, I suggest some research lines that may result relevant to follow on amphibian field radioecology in the near future. Among them, it would be highly relevant to examine the effects of radiation on the lifespan and ageing of amphibians living in contaminated areas; the ecophysiology of individuals with different levels of exposure to radiation, or the use that amphibian make of contaminated landscapes. At a genomic level, using state-of-the-art techniques should allow to evaluate demographic changes, effective population sizes, local adaption processes, as well as signs of adaptive evolution. At a more methodological level, focussing on the embryonic and larval part of the amphibian life cycle, and the use of mesocosms

G. Orizaola (✉)
IMIB-Biodiversity Research Institute (Univ. Oviedo-CSIC-Princip. Asturias), Mieres, Asturias, Spain

Zoology Unit, Department of Biology of Organisms and Systems, University of Oviedo, Oviedo, Asturias, Spain

Animal Ecology Unit, Department of Ecology and Genetics, Uppsala University, Uppsala, Sweden
e-mail: orizaolagerman@uniovi.es

M. D. Wood et al. (eds.), *Biomarkers of Radiation in the Environment*, NATO Science for Peace and Security Series A: Chemistry and Biology,
https://doi.org/10.1007/978-94-024-2101-9_12

and reciprocal transplants experiment would help to have a more complete evaluation of the effects of, and adaptive responses to, ionizing radiation on amphibians under field conditions.

12.1 Introduction

Ionizing radiation is harmful for organisms since the free radicals it generates can directly damage DNA and other organic molecules (UNSCEAR, 2008). Ionizing radiation is naturally present in the environment in highly variable levels, although the main concern over radiation effects on living organisms came as a consequence of radioactive materials released by human activity, from radioactive mining and medical use, to nuclear weapons tests (Beresford & Copplestone, 2011). Accidental releases of radionuclides from nuclear power plants are the most evident case of concern for the impact of artificial radiation on living organisms. The accidents in the nuclear power plants of Chernobyl (Ukraine) in 1986 and Fukushima (Japan) in 2011, represent the more obvious (and more dramatic) examples of radioactive contamination due to human action. These accidents forced the evacuation of human populations from vast areas and had an immediate effect on human lives and on natural ecosystems (Møller & Mousseau, 2006; Beresford & Copplestone, 2011; Mousseau & Møller, 2014; Beredsford et al., 2016). However, besides the direct effects of radiation immediately after the accidents, there is no clear consensus about the long-term effects of the chronic exposure to the remaining low-dose radiation levels in areas affected by radioactive pollution (Møller & Mousseau, 2006; 2016; Beresford & Copplestone, 2011; Beredsford et al., 2016). A good understanding, based on solid scientific studies, of the long-term effects of ionizing radiation on the environment is crucial for the correct management of radioactive pollution for humans and for wildlife.

Susceptibility to artificial ionizing radiation varies greatly among taxa, although vertebrates are the more sensitive group (Whicker & Schultz, 1982, Adam-Guillermin et al., 2018). Among vertebrates, amphibians, are a particularly interesting group for studying the impact of radiation in wildlife since they are highly sensitive to radiation, and occupy a wide diversity of habitats during their complex-life cycles. The classic amphibian life cycle is characterised by the maintenance of reproduction activities linked to the aquatic environment, and by the existence of a larval aquatic stage that after growing and developing in water experiences a process of complete morphological and physiological reorganization, i.e. metamorphosis, before reaching a juvenile terrestrial stage (Wilbur, 1980). In temperate areas, juvenile growth and development generally occurs completely in the terrestrial environment, as well as hibernation. Therefore, most amphibians occupy both the aquatic and the terrestrial environments during crucial parts of their life cycle and, thus, can be exposed to a great variety of radioactive contamination on both types of substrate. Furthermore, amphibians, either at the adult, juvenile or larval stages, have moist, permeable skin that can readily absorb toxic substances, making them highly vulnerable to general pollution. Interestingly, amphibians are also characterised for having low dispersal capacities and high philopatry, i.e. they often use for breeding the same area and even the same pond in which they were born (Berven & Grudzien, 1990; Reading et al., 1991; Sinsch, 1991; Blaustein et al., 1994). This restriction in movement and their dependence of both the aquatic and the terrestrial environment makes amphibians highly sensitive to environmental alterations, in particular to habitat fragmentation, the emergence of new diseases, and pollution (Kiesecker et al., 2001). In fact, amphibians are currently the most endangered group of vertebrates, with about a third of the species considered at high risk of disappearance (IUCN, 2008), and most experiencing significant population declines (Stuart et al., 2004).

Amphibians are sensitive to different types of radiation. Actually, one of the factors that may have contributed to the decline of amphibian populations in recent times is the increase on the levels of non-ionizing, UV radiation (Blaustein et al., 1998; Kiesecker et al., 2001). UV radiation can cause lethal and sublethal effects in embryos, larvae and adult amphibians (Licht & Grant, 1997; Blaustein & Belden, 2003). Direct exposure to UV radiation often causes a significant increase in embryonic mortality in frogs, toads, and salamanders (Blaustein et al., 1998). UV radiation can also alter amphibian behaviour, reduce larval growth and development, and induce different types of malformations (review in Blaustein et al., 2003). Amphibians are also affected by ionizing radiation, and they have been, for many decades, the subject of detailed radiobiology studies in the lab (e.g. Luther, 1939; Rugh, 1954; Brunst, 1965). These studies focussed on a vast variety of traits and radiation levels and demonstrated the high sensitivity of amphibians, at the embryonic, larval and adult stages, to high doses of ionizing radiation. For example, Brunst (1965) described in great detail severe alterations in cell division and morphology when amphibians were exposed to high radiation in the lab. On the same line, Masui (1973) later reported how amphibian oocytes responded to radiation; with oocytes irradiated with high doses before the initiation of maturation usually failing to be fertilized, and showing clear damage of the jelly surrounding the eggs, whereas oocytes irradiated after maturation normally showed an increase of developmental abnormalities. Studies focussed on determining lethal doses (LD_{50}) showed that the fertilised egg was the more sensitive stage of the amphibian life cycle, whereas medium to large tadpoles were the most resistant stages (e.g. for *Limnodynastes tasmaniensis*; Panter, 1986). Other studies examining the survival of early amphibian stages exposed to high doses of ionizing radiation in the lab reported that early larval stages were more resistant than larvae approaching metamorphosis (*Rana catesbeiana*; Just & Sperka, 1977), and even that in some species irradiated tadpoles can fail to metamorphose completely (*Bufo woodhousei foleri*; Landreth et al., 1974).

Most of the studies examining the effects of ionizing radiation on living organisms conducted under laboratory conditions have in common the use of very high dose rates rarely encountered in nature, as well as the use of laboratory setups lacking in ecological context and complexity. Additionally, many of the traits examined in these studies were connected with very specific steps of embryonic and larval development, and focussed mostly on end points in the form of mortality or physical malformations (e.g. Bondar'kov et al., 2002). As a consequence of the acute exposure used in most of the experiments, it is often not possible to evaluate mid- or long-term effects of radiation on individuals. However, mostly after the accidents in the Chernobyl and Fukushima power plants, some studies on the effects of ionizing radiation have been also conducted in the field. Here, I present a review of these radioecology studies conducted on amphibians living under natural, i.e. field conditions. The aim of this review is to summarise the diversity of studies conducted on amphibians, the species involved in these studies, and the different traits examined. Finally, I suggest some research lines that may be interesting to follow for the future development of radioecology studies on amphibians within and eco-evolutionary framework. This review should help to identify gaps in field radioecology studies with amphibians, as well promote the use of this group of vertebrates as a very pertinent model for the evaluation of the effects of, and the adaptive responses to, chronic exposure to ionizing radiation in wildlife.

12.2 Amphibians in Field Radioecology: A Review

After the accidents occurred in the nuclear power plants of Chernobyl, Ukraine, in 1986 and Fukushima, Japan, in 2011, many studies have been conducted in order to evaluate the impact that the accidental release of large amounts of radiation to the environment has on the wildlife inhabiting the contaminated areas. Studies moved from measuring the concentration of radionuclides and the immediate effects of the releases during the acute phase of the accidents (i.e. first years), to examine the effects that the exposure to chronic low-dose ionizing radiation may have on organisms living in areas that are still contaminated years after the accidents. Among these studies, the effects of radiation on plants, birds and mammals have focussed most of the research efforts, generating a fairly good amount of scientific literature (e.g. Mousseau & Møller, 2014; see also Chap. 11). However, studies on other organisms are much scarcer (Møller & Mousseau, 2006; Geras'kin et al., 2008; Yablokov, 2011; Mousseau & Møller, 2014). In particular, the effects of radioactive contamination on wild amphibians have been only superficially examined.

In order to summarise the different studies reporting how amphibians respond to ionizing radiation in the wild, I performed a search on October 2018 using ISI Web of Knowledge database and the keywords "radioact* or radiation", in combination with "frog* or toad* or salamander* or amphibian*". I conducted additional specific searchers using "Chernobyl" or "Fukushima" in combination with "radioact* or radiation". Afterwards, I filtered the list of articles to pick only those ones including information from amphibians exposed to radioactive contamination in the wild. In the final group of publications considered, I also included an additional set of studies, written in Russian, Ukrainian, and Belarussian, not accessible through ISI Web of Knowledge, and detected searching through the references of the previous papers, mostly Yablokov (2011).

A total of 24 studies matched the different searching criteria, covering 16 species (13 anurans and 3 urodeles), and including studies conducted around the three major nuclear accidents: Chernobyl, Fukushima, and Mayak (1957, former Soviet Union).

12.2.1 Radionuclide Concentrations in Wild Amphibians

A total of 13 studies have reported levels of radioactive concentration in amphibians collected in the field, covering 16 species (13 anurans and 3 urodeles, Tables 12.1 and 12.2). Studies mostly reported internal concentrations of strontium (^{90}Sr) and/or caesium (^{134}Cs, ^{137}Cs) expressed here in Bq/kg (Tables 12.1 and 12.2). Most of these studies have been focused on determining the concentration levels of the different radionuclides on the bodies of adult amphibians, and only four times in amphibian larvae (Sakai et al., 2014; Fuma et al., 2015; Tagami et al., 2018). A drawback of many of the studies is the low sample size used to analysed radionuclide concentration; in about a half of the cases sample size was lower than 10 individuals (17 out of 35 cases; Tables 12.1 and 12.2). Additional problems of these studies are the mixing of males and females (e.g. Matsushima et al., 2015), adults and juveniles (e.g. Takahara et al., 2015), and even different species (Tagami et al., 2018; Table 12.1).

The highest concentration of caesium in amphibians was found in juveniles of the Japanese brown frog (*Rana japonica*) collected inside Fukushima Exclusion Zone in July 2011 (2 years after the accident), with maximum concentrations of 188 and 205 kBq/kg for ^{134}Cs and ^{137}Cs respectively (Fuma et al., 2017). For strontium, the highest levels were found in adult common spadefoot toads (*Pelobates fuscus*) collected inside Chernobyl Exclusion Zone in April 2003, with levels of 3090 kBq/kg ^{90}Sr (Gaschak et al., 2011). Only two studies evaluated sexual differences in radionuclide concentration, finding much higher levels of ^{90}Sr in females than males for moor frogs (*Rana arvalis*) living in the Eastern Ural Radioactive Trace (Pyastolova & Vershinin 1999), but no differences between sexes for Eurasian marsh frogs (*Pelophylax ridibundus*) living in the same area, neither for ^{90}Sr or ^{137}Cs (Guseva et al., 2017). Again, there seems to be not a clear pattern about how radioactive concentration changes across the different life stages on the amphibian cycle; some studies suggested that concentrations are higher in juveniles (Pyastolova & Vershinin, 1999, in *Rana arvalis*), whereas other studies find higher levels in adults or embryos (Fuma et al., 2015, in *Hynobius lichenatus*), higher in adults than larvae (Tagami et al., 2018, in *Buergeria buergeri*), or no clear pattern (Guseva et al., 2017, in *Pelophylax ridibundus*).

Clearly, more studies, with larger samples sizes and standardized sampling protocols, are needed in order to understand how radionuclide concentration varies among species and across time in amphibians. Other aspects that remain open are how radionuclide concentration can vary across life stages in amphibians, how it may change across the life of an individual, or the putative effects that mortality in

Table 12.1 Radionuclide concentration (Bq/kg) in urodeles collected in the wild

Species	Area	n	Stage	Radionuclide concentration (Bq/kg)						References
				^{134}Cs		^{137}Cs		^{90}Sr		
				Max.	Min.	Max.	Min.	Max.	Min.	
Amphibians	Chernobyl	6	Adult	–	–	4.15	0.55	55.27	2.027	Oskolkov et al. (2010)
Caudata										
Cynops pyrrhogaster	Fukushima	2	Adult	84.6E+3	0.15E+3	99.9E+3	0.18E+3	–	–	Fuma et al. (2017)
Hynobius lichenatus	Fukushima	12	Adult	96.5E+3	150	126.0 E+3	293	–	–	Fuma et al. (2015)
Hynobius lichenatus	Fukushima	17	Larvae	20.0E+3	<149	19.6E+3	141	–	–	Fuma et al. (2015)
Hynobius lichenatus	Fukushima	153	Embryo	58.5E+3	265	83.1E+3	320	–	–	Fuma et al. (2015)
Triturus cristatus	Chernobyl	1	Adult	–	–	–	–	6.53E+3	–	Oskolkov et al. (2011)

Table 12.2 Radionuclide concentration (Bq/kg) in anurans collected in the wild

Species	Area	n	Stage	Radionuclide concentration (Bq/kg)						References
				^{134}Cs		^{137}Cs		^{90}Sr		
				Max.	Min.	Max.	Min.	Max.	Min.	
Pelobates fuscus	Chernobyl	46	Adult	–	–	718E+3	36.4E+3	309E+4	49.9E+3	Gaschak et al. (2011)
Bufo japonicus	Fukushima	1	Adult		0.01		0.01	–	–	Tagami et al. (2018)
Hyla japonica	Fukushima	2	Adult	18.0E+3	15.9E+3	17.7E+3	16.9E+3	–	–	Fuma et al. (2017)
Hyla japonica	Fukushima	27	Adult	106.4*	46.2*	–	–	–	–	Matsushima et al. (2015)
Hyla japonica	Fukushima	142	Adult	42.8E+3	0	56.2E+3	53	–	–	Giraudeau et al. (2018)
Hyla japonica	Fukushima	42	Adult	4.5**		11**		–	–	Tagami et al. (2018)
Pelophylax p. porosus	Fukushima	3	Adult	1.7E+3*	43.1*	–	–	–	–	Matsushima et al. (2015)
Pelophylax p. porosus	Fukushima	50	Larvae	3000	600	4500	0890	–	–	Sakai et al. (2014)
Pelophylax p. porosus	Fukushima	85	Adult	27	0.78	54	1.9	–	–	Tagami et al. (2018)
Pelophylax ridibundus	Urals	56	Adult	–	–	53.0E+3	8.0	24.9	1.4	Guseva et al. (2017)
Pelophylax esculentus	Chernobyl	5	Adult	–	–	4.15E+3	0.55E+3	55.3E+3	2.27E+3	Oskolkov et al. (2011)
Rana arvalis	Chernobyl	1	Adult	–	–	20.5E+3		32.1E+3		Oskolkov et al. (2011)
Rana arvalis	Sweden	5	Adult	–	–	3.5E+3		–	–	Stark et al. (2004)
Rana arvalis	Urals	22	Adult	–	–	–	–	15.9E+5	0	Pyastolova and Vershinin (1999)
Rana catesbeiana	Fukushima	8	Adult	270	1.2	66	290	–	–	Tagami et al. (2018)
Rana japonica	Fukushima	111	Adult	39.5E+3*	36.8*	–	–	–	–	Matsushima et al. (2015)
Rana japonica	Fukushima	1	Juv	188E+3		205E+3		–	–	Fuma et al. (2017)
Rana japonica	Fukushima	24	Adult	5.7	3.3	12	8.4	–	–	Tagami et al. (2018)
Rana ornativentris	Fukushima	3	Adult	220.6*		–	–	–	–	Matsushima et al. (2015)
Rana ornativentris	Fukushima	1	Adult	93.3E+3		110E+3		–	–	Fuma et al. (2017)
Rana ornativentris	Fukushima	5	Juv	52.1E+3	8.29E+3	61.7E+3	9.78E+3	–	–	Fuma et al. (2017)
Rana ornativentris	Fukushima	42	Adult	160	7.7	670	22	–	–	Tagami et al. (2018)
Rana ornativentris	Fukushima	?	Larvae		220**		320**	–	–	Tagami et al. (2018)

(continued)

Table 12.2 (continued)

Species	Area	n	Stage	Radionuclide concentration (Bq/kg)						References
				^{134}Cs		^{137}Cs		^{90}Sr		
				Max.	Min.	Max.	Min.	Max.	Min.	
Rana rugosa	Fukushima	208	Adult	79	0.64	190	1.2	–	–	Tagami et al. (2018)
Rana t. tagoi	Fukushima	8	Adult	47.3E+3*		–	–	–	–	Matsushima et al. (2015)
Rana t. tagoi	Fukushima	66	Ad+Juv	27.7E+3*	2.6E+3*	–	–	–	–	Takahara et al. (2015)
Buergeria buergeri	Fukushima	1	Adult	16.0E+3*		–	–	–	–	in Matsushima et al. (2015)
Buergeria buergeri	Fukushima	2	Adult		74**		370**	–	–	Tagami et al. (2018)
Buergeria buergeri	Fukushima	293	Larvae		55**		81**	–	–	Tagami et al. (2018)

*Values represent the sum of ^{134}Cs and ^{137}Cs; **Values represent the average of the total sample size early stages may have on the patterns of radionuclide concentration observed when examining adult stages.

12.2.2 Biological Effects of Ionizing Radiation on Wild Amphibians

A total of 21 studies have examined the biological effects that the exposure to ionizing radiation has on wild amphibians. These studies vary deeply in the type of traits examined, from chromosome aberrations and the presence of micronuclei, to changes in immune cell counts.

In the common (*Rana temporaria*) and moor (*R. arvalis*) frogs, the percentage of bone marrow cells with chromosome aberrations was higher in radioactive contaminated areas of Belarus (1.8%) than in control ones (0.4%) during 1986–1992 (Yeliseeva et al., 1995). These authors also found a non-significant tendency to a reduction in chromosomal damage in some of the contaminated areas between 1990 and 1992 (Yeliseeva et al., 1995). Both *Rana temporaria* and *R. arvalis* presented significant cytogenetic damage in bone marrow cells and erythrocytes, as well as a change in the ratio of erythrocytes in peripheral blood when exposed to radiation in the wild (Voitovich, 2000). In Chernobyl, studies conducted on the moor frog (*Rana arvalis*) detected higher levels of cells with chromatin changes in individuals from populations located in areas that have been contaminated by radiation for 7 years, and once these frogs were experimentally exposed to additional radiation they showed higher levels of apoptosis in spleen and bone marrow cells (Afonin & Voitovich, 1998; Afonin et al., 1999).

Micronuclei are extra-nuclear bodies that contain damaged chromosome fragments and/or whole chromosomes that were not incorporated into the nucleus after cell division (Luzhna et al., 2013). The micronuclei test, consisted in the estimation of the frequency of cells harbouring micronucleus among normal cells, has been widely used in toxicological studies as an indicator of the adverse effects of chemicals on animals (Hayashi, 2016). Inside Chernobyl Exclusion Zone, adults of *Rana temporaria* and *R. arvalis* frogs from highly contaminated localities had higher number of micronuclei in blood cells than frogs from non-contaminated areas (Yeliseeva et al., 1996). On a different study, *Rana temporaria* frogs collected before 1991 from highly contaminated localities also presented higher number of micronuclei (Voitovich, 2000). On the same line, adult edible frogs (*Pelophylax esculentus*) collected in 1993 from contaminated areas in Bryansk Province (Russia), also presented higher proportion of micronuclei in erythrocytes than frogs from non-contaminated areas (from 0.22% to 1.55%; Chubanishvyli et al., 1996). Frogs that developed in water bodies with high concentrations of radionuclides and heavy metals, presented also higher frequency of single-strand breaks of DNA than individuals in control, non-contaminated localities, although no significant differences were found when estimating the level of double-strand breaks (Yushkova et al., 2018). Also, a reduction in genome size, estimated by DNA flow cytometry, was found in edible frogs (*Pelophylax esculentus*) living in areas with radioactive contamination from the Chernobyl fallout inside Belorussia (Vinogradov & Chubinishvili, 1999).

Lower body condition was found in moor frogs (*Rana arvalis*) breeding in areas contaminated by the Mayak accident, using the ratio of body weight to body length as a measure of individual condition (Versihin & Seredyuk, 2000). On the same species and area, two studies reported a shorter lifespan in individuals breeding in radioactive contaminated localities (Ushakov et al., 1982; Pyastolova & Vershinin, 1999, in Versihin & Seredyuk, 2000). Smaller egg size was also found for this species in areas affected by radioactive contamination, both around Mayak (Pyastolova et al., 1996), and in the Chernobyl Exclusion Zone (Cherdantsev et al., 1993), as well as a higher proportion of clutches containing small embryos (Cherdantsev et al., 1993). Moor frogs (*R. arvalis*) living in localities within the Eastern Ural Radioactive Trace had lower metabolism, indicated by lower oxygen consumption, than frogs from non-contaminated areas (Versihin & Seredyuk, 2000).

The amount of nutrients stored as fat in organs as the liver is essential for amphibian survival, especially during hibernation. One study found that both adult and juvenile moor frogs (*Rana arvalis*) from radioactive contaminated sites at the Eastern Ural Radioactive Trace had lower amounts of nutrients in liver cells than frogs from clean localities, as well as an increase in the size of hepatocytes that may be linked to chronic load of protective mechanisms, and ca. 13% of juveniles showed chromosome damage (Pyastolova & Vershinin, 1999). Liver with signs of hepatocyte destruction were also observed in Chernobyl frogs, as well as hypertropia in other cells of the hemopoietic system (Nosova et al., 1994). Other study reported smaller size at metamorphosis in juveniles from contaminated areas (Vershinin & Tereshin, 1996, cited in Pyastolova & Vershinin, 1999), which may reduce their survival at the terrestrial environment. The rate of morphological anomalies in *R. arvalis* metamorphs was also higher in radioactive contaminated localities of the Eastern Ural Radioactive Trace than in non-contaminated localities (17% higher; Pyastolova et al., 1996). In another area, place of nuclear tests at Semipalatinsk (Kazakhstan) 16.3% of marsh frogs (*Pelophylax ridibundus*) showed different morphological anomalies: ectrodactyly, syndactyly, asymmetrical limbs, melanism and dorsal pattern anomalies (Rakhimzhanova & Khromov, 1998, in Henle et al., 2017).

The effects of the chronic exposure to radiation in wild amphibians have been also examined at the level of the immune system, reporting that adult edible frogs (*Pelophylax esculentus*) inhabiting contaminated areas inside Chernobyl Exclusion Zone had higher leukocyte and lymphocyte count, and lower percentage of neutrophils, T and B lymphocytes than frogs from clean localities (Isaeva & Vyazov, 1996).

The only study published, so far, examining the biological effects of the nuclear accident at the Fukushima power plant on amphibians showed no effects of radiation levels on the amount of carotenoids on the blood, liver and vocal sac skin in adult Japanese tree frogs (*Hyla japonica*), which suggests that carotenoid distribution in amphibians might be less sensitive to ionizing radiation exposure than in other organisms (Giraudeau et al., 2018).

One crucial aspect when evaluating the impact of a nuclear accident is to properly estimate the changes in the abundance and diversity of organisms living inside the contaminated areas. Only one study has aimed at examining the effects of radioactive contamination on the abundance of amphibians. Møller and Mousseau (2011) evaluated the abundance of different groups of animals across a radiation gradient inside Chernobyl Exclusion Zone, and reported a lower abundance of amphibians across that gradient. However, this result should be taken with great caution since the methodology to estimate amphibian abundance was far from ideal, as authors acknowledge in the paper. Amphibian sampling was conducted on a semi-opportunistic way over transects designed mostly for examining bird abundance, and during daylight, whereas amphibian abundance should be estimated around breeding localities, preferentially by night and during repeated visits to the same locality across the breeding season (Heyer et al., 1994). Therefore, further researcher is needed to evaluate the effects of radioactive contamination on the abundance and distribution of amphibians both at Chernobyl and Fukushima, where otherwise abundant (and diverse) amphibian populations can be seen using localities within highly contaminated areas (G. Orizaola, P. Burraco pers. obs).

In summary, few studies have examined the biological effects of the chronic exposure to ionizing radiation on amphibians, and these have addressed only a small range of traits, too often using very low sample sizes and lacking, for example, in control by age in most cases (i.e. age effects were not included in the analyses). Overall, there is also a clear lack of studies reporting negative results, i.e. lack of effects of the exposure to ionizing radiation. Undoubtedly, more detailed studies with carefully thought sampling design, both in terms of examined traits and sample sizes, are needed in order to establish amphibians as model organisms in which to understand the effects of chronic exposure to ionizing radiation on wildlife.

12.3 Future Research Directions in Field Radioecology with Amphibians

Amphibians are an ideal study system to examine the impact of the chronic exposure to ionizing radiation on wildlife, as they are vertebrates with reduced dispersion capacities, which use both the aquatic and terrestrial environment, and are also crucial elements of the food chain (Wells, 2007). However, as it is summarised above, the study of the effects of ionizing radiation on wild amphibians has been very descriptive until now, focussed on evaluating absorbed dose rates, and only marginally on understanding biological effects with ecological or evolutionary relevance. Research on amphibian field radioecology should move forward and identify traits directly linked with individual survival and fitness, as well as their genetic and genomic underpinning. This will allow to properly evaluate the impact of radioactive contamination on these organisms, as well as the adaptive responses that may have emerged as a consequence of the chronic exposure to radiation. Below, I will highlight some aspects that should be part of a research program in field radioecology with amphibians in the near future.

Obvious points to be implemented in future work on amphibian field radioecology are the use of adequate samples sizes, as well as a better focus on the type of organisms used in the studies. As mentioned before, many studies on the effects of radiation on amphibians have been conducted using extremely low samples sizes (see Tables 12.1 and 12.2), which compromise the validity and generalization of the results. On the other size, studies often did not differentiate between males and females, or adults and juveniles when measuring dose rates or biological responses, which adds unnecessary uncertainty. A precise and justified use of a particular sex and life cycle stage is, thus, needed in the future in order to standardize the observed responses.

One interesting aspect that warrants further investigation in field radioecology is the possible cost of radiation exposure in terms of the lifespan of exposed organisms. The exposure to high levels of environmental stress can shorten the lifespan of wildlife (e.g. Hayward et al., 2009, Monaghan et al., 2012). In this context, the possible effects of radiation would not be directly observable in the field, but would result in a reduction of the lifespan of individuals exposed to higher levels of radiation. In order to detect this effect, it will be needed either to follow individuals over time (using e.g. capture-mark-recapture techniques) or apply some methods to estimate the age of the different individuals in a population. In the case of amphibians, skeletochronological methods in which lines of arrested growth in stained bone sections are observable, are ideal to assess the age of the individuals (Sinsch, 2015). Using skeletochronology, it would be also possible to evaluate if the age structure of populations living in radioactive contaminated areas differ from that of non-contaminated areas. Furthermore, skeletochronological techniques would allow to estimate the effects of individual age on any other type of trait examined. An alternative approach to the study of lifespan would be the use biomarkers for the examination of aging processes in animals in order to test if aging (i.e. the process of getting old) differs in function of the level of radioactive contamination. A clear candidate for this task, is the use telomere length. Telomeres are repetitive DNA segments at the end cap of chromosomes, essential in maintaining cell stability during replication, and that shorten after each cell division (Dunshea et al., 2011). Exposure to various stressors is associated with reductions in telomere length (e. g. Epel et al., 2004, Haussmann & Marchetto, 2010). Thus, either the analysis of telomere length dynamics over time, or the examination of telomere length on individuals of known age will help us to understand the effects of the chronic exposure to ionizing radiation on aging mechanisms in wildlife.

Another trait that has been increasingly used in eco-evolutionary studies in response to the exposure to environmental stress is the study of oxidative stress responses (Beaulieu & Costantini, 2014). In radioactive environments, oxidative stress can be generated when reactive oxygen species (ROS) forms after ionizing radiation hits the cells. Antioxidant compounds can eliminate ROS and also repair

DNA damaged by radiation exposure (Imlay & Linn, 1988), but the use of antioxidants needs to be balanced with the maintenance of other biological functions in the organism. When antioxidant levels are too low, or ROS levels too high, a situation of oxidative stress is generated. The impact that the chronic exposure to ionizing radiation has on oxidative stress levels has been studied in birds (e.g. Galván et al., 2014; Einor et al., 2016; see also Chap. 11). However, until now, no studies have examined antioxidant responses on amphibians as a consequence of the chronic exposure to ionizing radiation. Oxidative stress has been studied in amphibians in response to different levels of environmental stress, such as desiccation or the presence of predators (Gómez-Mestre et al., 2013; Pinya et al., 2016; Burraco et al., 2017, 2018). A detailed assessment of oxidative stress responses in amphibians exposed to ionizing radiation would contribute to a better knowledge of the physiological effects of radiation on this group of organisms.

Apart from the study of telomeres and oxidative stress responses, more detailed eco-physiological studies are needed in order to fully understand non-lethal effects of ionizing radiation on wildlife. Over the last decades, precise and field-friendly techniques have been developed allowing the direct measurement of physiological parameters in natural settings, as well as the collection, storage and processing of a wide array of physiological samples outside big laboratories (Cooke et al., 2013; Madliger et al., 2018). The availability of these techniques should lead to the incorporation of a complete physiological toolbox to the studies on the effects of ionizing radiation on wildlife (Madliger et al., 2018).

Another move forward on amphibian field radioecology would be to increase the focus on the initial stages of the life cycle, i.e. embryonic and larval stages. The vast majority of studies conducted so far on the effects of ionizing radiation on wild amphibians have focussed on the adult stage (see Tables 12.1 and 12.2 for dose rates). Studying embryonic and larval stages is especially relevant in order to assess the potential early life effects of radiation, as well as the effect that mortality occurred during these stages may have on the response patterns observed in the wild when considering only the adult stage.

Most of the studies conducted on the biological effects of ionizing radiation on wildlife, and this is not only the case of amphibians, have used measurements on the genetics, physiology, abundance or behaviour of individuals sampled or observed under wild conditions, comparing effects across different radiation levels. This is a highly valuable approach, but in order to fully assess the adaptive potential of organisms exposed to chronic radiation in the wild, it is required to conduct controlled experiments under field conditions, either micro- or mesocosms experiments, or as already suggested by Møller and Mousseau (2016), experiments including reciprocal transplants. Mesocosms are widely used for the study of amphibian ecology (see e.g. Rowe & Dunson, 1998), and would allow the evaluation of radiation effects on species interactions under a controlled environment, but in organisms exposed to natural levels of radiation. Experiments using a reciprocal transplant design in which organisms from clean areas would be reared with and without exposure to radiation, and individuals from contaminated areas would be reared also under contaminated and non-contaminated conditions, would be highly interesting allowing to test for the existence of local adaptation process, the amount of plasticity of these responses, as well as the factors contributing to adaptive differentiation (Kawecki & Ebert, 2004).

Another aspect largely ignored in field radioecology is the examination of the genomic basis of the effects and responses of wildlife exposed to ionizing radiation. There have not even been studies looking at whole-genome estimates of mutation accumulation caused by chronic radiation exposure (Møller & Mousseau, 2016). The development of ecological genomics techniques makes these studies possible now, and considerable cheaper than a decade ago, even in non-model organisms as the amphibians living in radio-contaminated areas (Ellegren, 2014, Shaffer et al., 2015). Amphibians have extremely large genomes, and thus the use of genomic techniques is relatively more complex than in other organisms, but recent technical advances in this field (Shaffer et al., 2015), as well as the sequence of the first genomes of some amphibian groups (e.g. Hammond et al., 2017, Edwards et al., 2018) should act as good starting points for the development of a genomic research program in amphibian radioecology. By using a genomic approach, it would be possible to look for demographic changes, effective population sizes, local adaptation processes, as well as to detect signs of selection underlying adaptive evolution in populations exposed to ionizing radiation (Ellegren, 2014).

Recent studies examining the abundance of wildlife inside areas heavily affected by nuclear fallout after the Chernobyl accident have reported the existence of an abundant mammal community after nearly three decades of chronic radiation exposure (Deryabina et al., 2015). These results contrast with previous studies on different animal groups that reported a general reduction in abundance associated to increasing radiation level (Møller & Mousseau, 2011). Clearly, future studies on the responses of wildlife to ionizing radiation should be addressed at population level, and include the periodic census of breeding populations using adequate methods, capture-mark-recapture studies to infer individual survival and movement, as well as genetic analyses that could test for changes in effective population sizes and possible source-sink dynamics. This last point, the examination of potential source-sink dynamics on Chernobyl wildlife is highly relevant since Chernobyl Exclusion Zone has been suggested to act as a sink area for wildlife, at least for barn swallows (*Hirundo rustica*; Møller et al., 2006). The use of genetic tools at population level would solve this issue and it would be also highly useful in order to examine the connectivity between subpopulations, and the changes in genetic diversity that can be associated with potential source-sink dynamics.

The study of the use that individuals make of contaminated landscapes, as the ones in Chernobyl or Fukushima, is also highly interesting both from a theoretical and from a management point of view. The use of tracking devices should be implemented with amphibians as a tool for understanding how these organisms move across the contaminated landscape, how much are they exposed to radiation, and how they disperse across environments with different levels of radioactive contamination. Remote tracking would be a great tool to ask these questions, although the small size of most amphibians is challenging, as it forces to the use of small tracking devices with shorter battery life, although recent technological development (e.g. ICARUS, 2008) may solve most of these limitations in the near future.

Overall, more studies are needed in order to understand how amphibians are affected by the chronic exposure to ionizing radiation. Research on amphibians should also embrace a critical sift needed in field radioecology, which is to focus not only on detecting the negative effects of radiation, but also to start examining the potential adaptive responses that organism may have developed to cope with the chronic exposure to low-dose radiation in the wild.

Acknowledgements During the writing of this review, I was supported by Swedish Radiation Safety Authority projects SSM2017-269 and SSM2018-2038, and a Spanish Ministry of Education, Innovation and Universities "Ramón y Cajal" grant RYC-2016-20656.

References

Adam-Guillermin, C., Hertal-Aas, T., Oughton, D., Blanchard, L., Alonzo, F., Armant, O., & Horemans, N. (2018). Radiosensitivity and transgenerational effects in non-human species. *Annals of the ICRP, 47*, 327–341.

Afonin, V. Y., & Voitovich, A. M. (1998). Ionizing irradiation impact on cell destruction in frog spleens. *Herald National Belarus Academy of Sciences (Biology), 4*, 153–154. (in Belarussian).

Afonin, V. Y., Voitovich, A. M., & Yeliseeva, K. G. (1999). Additional γ-irradiation induced apoptosis of bone marrow cells in amphibians inhabiting radio-contaminated areas. *Herald National Belarus Academy Sciences (Biology), 4*, 131–132. (in Belarussian).

Beaulieu, M., & Costantini, D. (2014). Biomarkers of oxidative status: Missing tools in conservation physiology. *Conservation Physiology, 2*, cou014.

Beredsford, N. A., Fesenko, S., Konoplev, A., Skuterud, L., Smith, J. T., & Voigt, G. (2016). Thirty years after the Chernobyl accident: What lessons have we learnt? *Journal of Environmental Radioactivity, 157*, 77–89.

Beresford, N. A., & Copplestone, D. (2011). Effects of ionizing radiation on wildlife: What knowledge have we gained between the Chernobyl and Fukushima accidents? *Integrated Environmental Assessment and Management, 7*, 371–373.

Berven, K. A., & Grudzien, T. A. (1990). Dispersal in the wood frog (*Rana sylvatica*): Implications for genetic population structure. *Evolution, 44*, 2047–2056.

Blaustein, A. R., & Belden, L. K. (2003). Amphibian defenses against ultraviolet-B radiation. *Evolution & Development, 5*, 89–97.

Blaustein, A. R., Wake, D. B., & Sousa, W. P. (1994). Amphibian declines: Judging stability, persistence, and susceptibility of populations to local and global extinctions. *Conservation Biology, 8*, 60–71.

Blaustein, A. R., Kiesecker, J. M., Chivers, D. P., Hokit, D. G., Marco, A., Belden, L. K., & Hatch, A. (1998). Effects of ultraviolet radiation on amphibians: Field experiments. *Integrative and Comparative Biology, 38*, 799–812.

Blaustein, A. R., Romansic, J. M., Kiesecker, J. M., & Hatch, A. C. (2003). Ultraviolet radiation, toxic chemicals and amphibian population declines. *Diversity and Distributions, 9*, 123–140.

Bondar'kov, M. D., Gatshak, S. P., Goryanaya, Y. A., Maksymenko, A. M., Shul'ga, A. A., et al. (2002). Amphibian abnormalities from radioactive contamination in the Chernobyl zone. *Scientific and Technical Aspects of Chernobyl (Slavutich), 4*, 508–517. (in Ukrainian).

Brunst, V. V. (1965). Effects of ionizing radiation on the development of amphibians. *The Quarterly Review of Biology, 40*, 1–67.

Burraco, P., Díaz-Paniagua, C., & Gomez-Mestre, I. (2017). Different effects of accelerated development and enhanced growth on oxidative stress and telomere shortening in amphibian larvae. *Scientific Reports, 7*, 7494.

Burraco, P., Valdés, A. E., Johansson, F., & Gomez-Mestre, I. (2018). Physiological mechanisms of adaptive developmental plasticity in *Rana temporaria* island populations. *BMC Evolutionary Biology, 17*, 164.

Cherdantsev, V. G., Ljapkov, S. M., Cherdantseva, E. M., & Severtsov, A. S. (1993). The methods and data of ecological stability analysis of frog populations at radioactive contamination. In *Ecological consequences of radioactive contamination in the Southern Urals* (pp. 303–307). Nauka. (in Russian).

Chubanishvyli, A. T., Borisov, V. I., & Zakharov, V. M. (1996). Amphibians: Developmental stability. In V. M. Zakharov & E. Y. Krysanov (Eds.), *Consequences of the Chernobyl catastrophe: Environmental health* (pp. 47–51). Center for Russian Environmental Policy. (in Russian).

Cooke, S. J., Sack, L., Franklin, C. E., Farrell, A. P., Beardall, J., Wikelski, M., & Chown, S. L. (2013). What is conservation physiology? Perspectives on an increasingly integrated and essential science. *Conservation Physiology, 1*, cot001.

Deryabina, T. G., Kuchmel, S. V., Nagorskaya, L. L., Hinton, T. G., Beasley, J. C., Lerebours, A., & Smith, J. T. (2015). Long-term census data reveal abundant wildlife populations at Chernobyl. *Current Biology, 25*, 824–826.

Dunshea, G., Duffield, D., Gales, N., Hindell, M., Wells, R. S., & Jarman, S. N. (2011). Telomeres as age markers in vertebrate molecular ecology. *Molecular Ecology Resources, 11*, 225–235.

Edwards, R. J., Tuipulotu, D. E., Amos, T. G., O'Meally, D., Richardson, M. F., Russell, T. L., Vallinoto, M., Carneiro, M., Ferrand, N., Wilkins, M. R., Sequeira, F., Rollins, L. A., Holmes, E. C., Shine, R., & White, P. A. (2018). Draft genome assembly of the invasive cane toad, *Rhinella marina*. *GigaScience, 7*, giy095.

Einor, D., Bonisoli-Alquati, A., Costantini, D., Mousseau, T. A., & Møller, A. P. (2016). Ionizing radiation, antioxidant response and oxidative damage: a meta-analysis. *Science of the Total Environment, 548–549*, 463–471.

Ellegren, H. (2014). Genome sequencing and population genomics in non-model organisms. *Trends in Ecology & Evolution, 29*, 51–63.

Epel, E. S., Blackburn, E. H., Lin, J., Dhabhar, F. S., Adler, N. E., Morrow, J. D., & Cawthon, R. M. (2004). Accelerated telomere shortening in response to life stress. *Proceedings of the National Academy of Sciences USA, 101*, 17312–17315.

Fuma, S., Ihara, S., Kawaguchi, I., Ishikawa, T., Watanabe, Y., Kubota, Y., Sato, Y., Takahashi, H., Aono, T., Ishii, N., Soeda, H., Matsui, K., Une, Y., Minamiya, Y., & Yoshida, S. (2015). Dose rate estimation of the Tohoku hynobiid salamander, *Hynobius lichenatus*, in Fukushima. *Journal of Environmental Radioactivity, 143*, 123–134.

Fuma, S., Ihara, S., Takahashi, H., Inaba, O., Sato, Y., Kubota, Y., Watanabe, Y., Kawaguchi, I., Aono, T., Soeda, H., & Yoshida, S. (2017). Radiocaesium contamination and dose rate estimation of terrestrial and freshwater wildlife in the exclusion zone of the Fukushima Dai-ichi Nuclear Power Plant accident. *Journal of Environmental Radioactivity, 173*, 176–188.

Galván, I., Bonisoli-Alquati, A., Jenkinson, S., Ghanem, G., Wakamatsu, K., Mousseau, T. A., & Møller, A. P. (2014). Chronic exposure to low-dose radiation at Chernobyl favours adaptation to oxidative stress in birds. *Functional Ecology, 28*, 1387–1403.

Gaschak, S. P., Makliuk, Y. A., Maksimenko, A. M., Bondarkov, M. D., Chizhevsky, I., Caldwell, E. F., Jannik, G. T., & Farfán, E. B. (2011). Frequency distributions of ^{90}Sr and ^{137}Cs concentrations in an ecosystem of the "Red Forest" area in the Chernobyl exclusion zone. *Health Physics, 101*, 409–415.

Geras'kin, S. A., Fesenko, S. V., & Alexakhin, R. M. (2008). Effects of non-human species irradiation after the Chernobyl NPP accident. *Environment International, 34*, 880–897.

Giraudeau, M., Bonzom, J.-M., Ducatez, S., Beaugelin-Seiller, K., Deviche, P., Lengagne, T., Cavalie, I., Camilleri, V., Adam-Guillermin, C., & McGraw, K. J. (2018). Carotenoid distribution in wild Japanese tree frogs (*Hyla japonica*) exposed to ionizing radiation in Fukushima. *Scientific Reports, 8*, 7438.

Gomez-Mestre, I., Kulkarni, S., & Buchholz, D. R. (2013). Mechanisms and consequences of developmental acceleration in tadpoles responding to pond drying. *PLoS One, 8*, e84266.

Guseva, V. P., Chebotina, M. J., Ishchenko, V. G., & Berzin, D. L. (2017). Accumulation of radionuclids in amphibians (*Pelophylax ridibundus* Pall.) in the Middle Urals. *Contemporary Problems of Ecology, 10*, 84–89.

Hammond, S. A., Warren, R. L., Vandervalk, B. P., Kucuk, E., Khan, H., Gibb, E. A., Pandoh, P., Kirk, H., Zhao, Y., Jones, M., Mungall, A. J., Coope, R., Pleasance, S., Moore, R. A., Holt, R. A., Round, J. M., Ohora, S., Walle, B. V., Veldhoen, N., Helbing, C. C., & Birol, I. (2017). The North American bullfrog draft genome provides insight into hormonal regulation of long noncoding RNA. *Nature Communications, 8*, 1433.

Haussmann, M. F., & Marchetto, N. M. (2010). Telomeres: Linking stress and survival, ecology and evolution. *Current Zoology, 56*, 714–727.

Hayashi, M. (2016). The micronucleus test—Most widely used *in vivo* genotoxicity test. *Genes and Environment, 38*, 18.

Hayward, A. D., Wilson, A. J., Pilkington, J. G., Pemberton, J. M., & Kruuk, L. E. B. (2009). Ageing in a variable habitat: Environmental stress affects senescence in parasite resistance in St Kilda Soay sheep. *Proceedings of the Royal Society B, 276*, 3477–3485.

Henle, K., Dubois, A., & Vershinin, V. (2017). A review of anomalies in natural populations of amphibians and their potential causes. *Mertensiella, 25*, 56–164.

Heyer, W. R., Donnelly, M. A., McDiarmid, R. W., Hayek, L. A. C., & Foster, M. S. (1994). *Measuring and monitoring biological diversity: Standard methods for amphibians*. Smithsonian Institution.

ICARUS. (2008). *White paper: Global satellite tracking of (small) animals will revolutionize insight into migration, human health, and conservation*. ICARUS.

Imlay, J. A., & Linn, S. (1988). DNA damage and oxygen radical toxicity. *Science, 240*, 1302–1309.

Isaeva, E. I., & Vyazov, S. O. (1996). Amphibians: Immune status. In V. M. Zakharov & E. Y. Krysanov (Eds.), *Consequences of the Chernobyl catastrophe: Environmental health* (pp. 52–59). Center for Russian Environmental Policy. (in Russian).

IUCN. (2008). *An analysis of amphibians on the 2008 IUCN Red List*. Available at: http://www.iucnredlist.org/initiatives/amphibians

Just, J. J., & Sperka, R. (1977). The effects of whole body irradiation on survival and erythropoiesis of *Rana catesbeiana* tadpoles. *Radiation Research, 69*, 258–266.

Kawecki, T. J., & Ebert, D. (2004). Conceptual issues in local adaptation. *Ecology Letters, 7*, 1225–1241.

Kiesecker, J. M., Blaustein, A. R., & Belden, L. K. (2001). Complex causes of amphibian population declines. *Nature, 410*, 681–684.

Landreth, H. F., Dunaway, P. B., & Cosgrove, G. E. (1974). Effects of whole body gamma irradiation on various life stages of the toad, *Bufo woodhousei fowleri*. *Radiation Research, 58*, 432–438.

Licht, L. E., & Grant, K. P. (1997). The effects of ultraviolet radiation on the biology of amphibians. *Integrative and Comparative Biology, 37*, 137–145.

Luther, W. (1939). The radiation effect on amphibian skin before and after metamorphosis. *Naturwissenschaften, 27*, 713–720.

Luzhna, L., Kathiria, P., & Kovalchuk, O. (2013). Micronuclei in genotoxicity assessment: From genetics to epigenetics and beyond. *Frontiers in Genetics, 4*, 131.

Madliger, C. L., Love, O. P., Hultine, K. R., & Cooke, S. J. (2018). The conservation physiology toolbox: Status and opportunities. *Conservation Physiology, 6*, coy029.

Masui, Y. (1973). Effects of ionizing radiation on meiotic maturation of frog oocytes. *Journal of Embryology and Experimental Morphology, 29*, 87–104.

Matsushima, N., Ihara, S., Takase, M., & Horiguchi, T. (2015). Assessment of radiocesium contamination in frogs 18 months after the Fukushima Daiichi nuclear disaster. *Scientific Reports, 5*, 9712.
Møller, A. P., & Mousseau, T. A. (2006). Biological consequences of Chernobyl: 20 years after the disaster. *Trends in Ecology & Evolution, 21*, 200–207.
Møller, A. P., & Mousseau, T. A. (2011). Efficiency of bio-indicators for low-level radiation under field conditions. *Ecological Indicators, 11*, 424–430.
Møller, A. P., & Mousseau, T. A. (2016). Are organisms adapting to ionizing radiation at Chernobyl? *Trends in Ecology & Evolution, 31*, 281–289.
Møller, A. P., Hobson, K. A., Mousseau, T. A., & Peklo, A. M. (2006). Chernobyl as a population sink for barn swallows: Tracking dispersal using stable-isotope profiles. *Ecological Applications, 16*, 1696–1705.
Monaghan, P., Heidinger, B. J., D'Alba, L., Evans, N. P., & Spencer, K. A. (2012). For better or worse: Reduced adult lifespan following early-life stress is transmitted to breeding partners. *Proceedings of the Royal Society B, 279*, 709–714.
Mousseau, T. A., & Møller, A. P. (2014). Genetic and ecological studies of animals in Chernobyl and Fukushima. *The Journal of Heredity, 105*, 704–709.
Nosova, L. I., Shalimov, V. A., Ryasenko, V. I., & Shevchenko, Zh. T. (1994). Dynamics of hematological parameters in rodents from the Zone of Chernobyl Power Plant. *Chernobyl-94: 4th international conference on results of eight-year work on eliminating the consequences of Chernobyl disaster* (p. 253).
Oskolkov, B. Y., Bondarkov, M. D., Gaschak, S. P., Maksymenko, A. M., Maksymenko, V. M., Martynenko, V. I., Farfán, E. B., Jannik, G. T., & Marra, J. C. (2010). Environmental problems associated with decommissioning the Chernobyl nuclear power plant cooling pond. *Health Physics, 99*, 639–648.
Oskolkov, B. Y., Bondarkov, M. D., Gaschak, S. P., Maksimenko, A. M., Hinton, T. G., Coughlin, D., Jannik, G. T., & Farfán, E. B. (2011). Radiation dose assessment for the biota of terrestrial ecosystems in the shoreline zone of the Chernobyl nuclear power plant cooling pond. *Health Physics, 101*, 349–361.
Panter, H. C. (1986). Variations in radiosensitivity during development of the frog *Limnodynastes tasmaniensis*. *Developmental Cell Biology, 238*, 193–199.
Pinya, S., Tejada, S., Capó, X., & Sureda, A. (2016). Invasive predator snake induces oxidative stress responses in insular amphibian species. *Science of the Total Environment, 566–567*, 57–62.
Pyastolova, O. A., & Vershinin, V. L. (1999). Some cytological features of *Rana arvalis* Nilss. frogs on the territory of the Eastern Ural radioactive trace. *Russian Journal of Ecology, 30*, 26–31.
Pyastolova, O. A., Vershinin, V. L., Trubetskaya, E. A., & Gatiyatullina, E. Z. (1996). Utilization of amphibians in bioindication research on territories of the Eastern Urals radioactive trace. *Russian Journal of Ecology, 27*, 361–365.
Rakhimzhanova, A. K., & Khromov, V. A. (1998). Morphological study of the green frog (*Rana ridibunda* Pallas, 1771) from eastern Khazakstan province. *Asian Herpetological Meeting Almaty Abstracts, 3*, 31.
Reading, C. J., Loman, J., & Madsen, T. (1991). Breeding pond fidelity in the common toad, *Bufo bufo*. *Journal of Zoology, 225*, 201–211.
Rowe, C. L., & Dunson, W. A. (1998). The value of simulated pond communities in mesocosms for studies of amphibian ecology and ecotoxicology. *Journal of Herpetology, 28*, 346–356.
Rugh, R. (1954). The effect of ionizing radiations on amphibian development. *Journal of Cellular and Comparative Physiology, 43*, 39–75.
Sakai, M., Gomi, T., Nunokawa, M., Wakahara, T., & Onda, Y. (2014). Soil removal as a decontamination practice and radiocesium accumulation in tadpoles in rice paddies at Fukushima. *Environmental Pollution, 187*, 112–115.
Shaffer, H. B., Gidiş, M., McCartney-Melstad, E., Neal, K. M., Oyamaguchi, H. M., Tellez, M., & Toffelmier, E. M. (2015). Conservation genetics and genomics of amphibians and reptiles. *Annual Review of Animal Biosciences, 3*, 113–1138.

Sinsch, U. (1991). Mini-review: The orientation behaviour of amphibians. *Herpetological Journal, 1*, 541–544.
Sinsch, U. (2015). Skeletochronological assessment of demographic life-history traits in amphibians. *Herpetological Journal, 25*, 5–13.
Stark, K., Avila, R., & Wallberg, P. (2004). Estimation of radiation doses from ^{137}Cs to frogs in a wetland ecosystem. *Journal of Environmental Radioactivity, 75*, 1–14.
Stuart, S. N., Chanson, J. S., Cox, N. A., Young, B. E., Rodrigues, A. S. L., Fischman, D. L., & Waller, R. W. (2004). Status and trends of amphibian declines and extinctions worldwide. *Science, 306*, 1783–1786.
Tagami, K., Uchida, S., Wood, M. D., & Beresford, N. A. (2018). Radiocaesium transfer and radiation exposure of frogs in Fukushima prefecture. *Scientific Reports, 8*, 10662.
Takahara, T., Endo, S., Takada, M., Oba, Y., Nursal, W. Y., Igawa, T., Doi, H., Yamada, T., & Okuda, T. (2015). Radiocesium accumulation in the anuran frog, *Rana tagoi tagoi*, in forest ecosystems after the Fukushima Nuclear Power Plant accident. *Environmental Pollution, 199*, 89–94.
UNSCEAR. (2008). *Sources and effects of ionizing radiation*. UNSCEAR 2008 Report, Vol. II.
Ushakov, V. A., Lebedinskii, A. A., & Grefner, N. M. (1982). The analysis of size and age structure of grass frog populations in urbanized areas. *Vestnik Zoologii, 2*, 67–68.
Versihin, V. L., & Seredyuk, S. D. (2000). Trophic specificity of the *Rana arvalis* population from the Eastern Ural radioactive trace in relation to the state of the soil mesofauna. *Russian Journal of Ecology, 31*, 330–335.
Vinogradov, A. E., & Chubinishvili, A. T. (1999). Genome reduction in a hemiclonal frog *Rana esculenta* from radioactively contaminated areas. *Genetics, 151*, 1123–1125.
Voitovich, A. M. (2000). Micronuclei frequency in erythrocytes and disturbance in differentiation process in brown-frog under chronic radiation exposure. *Herald National Belarus Academy of Sciences (Biology), 3*, 60–63. (in Belarussian).
Wells, K. D. (2007). *The ecology and behavior of amphibians*. Chicago Univesity Press.
Whicker, F. W., & Schultz, V. (1982). *Radioecology: Nuclear energy and the environment I*. CRC Press.
Wilbur, H. M. (1980). Complex life cycles. *Annual Review of Ecology, Evolution, and Systematics, 11*, 67–93.
Yablokov, A. V. (2011). Chernobyl's radioactive impact on fauna. *Annals of the New York Academy of Sciences, 1181*, 255–280.
Yeliseeva, K. G., Mikhalevich, L., & Kartel, N. (1995). Cytogenetic effects of radiation from Chernobyl nuclear accident on humans and animals in the contaminated area of Belarus. *Revista Internacional de Contaminacion Ambiental, 11*, 77–85.
Yeliseeva, K. G., Kartek, N. A., Voitovich, A. M., Trusova, V. D., Ogurtsova, S. E., & Krupnova, E. V. (1996). Chromosome aberrations in some tissues of murine rodents and amphibians from Belarussian radionuclide contaminated territories. *Cytology and Genetics, 30*, 20–25. (in Russian).
Yushkova, E. A., Bodnar, I. S., Shadrin, D. M., Pylina, Y. I., & Zainullin, V. G. (2018). Cytogenetic and molecular genetic indexes in populations of Anura (*Rana arvalis* Nilsson) under conditions of radioactive and chemical pollution of an aquatic environment. *Inland Water Biology, 11*, 349–358.

Chapter 13
Measuring Adaptive Responses Following Chronic and Low Dose Exposure in Amphibians

Marilyne Audette-Stuart, Sang-Bog Kim, Danielle McMullin, Amy Festarini, Tamara L. Yankovich, Carla Ferreri, James Carr, and Sunita Mulpuru

Abstract Adaptive responses can be used as biological indicators of chronic low dose exposure to ionizing radiation. The micronucleus assay has been used to demonstrate this, in vivo and in vitro, in liver cells obtained from amphibians collected in radiologically-contaminated locations (with tritium being the main contaminant) and in locations that had significantly lower radiological contamination. In addition, under controlled laboratory settings, evidence that bystander effects can prompt adaptive responses were also observed. This later finding indicates that membranes may play a role in the induction of adaptive responses. In support of this hypothesis, muscle fatty acid composition analysis indicated a partial remodeling of muscle lipids in response to a chronic low dose tritium exposure and different responses to a subsequent higher dose between control amphibians and amphibians that have been chronically exposed to ionizing radiation.

M. Audette-Stuart (✉) · S.-B. Kim · D. McMullin · A. Festarini · J. Carr
Canadian Nuclear Laboratories, Chalk River, ON, Canada
e-mail: marilyne.stuart@cnl.ca

T. L. Yankovich
International Atomic Energy Agency (IAEA), Vienna International Centre, Vienna, Austria

C. Ferreri
Institute for Organic Synthesis and Photoreactivity, Consiglio Nazionale delle Ricerche, Bologna, Italy

S. Mulpuru
University of Ottawa, Ottawa, ON, Canada

M. D. Wood et al. (eds.), *Biomarkers of Radiation in the Environment*, NATO Science for Peace and Security Series A: Chemistry and Biology,
https://doi.org/10.1007/978-94-024-2101-9_13

13.1 Introduction

Environmental radiation doses and dose rates are generally well below those expected to produce mortality or significant measurable detrimental effects. Although environmental radiation exposures may modulate the frequency of micronucleus formation, fatty acid composition or elicit bystander effects, a multitude of other environmental and biological factors may contribute to inter-individual variability (see for example Almássy et al., 1987). Therefore, attempting to establish a relationship between exposure and effect when there is a low signal-to-noise ratio can translate into the need to gather a large number of measurements. To reduce the number of animals required, the studies presented in this chapter considered measurements in sampled biota subsequently exposed to a high dose of ionizing radiation, as a higher radiation dose results in higher signal-to-noise ratio. If exposure to above-background radiation affects the response to subsequent high radiation doses, an adaptive response could be a sensitive parameter to use as biomarker. The information presented in this chapter has been taken from Audette-Stuart et al. (2011), Audette-Stuart and Yankovich (2011) and Audette-Stuart et al. (2012).

The first objective of this preliminary work was, hence, to evaluate the feasibility of using the micronucleus assay for the detection of adaptive responses triggered by chronic exposure to environmental stressors (including low levels of ionizing radiation). If measurable, such induced competence could provide a biomarker for monitoring biological effects in natural populations. This was tested in two species (the northern leopard frog, *Rana pipiens*, and the mink frog. *Rana septentrionalis*).

Research at Chalk River Laboratories showed that, for the same activity concentration of tritium, more cells died when the tritium was delivered in the form of tritiated fatty acids compared to when tritium was delivered in the form of tritiated bases. Delivering tritium in the form of tritiated amino acids did not affect cell survival as much as the other two forms (data not shown). Knowing that lipids could be more sensitive to ionizing radiation than DNA, our secondary objective was to determine whether low levels of radiation exposure could produce measureable changes in fatty acid composition. We also wanted to investigate whether exposure to a high dose of ionizing radiation would lead to a different remodeling of the lipid asset in control amphibians compared to amphibians that have been chronically exposed to low levels of ionizing radiation. This was investigated in one frog species (the American bullfrog, *Lithobates catesbeiana*).

Finally, a study was designed to test if adaptive responses, expressed by an individual, could be passed on to another individual. This study was conducted because evidence of such types of effects were being detected in fish, as documented by O'Neill-Mehlenbacher et al. (2007), Mothersill et al. (2014) and Smith and Moccia (2016) and references therein.

13.2 Methodology

Frogs (northern leopard frog *Rana pipiens* and mink frog *Rana septentrionalis*) were chosen for these studies as they are common across North America, utilize both terrestrial and aquatic environments and could be used as indicator species for environmental assessments. *Rana pipiens* and *Rana septentrionalis* were used to evaluate micronucleus frequencies following in vivo and in vitro exposures to ionizing radiation. For both species, field-based studies were conducted. In the case of *Rana septentrionalis*, information regarding fatty acid composition was also collected. The American bullfrog (*Lithobates catesbeiana*) was used to evaluate the role of bystanders using the micronucleus assay.

All animals were handled in accordance with an approved animal care protocol. A list of the leopard frogs (*Rana pipiens*) used is provided in Table 13.1, the mink frogs (*Rana septentrionalis*) are listed in Table 13.2 and the American bullfrog (*Lithobates catesbeiana*) tadpoles in Table 13.3. As the micronucleus frequency in frogs tends to increase with increasing temperatures, care was taken to collect frogs in such a way that the distribution of ambient temperature at the time of collection was the same among treatment groups. This was achieved by collecting a control frog every time that a test frog was collected and/or ensuring that the distribution of temperature was the same in each group. In addition, MS-222 (tricaine methanesulfonate) was not used to euthanize the animals, as a pilot study previously conducted on frogs, at the Chalk River Laboratories, showed that MS-222 affected the outcome of the micronucleus assay (data not shown).

Table 13.1 Origin of the primary leopard frog (*Rana pipiens*) liver cells cultures used to conduct the micronucleus assay

Type of experiment, frog#	Site of capture	Date	Gender	Length (nose to fork)
In vitro, 1	Dew Drop Lake (B)	2003 July 12	Female	65 mm
In vitro, 2	Duke Swamp (AB)	2003 July 24	Female	77 mm
In vitro, 3	Twin Lake (B)	2004 October 05	Male	52 mm
In vivo, 1	Twin Lake (B)	2003 July 08	Male	70 mm
In vivo, 2	Dew Drop Lake (B)	2003 July 23	Female	82 mm
In vivo, 3	Twin Lake (B)	2003 July 29	Female	77 mm
In vivo, 4	Duke Swamp (AB)	2003 July 08	Female	73 mm
In vivo, 5	Duke Swamp (AB)	2003 July 22	Female	71 mm
In vivo, 6	Duke Swamp (AB)	2003 July 23	Female	81 mm

B Background, *AB* Above-background ionizing radiation levels

Table 13.2 Origin of the primary mink frog (*Rana septentrionalis*) liver cells cultures used to conduct the micronucleus assay and the muscle samples used to determine the fatty acid composition

Type of experiment, frog#	Site of capture	Date	Gender	Length (nose to fork)
Pilot, 1	Twin Lake (B)	2008 July 07	Male	10 cm
FA, 1	Twin Lake (B)	2008 July 07	Female	10 cm
Pilot, 2*	Twin Lake (B)	2008 July 07	Male	10 cm
FA, 2*	Twin Lake (B)	2008 July 07	Female	10 cm
FA, 3*	Twin Lake (B)	2008 July 21	Female	6 cm
In vitro, 1	Twin Lake (B)	2008 July 14	Male	6 cm
In vitro, 2	Twin Lake (B)	2008 July 25	Female	7 cm
In vitro, 3	Twin Lake (B)	2008 July 25	Female	6 cm
In vitro, 4	Twin Lake (B)	2008 July 28	Male	14 cm
In vitro, 5	Twin Lake (B)	2008 July 28	Female	12 cm
In vitro, 6	Duke Swamp (AB)	2008 July 25	Female	5 cm
In vitro, 7	Duke Swamp (AB)	2008 July 25	Female	8 cm
In vitro, 8	Duke Swamp (AB)	2008 July 25	Male	15 cm
In vitro, 9	Duke Swamp (AB)	2008 July 28	Male	16 cm
In vitro, 10	Duke Swamp (AB)	2008 July 28	Female	7 cm
In vivo, 1	Twin Lake (B)	2008 July 14	Female	10 cm
In vivo, 2	Twin Lake (B)	2008 July 14	Female	10 cm
In vivo, 3	Twin Lake (B)	2008 July 21	Female	7 cm
In vivo, 4	Twin Lake (B)	2008 July 21	Female	7 cm
In vivo, 5	Duke Swamp (AB)	2008 July 14	Male	15 cm
In vivo, 6	Duke Swamp (AB)	2008 July 14	Female	10 cm
In vivo, 7	Duke Swamp (AB)	2008 July 21	Male	15 cm
In vivo, 8	Duke Swamp (AB)	2008 July 21	Male	15 cm
In vivo, 9	Duke Swamp (AB)	2008 July 21	Male	15 cm

FA Fatty Acid analysis only, *B* Background, *AB* Above-background ionizing radiation levels. *Received 4 Gy in vivo. The frogs labelled "Pilot" were used to test methodologies and were not contributing to the data set

Table 13.3 Origin of the primary bullfrog (*Lithobates catesbeiana*) liver cells cultures used to conduct the micronucleus assay

Type of experiment	Site of capture	Date	Number of frogs in the control group	*Number of frogs in the exposed group
Laboratory	Twin Lake (B)	2005 June 02	15	15

B Background. *Over the 57-day exposure, the tritium concentrations fluctuated between 25,000 and 45,000 Bq/L. Note that, contrary to the tritium exposures, the gamma irradiations were performed on cell cultures (not live animals)

13.2.1 The Study Sites

Leopard and mink frogs were captured in Duke Swamp at Chalk River Laboratories, an area with above-background tritium and carbon-14 levels, and from two background areas (Dew Drop Lake and Twin Lake). The bullfrog tadpoles were collected from Twin Lake. These areas are all found on the Chalk River Laboratories property (Fig. 13.1).

When the leopard frogs were collected, the tritium levels measured in Dew Drop Lake and Twin Lake were about 475 and 700 Bq/L, respectively. Within Duke Swamp, the frogs were captured in an area where the tritium porewater concentrations ranged from 5000 to 35,000 Bq/L. Within Duke Swamp, the mink frogs were captured from a pond with a tritium activity concentration of about 2800 Bq/L. Duke Swamp is receiving tritium and carbon-14 through groundwater. In Duke Swamp, the carbon-14 concentrations in mosses were estimated to be between 237 and

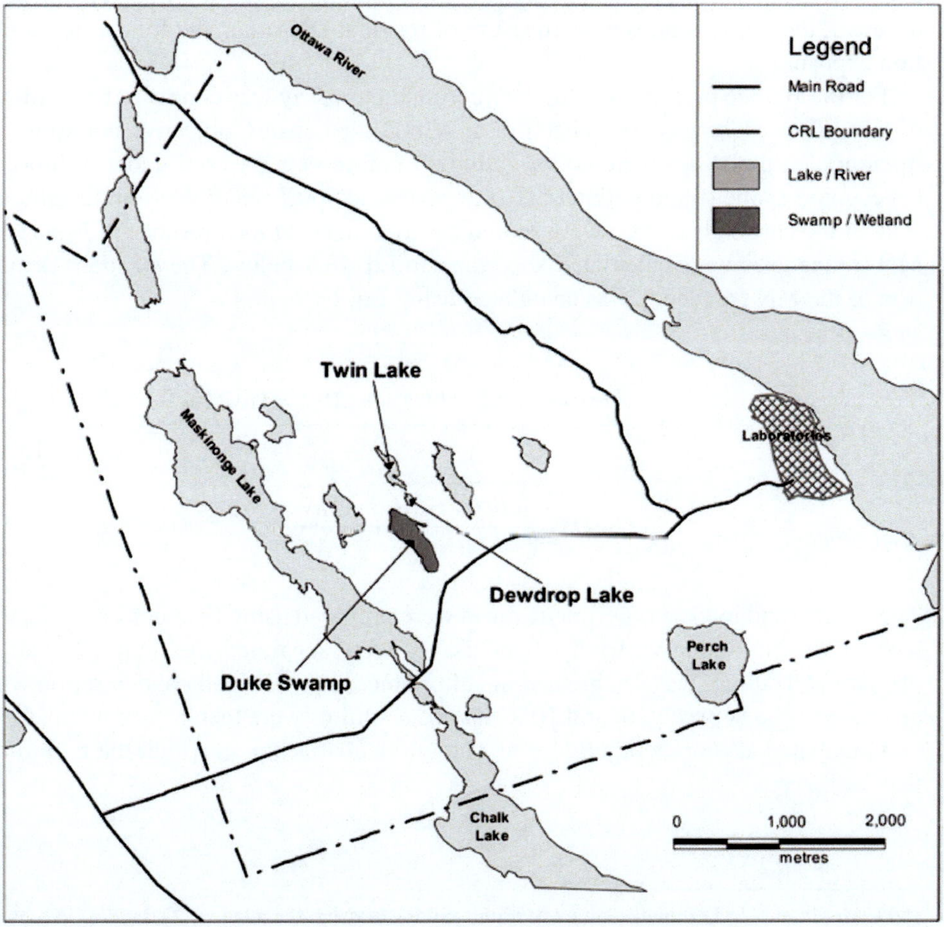

Fig. 13.1 Sites where amphibians were collected at Chalk River Laboratories

1000 Bq/kg C (Yankovich et al., 2013 and 2014). The tritium levels measured in Twin Lake when the mink frogs were collected (see Table 13.2) were about 215 Bq/L. Tritium levels measured in Twin Lake water were between 76 and 511 Bq/L throughout the bullfrog study.

13.2.2 Micronucleus Assay

Ionizing radiation can induce DNA breaks in chromosomes. If unrepaired, these breaks may lead to improper segregation of chromosomal fragments into the two daughter nuclei during cell division, and the fragments may subsequently be packaged into micronuclei (MN). It follows that micronuclei contain chromosomal fragments or whole chromosomes that were not incorporated into the main cell nucleus during cell division (Fenech et al., 2003). The frequency of micronuclei formation in cells that have been exposed to high doses of radiation and, subsequently, allowed to repair, therefore, represents a measure of residual DNA damage following radiation exposure.

For the studies presented here, the micronucleus assay was conducted according to the criteria described in Fenech et al. (2003). To ensure objective counting of micronuclei, the slides were coded (blinded) before being scored under a fluorescence microscope (fitted with a FITC filter cube) at 400X magnification. One thousand binucleate cells (BNC) were scored per treatment for the presence of MN. The MN frequencies were calculated according to Eq. 13.1 below. The standard deviation of the MN frequency was calculated using Eq. 13.2.

$$MN\ frequency\left(MN/100\,BNC\right)=\frac{\left(\left(BNC\ with\,1MN\right)+2\left(BNC\ with\,2MN\right)+3\left(BNC\ with\,3MN\right)\right)}{BNC} \tag{13.1}$$

$$STD=\sqrt{\frac{\left(BNC+MC\right)}{\left(BNC\right)}\cdot\frac{MN}{BNC}} \tag{13.2}$$

The in vitro and in vivo experiment slides were prepared using liver cells aseptically harvested from animals. Briefly, liver tissues were separated mechanically and a primary cell culture was established in culture medium.[1] The cell concentration was estimated to be between 10^7 and 10^8 cells/mL. Cultures were then placed at an angle and incubated at room temperature in a small incubator box in which the humidity was high.

[1] 50% Modified L-15 Leibovitz Culture Medium (Sigma-Aldrich Canada Ltd., Oakville, ON), 39% deionized, distilled, filter-sterilized water, 10% Fetal Calf Serum (Invitrogen Canada Ltd.) supplemented with 1% of an antibiotic solution.

13.2.2.1 In Vitro Experiments (Gamma Irradiations Performed on the Cell Cultures)

For the in vitro experiments, the cells were maintained in culture for approximately 24 h prior to being irradiated. Some samples were subjected to an "adapting" irradiation dose (50 to 100 mGy) delivered[2] at a dose rate of 5–10 mGy/min. This could be seen as a low dose exposure. The cultures were incubated at room temperature for 3 h, after which some received a challenge dose of 4 Gy delivered[3] at a dose rate of 8 or 16 Gy/min. This could be seen as an acute high dose exposure. Apart from the actual exposures to ionizing radiation, care was taken to ensure that all samples (including controls) were subjected to the same treatments, including transport and handling. In preparation for the micronucleus assay, 10μL of cytochalasin B (0.2 mg/mL) was added per 1.0 mL of culture suspension to arrest cells at cytokinesis, 48 h before the cells were harvested and fixed onto slides.

13.2.2.2 In Vivo Experiments (Live Animals Exposed to Gamma Irradiation)

During the in vivo experiments, the amphibians received a 4 Gy dose (at a dose rate of 8 or 16 Gy/min). The frogs were kept in a cool environment for 30 to 60 min after irradiation before the liver tissue was harvested. Primary cell cultures were then established, as stated earlier, and cytochalasin B was added. Forty-eight hours later, the cells were harvested and fixed onto slides.

13.2.2.3 Slide Preparation

The cultures were gently aspirated to dislodge the cells from the culture flask walls. The suspension was then transferred to a 15 mL polypropylene centrifuge tube and the cultures were centrifuged. Most of the supernatant was removed and the pellet was gently re-suspended in the remaining supernatant. Soft fixative solution (0.75 mL of 10% acetic acid) was added to the cells. After 5 min, 0.75 mL of fixative solution (3:1 methanol/acetic acid containing 10% formaldehyde) was also added. Five minutes later, the cells were centrifuged. Most of the supernatant was removed and the pellet was re-suspended in 1.5 mL of fixative solution (3:1 methanol/acetic acid). Following a standing time of 5 min, the cells were centrifuged again and the supernatant was replaced with some fresh fixative. The slides were immediately prepared by dropping 12μL of the cell suspension onto each slide. The slides were flushed three times with fixative (3:1 methanol/acetic acid) and then dried over a hot water bath (80 °C). Slides were stained in acridine orange dye

---

[2] Gamma Beam 150 C ^{60}Co irradiator, AECL

[3] Gammacell 220, AECL

(50μg/mL) and then rinsed twice with deionized, distilled water. A cover slip was placed over the cells and the slides were stored in the dark at room temperature until evaluation.

13.2.3 Fatty Acid Composition

Several reports have highlighted the importance of lipids (type and quantity) in the regulation of cellular responses, including modulating the ability to cope with various stresses (e.g., Willmer et al., 2002; Csermely & Vigh, 2007). Fatty acids are constituents of cellular membranes, and they regulate protein and channel functioning. Additionally, fatty acid composition appear to relate to the metabolic rates in small animals, and muscle tissue lipid content is often considered to be correlated with mitochondrial activity and stress response (Hulbert et al., 2007).

Experimentally, muscle fatty acid composition was assessed after part of a thigh muscle of each adult frog was carefully and aseptically removed and placed inside a sterile culture plate. This was done an hour after the in vivo irradiations. Tissues were covered with sterile phosphate-buffered saline solution (PBS). Each sample was aseptically homogenized in PBS, frozen at −20 °C and shipped frozen to Italy (ISOF, Consiglio Nazionale delle Ricerche) for lipid composition analysis. Each sample was analyzed separately. The analyses were performed following the methodology described in Ferreri et al. (2002). In this way, a quantitative estimation of the fatty acid changes could be made.

The groups of frogs examined in this study include 6 collected from the background site, 5 from the above-background site, 6 from the background site exposed to 4 Gy in vivo (6 samples were analyzed, but 5 provided valid data), and 5 frogs from the above-background site exposed to 4 Gy in vivo. The fatty acids that were considered in the analysis were: 14:0; 16:0; 17:0; 18:0; 19:0; 20:0; 22:0; 24:0; 16:1; 18:1 (9c); 18:1 (11c); 20:1; 20:4; 18:3 omega-3; 20:3 omega-3; 20:5 omega-3; 22:6 omega-3, 18:2 omega-6; 18:3 omega-6, 20:2 omega-6; 20:3 omega-6, 18:1 trans, 18:2 trans, and 20:4 trans. SPSS (Version 11, SPSS Inc., Chicago, IL) was used for all statistical analyses. The averages and the standard deviations were calculated for each experimental group. When data were normally distributed and variances were equal, an unpaired t-test was used for group comparisons. When normality and equal variance analyses failed, a Mann-Whitney ranks sum test was used instead of an unpaired t-test. As for previous statistical analysis, statistical significance was based on 95% confidence limits ($p = 0.05$).

13.2.4 Bystander Effect

For the study of bystander effects, the Chalk River Laboratories Biological Research Facility was used. This facility offers a controlled environment where low dose exposure is possible. The bullfrog tadpoles were placed in containers filled with 10 L of Twin Lake water, with five tadpoles placed in each container. This translated to two liters per tadpole. In 3 of the 6 containers, the tritium activity concentration was increased to about 30,000 Bq/L (between 25,000 and 45,000 Bq/L). The tadpoles were kept in the tritiated water for 57 days. Five tadpoles that were in tritiated water were transferred to a container filled with background water in preparation to conduct the bystander effect study (this served to clear the tritium and facilitate the handling of the animals). Out of these 5 tadpoles, 2 were placed in a background container with 3 control tadpoles and 3 were placed in a background container with 2 control tadpoles. Physiological features were used to distinguish the tadpoles when the previously exposed tadpoles were placed with control tadpoles. The tadpoles were kept together for a week. Amongst the control tadpoles there were three males, one female and one for which the gender could not be determined. Amongst the tadpoles previously exposed to tritium, there were four males and one female. For all of the treatment groups, the liver micronucleus frequencies following various ^{60}Co gamma radiation treatments (0 Gy, 100 mGy, 4 Gy, and 100 mGy 3 h prior to 4 Gy) were determined.

13.3 Results and Discussion

In vitro exposure to a low adaptive dose (100 mGy for leopard frog liver primary cultures) did not contribute to an increase in micronucleus frequency. For background frogs, exposure to a 4 Gy challenge dose caused a significant increase in micronucleus frequency ($p = 0.001$, Chi-square tests). For frogs that had been chronically exposed to ionizing radiation (mainly tritium) in the field, this increase was not observed. This suggested that cells harvested from radiologically exposed frogs would be better equipped to respond to an acute high dose of ionising radiation. The same effect was seen when cells from control frogs were pre-exposed to a low adaptive dose of ionizing radiation before being given the 4 Gy challenge dose (see Fig. 13.2).

Similar results were found for mink frogs compared to leopard frogs and using 50 mGy instead of 100 mGy as an adaptive dose (see Fig. 13.3). In this case, each treatment group included 5 frogs and the group of background cells exposed to 4 Gy was statistically different from the other groups ($p < 0.01$, Student t-test, unpaired).

The results presented in Fig. 13.4 confirmed that cells harvested from radiologically exposed leopard frogs were better equipped to respond to an acute high dose of ionising radiation *in vivo* than those obtained from control frogs. The difference between the mean values of the two groups (background and above-background)

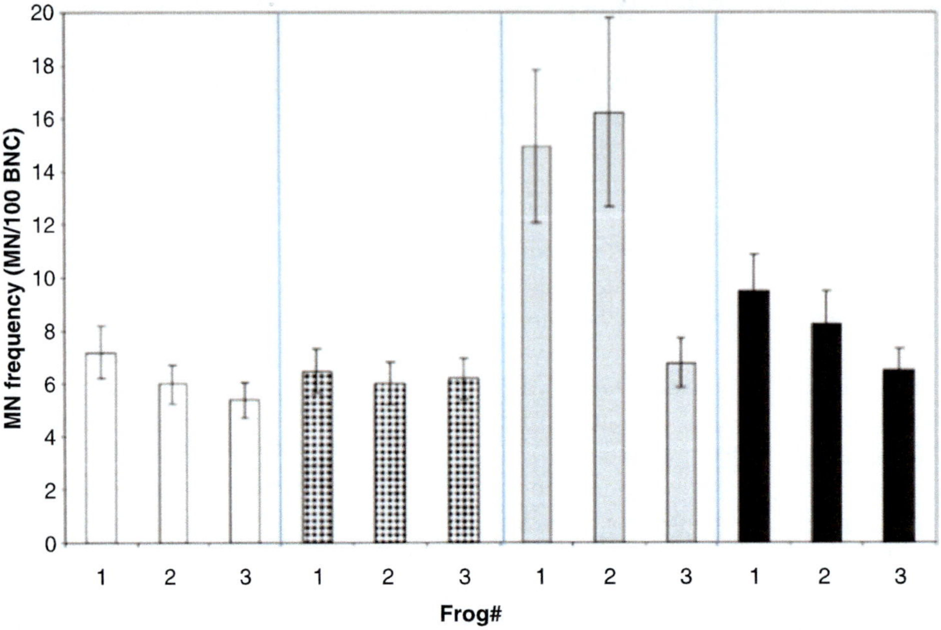

Fig. 13.2 Micronucleus (MN) frequency in binucleated cells (BNC) observed for the leopard frog in vitro experiment. In vitro frog information is shown in Table 13.1. White bars represent 0 mGy (adaptive dose) and 0 Gy (challenge dose), patterned bars represent 100 mGy (adaptive dose) and 0 Gy (challenge dose), grey bars represent 0 mGy (adaptive dose) and 4 Gy (challenge dose), and the black bars represent 100 mGy (adaptive dose) and 4 Gy (challenge dose). The error bars represent ±1σ

was found to be statistically significant ($p < 0.001$). Figure 13.5 provides a confirmation that the same holds true for mink frogs. In this figure, background area frogs ($n = 4$) and above-background area frogs ($n = 5$) are shown. The two groups are statistically different ($p < 0.01$, Student t-test, unpaired).

It should be noted that the adaptive responses are not radiation-specific, but is an example of a more general stress response (Boreham et al., 1991). Cross-adaptation has been observed between radiation and metals (Cai & Cherian, 1996), chemicals (Flores et al., 1996; Mitchel et al., 1990) and hyperthermia (Boreham et al., 1997; Cregan et al., 1994).

The data described here gives no indication that frogs are harmed by chronic low-level environmental ionizing radiation exposure. However, a decreased sensitivity to radiation damage (as measured using the micronucleus assay) was seen in liver cells from frogs collected in Duke Swamp, both when the radiation dose was delivered to live frogs and when it was delivered to cultured cells. This suggests that some stressor in Duke Swamp was acting to induce protection of DNA from the effects of high dose radiation exposure. It is suspected that the elevated radiation levels induces some level of DNA protection, although other environmental factors (temperature variations, diet or microbiological factors are examples) may also be

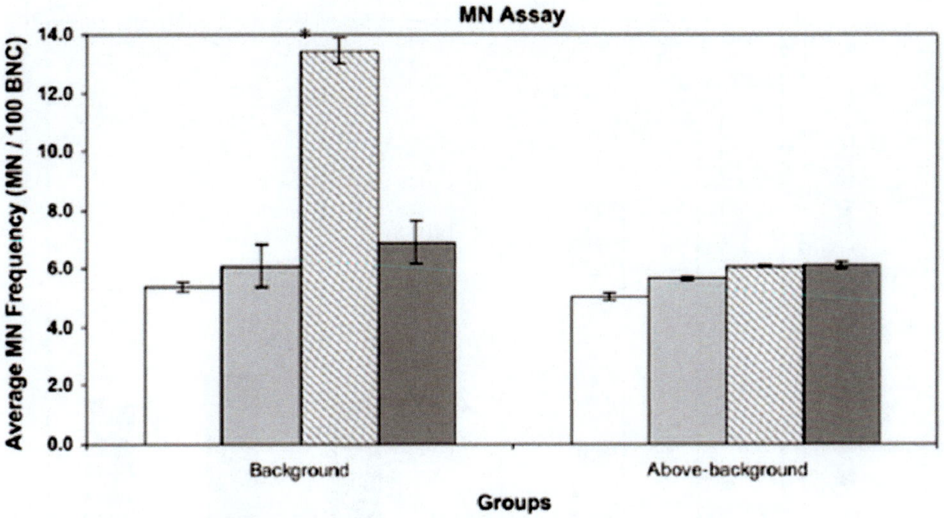

Fig. 13.3 Micronucleus (MN) frequency in binucleated cells (BNC) observed for the mink frog in vitro experiment. Average values were obtained from the in vitro frogs listed in Table 13.2. The white bars represent control cells, the light gray bars represent the cells exposed to 50 mGy (delivered at a dose rate of 10 mGy/min), the striped bars represent cells exposed to 4 Gy of cobalt-60 gamma radiation delivered at a dose rate of about 8 Gy/min, and the dark gray bars represent cells exposed to 50 mGy (delivered at a dose rate of 10 mGy/min) followed, 3 h later, by an exposure to 4 Gy of cobalt-60 gamma radiation delivered at a dose rate of about 8 Gy/min. The error bars represent $\pm 1\sigma$

responsible. Stress present in the above-background area seem to enhance cellular defense mechanisms and this resistance may provide a biomarker for biological effects for use in monitoring of natural populations.

Changes in lipids are known to occur in response to ionizing radiation (e.g., Nawar, 1973; Melchiorre et al., 2011; Kiang et al., 2012). As lipids can be more sensitive to ionizing radiation than DNA, fatty acid analyses were conducted on muscle tissues harvested from the mink frogs for which the micronucleus frequencies were measured (data shown above). The fatty acids that showed significant differences between groups are presented in Table 13.4.

Statistically significant differences for an increase of two polyunsaturated omega-3 fatty acid residues and a decrease of a polyunsaturated omega-6 fatty acid residue were observed in radiologically exposed frogs compared to background frogs. Changes in fatty acid composition did not seems to be attributable to diet. Furthermore, the effects of an acute high-dose exposure to ^{60}Co γ radiation indicated fast post-irradiation fatty acid changes with an increase of polyunsaturated fatty acids and decrease of saturated fatty acid contents in muscle tissues. As results indicated a partial remodeling of muscle lipids in response to a chronic low dose tritium exposure, it is postulated that fatty acid composition can play a role in low chronic and high acute stress responses. Fatty acid composition was found to be a sensitive marker that may be useful to study and monitor health of biota in

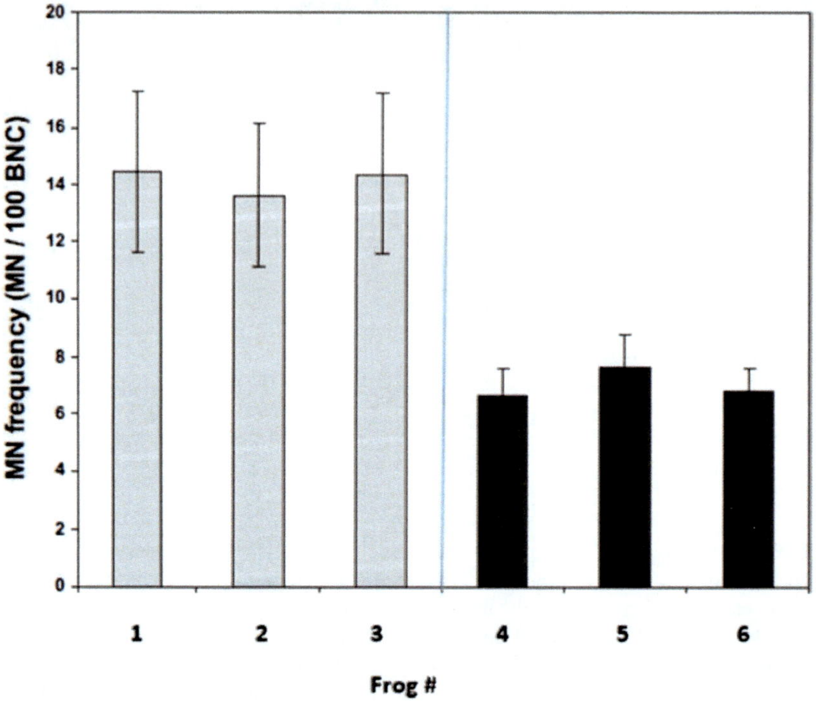

Fig. 13.4 Preliminary field experiments: Measure of the micronucleus (MN) frequency in binucleated cells (BNC) observed for the leopard frogs exposed to 4 Gy in vivo. The in vivo frog information is shown in Table 13.1. The grey bars represent frogs captured in the background areas of the Chalk River site and the black bars represent frogs collected in an area with above-background ionizing radiation levels. The error bars represent $\pm 1\sigma$

environments that are radiologically exposed, as well as for understanding the differences between low chronic and high acute stress responses. At this point, it is unknown if such changes in the muscles could be linked to adaptive responses that were observed using the micronucleus assay in liver cells harvested from the same animals.

These observations led to other questions. For example, at low doses and dose rate, is the DNA the primary target? The following experiment was designed to indirectly answer this type of question and to assess if being exposed to a stressor such as ionizing radiation, was necessary to exhibit an adaptive response, or if being in contact with other individuals that have previously been exposed is sufficient to acquire the ability. Tadpoles were exposed to about 30,000 Bq/L tritium for 57 days. After having been allowed to depurate for about an hour in Twin Lake water, test tadpoles were placed with control tadpoles (now called bystander tadpoles) for a period of 1 week. The cellular response to 0 Gy, 100 mGy, 4 Gy and 100 mGy 3 h prior to a 4 Gy exposure was then assessed using the micronucleus assay. The experimental design is presented in Fig. 13.6.

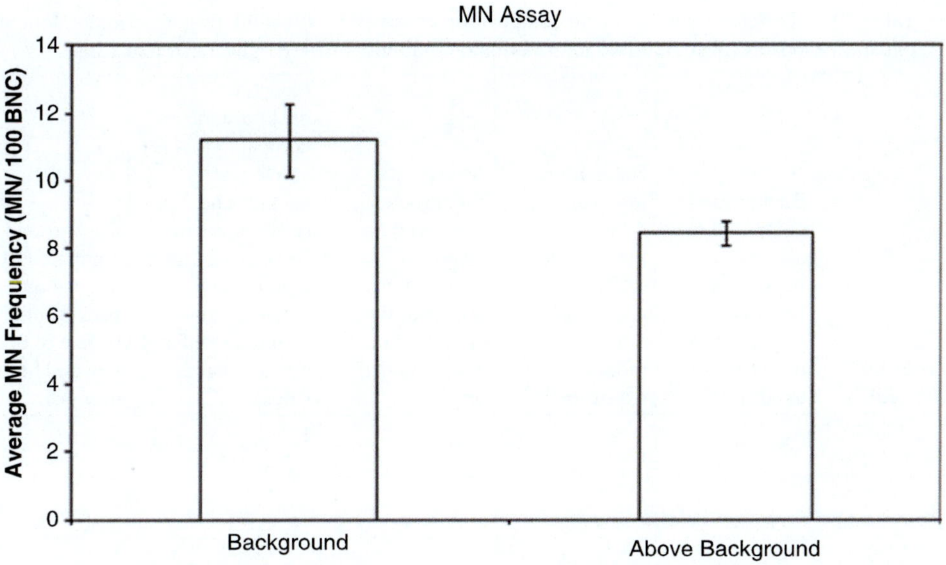

Fig. 13.5 In vivo field experiments: Average measure of the micronucleus (MN) frequencies in binucleated cells (BNC) observed for the mink frogs exposed to 4 Gy in vivo. The 4 Gy dose of Cobalt-60 Gamma radiation was delivered at about 8 Gy/min. The in vivo frog information is shown in Table 13.2. The first bar represents frogs captured in the background areas of the Chalk River site and the second bar represents frogs collected in an area with above-background ionizing radiation levels. The error bars represent $\pm 1\sigma$

In this case, control tadpoles did not show an adaptive response, while both exposed and bystander tadpoles did (n = 5) (see Fig. 13.7). Significant differences are noted with an asterisk (*). The data obtained suggest that bystander effects do play a role in wild populations, since bullfrog tadpoles that were not exposed to tritium responded in a similar manner to the tadpoles that were directly exposed to tritium after being placed in contact with them.

13.4 Concluding Remarks

The results reported in Audette-Stuart et al. (2011), Audette-Stuart and Yankovich (2011) and Audette-Stuart et al. (2012) indicate that the identification of the presence of adaptive responses can be used as a strategy to gain information regarding the effects of chronic low dose exposure to ionizing radiation. In leopard frog, mink frog and bullfrog liver cells, the method described here was effective both in vivo and in vitro and worked for both gamma and beta radiation. This means that cell culturing techniques can be used instead of collection of live animals for future assessments. Although adaptive responses have been shown to be triggered by other types of stressors, we have observed the same responses in field and lab settings.

Table 13.4 Differences in fatty acid composition observed for the mink frog experiment. Ionizing radiation exposures were in vivo. Information on individual frogs is given in Table 13.2

Fatty acid	Background compared to above background (effect of chronic low dose tritium exposure)	Background compared to 4 Gy background (effect of acute high dose γ radiation exposure)	Above background compared to 4 Gy above background (effect of the high dose on contaminated frogs)	Four Gy background compared to 4 Gy above background (differences in response to the high dose between non contaminated and contaminated frogs)	Background compared to 4 Gy above background (effect of both low and high doses)
14:0 (~0.7%)				↓	↓
16:0 (~34%)		↓			
17:0 (~1.7%)					↓
20:0 (~0.5%)		↓			↓
22:0 (~0.8%)			↓		↓
24:0 (~0.4%)		↓			↓
18:3 omega-3 (~2.3%)	↑	↑			↑
20:3 omega-3 (~0.1%)	↑	↑			↑
18:2 omega-6 (~0.3%)					↑
20:2 omega-6 (~0.3%)	↓	↑			↑
18:2 *trans* (~0.1%)			↑	↑	↑

Notes. ↑ Indicates a statistically significant increase, ↓ indicates a statistically significant decrease and the numbers in brackets indicate the proportions of each fatty acid in the control (background) muscle tissue. All fatty acids for which statistically significant differences were noted between any of the experimental groups are listed in this table

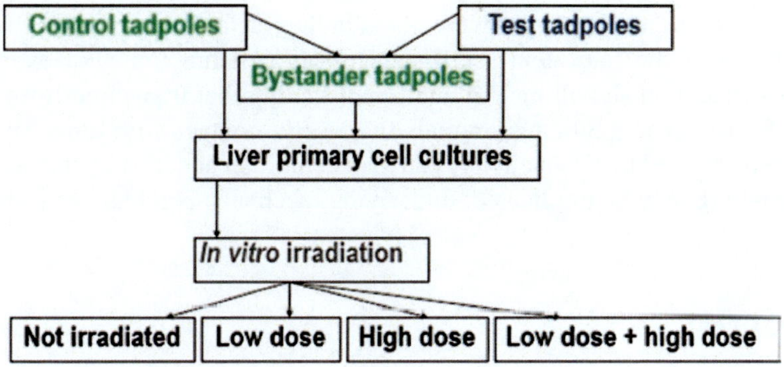

Fig. 13.6 Bystander effect experimental design

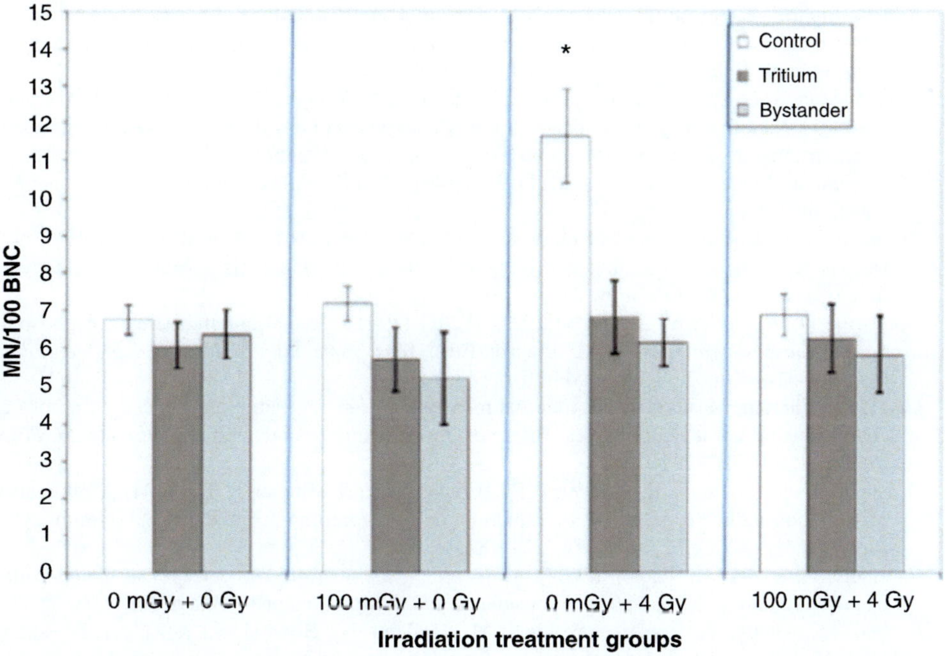

Fig. 13.7 Micronucleus (MN) frequencies in binucleated cells (BNC) for different irradiation treatments. The error bars represent $\pm 1\sigma$

This indicates that, for the type of endpoint used (the micronucleus frequency), the changes observed were likely due to the radiological exposure.

Differences in muscle fatty acids were also noted in response to ionizing radiation exposure. Specifically, low and high dose exposures seemed to cause different changes, and control and test frogs responded differently to an acute high dose of gamma radiation. Fatty acid composition seemed to be a more sensitive marker than the micronucleus frequency. For example, no change was noted between control and animal exposed to low doses using the micronucleus frequency, whereas statistically significant changes were noted in the relative proportions of fatty acids. Remodeling of the fatty acid assets can impact cell function. For example, it is likely to affect cell signalling. An additional finding that transpired from this work is that bystander amphibians (animals that are in contact with other animals that have been exposed to radioactivity) can also exhibit an adaptive response. Although cell signalling may be implicated, underlying mechanisms remain unknown.

References

Almássy, A., Krepinsky, A. B., Bianco, A., & Köteles, G. J. (1987). The present state and perspectives of micronucleus assay in radiation protection. A review. *International Journal of Radiation Applications and Instrumentation, 38*, 241–249.

Audette-Stuart, M., Ferreri, C., Festarini, A., & Carr, J. (2012). Fatty acid composition of muscle tissue measured in amphibians living in radiologically contaminated and non-contaminated environments. *Radiation Research, 178*, 173–181.

Audette-Stuart, M., Kim, S. B., McMullin, D., Festarini, A., Yankovich, T. L., Carr, J., & Mulpuru, S. (2011). Adaptive response in frogs chronically exposed to low doses of ionizing radiation in the environment. *Journal of Environmental Radioactivity, 102*, 566–573.

Audette-Stuart, M., & Yankovich, T. (2011). Bystander effects in bullfrog tadpoles. *Radioprotection, 46*(6), S497–S502.

Boreham, D. R., Dolling, J. A., Maves, S. R., Miller, S., Morrison, D. P., & Mitchel, R. E. (1997). Heat-induced thermal tolerance and radiation resistance to apoptosis in human lymphocytes. *Biochemistry and Cell Biology, 75*, 393–397.

Boreham, D. R., Trivedi, A., & Mitchel, R. E. (1991). Radiation and stress response in *Saccharomyces cerevisiae*. In R. Prasad (Ed.), *Yeast molecular biology and biotechnology* (pp. 294–314). Omega Scientific Publishers.

Cai, L., & Cherian, M. G. (1996). Adaptive response to ionizing radiation-induced chromosome aberrations in rabbit lymphocytes: Effect of pre-exposure to zinc, and copper salts. *Mutation Research, 369*, 233–241.

Cregan, S. P., Boreham, D. R., Walker, P. R., Brown, D. L., & Mitchel, R. E. (1994). Modification of radiation-induced apoptosis in radiation or hyperthermia-adapted human lymphocytes. *Biochemistry and Cell Biology, 72*, 475–482.

Csermely, P., & Vigh, L. (Eds.). (2007). *Molecular aspects of the stress response: Chaperones, membranes and networks*. Landes Bioscience and Spinger Science þ Business Media.

Fenech, M., Chang, W. P., Kirsch-Volders, M., Holland, N., Bonassi, S., & Zeiger, E. (2003). HUMN project: Detailed description of the scoring criteria for the cutokinesisblock micronucleus assay using isolated human lymphocyte cultures. *Mutation Research, 534*, 65–75.

Ferreri, C., Faraone Mennella, M. R., Formisano, C., Landi, L., & Chatgilialoglu, C. (2002). Arachidonate geometrical isomers generated by thiyl radicals: The relationship with trans lipids detected in biological samples. *Free Radical Biology & Medicine, 33*, 1516–1526.

Flores, M. J., Pinero, J., Ortiz, T., Pastor, N., Mateos, J. C., & Cortes, F. (1996). Both bovine and rabbit lymphocytes conditioned with hydrogen peroxide show an adaptive response to radiation damage. *Mutation Research, 372*, 9–15.

Hulbert, A. J., Pamplona, R., Buffenstein, R., & Buttemer, W. A. (2007). Life and death: Metabolic rate, membrane composition and life span of animals. *Physiological Reviews, 87*, 1175–1213.

Kiang, J. G., Fukumoto, R., & Gorbunov, N. V. (2012). Chapter 13: Lipid peroxidation after ionizing irradiation leads to apoptosis and autophagy. In *Biochemistry, genetics and molecular biology, "lipid Proxidation"*. Angel Catala. ISBN 978-953-51-0716-3.

Melchiorre, M., Torreggiani, A., Chatgilialoglu, C., & Ferreri, C. (2011). Lipid markers of "geometrical" radical stress: Synthesis of monotrans cholesteryl Ester isomers and detection in human plasma. *Journal of the American Chemical Society, 133*, 15184–15190.

Mitchel, R. E., Gragtmans, N. J., & Morrison, D. P. (1990). Beta-radiation-induced resistance to MNNG initiation of papilloma but not carcinoma formation in mouse skin. *Radiation Research, 121*, 180–186.

Mothersill, C., Smith, R. W., Heier, L. S., Teien, H.-C., Land, O. C., Seymour, C. B., Oughton, D., & Salbu, B. (2014). Radiation-induced bystander effects in the Atlantic salmon (*salmo salar* L.) following mixed exposure to copper and aluminum combined with low-dose gamma radiation. *Radiation and Environmental Biophysics, 53*, 103–114.

Nawar, N. N. (1973). The effects of ionizing radiation on lipids. *Progress in the Chemistry of Fats and Other Lipids, 13*, 89–118.

O'Neill-Mehlenbacher, A., Kilemade, M., Elliott, A., Mothersill, C., & Seymour, C. (2007). Comparison of direct and bystander effects induced by ionizing radiation in eight fish cell lines. *International Journal of Radiation Biology, 83*, 593–602.

Smith, R. W., & Moccia, R. D. (2016). The radiation induced bystander effect: Is there relevance for aquaculture? *Annals of Aquaculture and Research, 3*, 1026–1029.

Willmer, P., Stone, G., & Johnston, I. (2002). *Environmental physiology of animals*. Blackwell Science.

Yankovich, T., King-Sharp, K. J., Benz, M. L., Carr, J., Killey, R. W., Beresford, N. A., & Wood, M. D. (2013). Do site-specific radiocarbon measurements reflect localized distributions of ^{14}C in biota inhabiting a wetland with point contamination sources? *Journal of Environmental Radioactivity*, 352–366.

Yankovich, T. L., King-Sharp, K. J., Carr, J., Robertson, E., Killey, R. W. D., Beresford, N. A., & Wood, M. D. (2014). Spatial analysis of Carbon-14 dynamics in a wetland ecosystem (Duke Swamp, Chalk River Laboratories, Canada). *Journal of Environmental Radioactivity*, 173–180.

Chapter 14
Are There Ecosystem-Relevant Endpoints for Measuring Radiation Impacts?

Clare Bradshaw

Abstract Ecosystem-level effects of stress are the net result of the direct effects of the stressor and indirect effects caused by altered interactions between organisms and between biotic and abiotic components of the environment. Measuring impacts of any single stressor at the ecosystem level means acknowledging, and accounting for, a number of simultaneously acting processes.

It is therefore highly unlikely that there are radiation-specific responses at an ecosystem level. In the field, ecosystem-level responses will be the net result of the effects of the various radionuclides, other contaminants, environmental conditions and species interactions. However, this does not mean that we should not attempt to measure ecosystem-level endpoints, but in order to pinpoint the relative contribution of radiation in this multi-factorial situation, robust sampling and statistics and a sound ecological foundation is required.

What, then, are ecosystem-relevant endpoints? They include those that describe ecosystem structure and function and sometimes also the services an ecosystem provides to humans. Quantifying ecosystem effects may thus include measuring species composition, abundance, biodiversity, food web complexity and connectivity, habitat complexity (ie. aspects of ecosystem structure) and production, decomposition, pollination, functional or trait diversity (ie. ecosystem function). Many of these may also be used to estimate impacts on ecosystem services (e.g., provision of food, carbon storage). Many are by necessity proxy measures which we assume reflect the state of the ecosystem, since overall ecosystem condition is very hard to quantify. However, in risk assessment these proxies are sometimes combined into general measures of 'ecological status' and in ecosystem science, various metrics are used as integrative measures of the vulnerability or resilience of the ecosystem.

In this paper, I give an overview of the ways we can address and measure ecosystem-relevant and ecosystem-level endpoints in science and risk assessment, including community structure and function, ecological traits, ecosystem processes,

C. Bradshaw (✉)
Department of Ecology, Environment and Plant Sciences, Stockholm University, Stockholm, Sweden
e-mail: clare.bradshaw@su.se

M. D. Wood et al. (eds.), *Biomarkers of Radiation in the Environment*, NATO Science for Peace and Security Series A: Chemistry and Biology,
https://doi.org/10.1007/978-94-024-2101-9_14

ecological network metrics, ecological integrity and ecosystem services. I draw on examples from ecosystem science, stress ecology, ecotoxicology and environmental impact and risk assessment and discuss how these can inform assessment of radiation impacts.

14.1 Introduction and Terminology

Radiation protection and risk assessment are based on information about effects measured at the level of the individual organism or species. However, when an environmental assessment is conducted, we are rarely concerned with the individual organism or species, but rather the ecosystem as a whole. Indeed, radiation protection frameworks specify this: the ICRP states an explicit environmental protection goal of preventing or reducing deleterious radiation effects to a level where they would have "a negligible impact on the maintenance of biological diversity, the conservation of species, or the health and status of natural habitats, communities, and ecosystems" (ICRP, 2007). Such ecosystem-level protection goals arise from the ethical considerations for non-human species and also the recognition that ecosystems support human life. The individual organisms or species present in an ecosystem do not exist in isolation but interact with each other and with their environment. Thus, by basing assessments on individual- or species-level effects we may miss effects that might be manifested or modified as a result of these interactions. In recent decades, several scientists have argued that the current radiation protection system is therefore not appropriate and that 'higher' level effects should be considered to a greater extent (see Bradshaw et al., 2014; Bréchignac et al., 2016 and Rhodes et al., 2020 for more details).

This chapter addresses these higher level effects: what can and should be measured? First, some basic concepts and terminology need clarifying (Box 14.1: Definitions). In this chapter I draw on literature from radioecology, ecotoxicology, ecology and environmental and chemical risk assessment. These fields sometimes use different words to refer to the same, or similar, concepts (Table 14.1).

Table 14.1 Commonly-used, and often synonymous, terms for describing ways of measuring ecological or environmental effects of stressors

Things we can measure	The level at which we measure them	Integrative ways of describing or categorising the ecosystem/environment, usually including a degree of judgement
Endpoint	Ecosystem	Status
Indicator	Environment	Condition
Parameter	Ecological	Health
Descriptor		State
Biomarker (usually implies at sub-organism level)		Integrity
		Quality

Finally, the distinction should be made between ecosystem-relevant endpoints and ecosystem-level endpoints. The first are those that might be measured at lower levels (e.g., reproductive endpoints of a population) but that strongly influence a species' ecological competitiveness and thus have an effect at the ecosystem level. The second are those that are actually measured at the ecosystem level, such as net ecosystem production.

Box 14.1: Definitions

Organism: a single animal or plant

Population: a group of animals or organisms of one species that interbreed

Community: an assemblage of different species of animals and plants living in a particular area

Ecosystem: the biological community plus the abiotic (e.g., physical, chemical) parts of the environment, including the interactions between these

Environment: this term is sometimes used interchangeably with ecosystem and sometimes to mean the abiotic parts of an ecosystem

Environmental/Ecological Risk Assessment (ERA): an assessment of the biotic and abiotic parts of the environment to determine the likelihood and severity of detrimental effects to the ecosystem. These terms are often used interchangeably

Environmental Impact Assessment (EIA): an assessment of actual impacts to the ecosystem after a human intervention or contamination.

14.2 Why Can't We Just Extrapolate from 'Lower' Level to Ecosystem-Level Effects?

Biomarkers (a term used here in the usual sense, for endpoints at the sub-organismal level) have been advocated as a way of obtaining an early warning of effects at the individual- and population-level; sub-lethal changes to molecular processes may happen before the whole organism shows measureable effects (e.g., in growth, reproduction) (Mouneyrac & Amiard-Triquet, 2013). Molecular, immunological, physiological and behavioural endpoints have sometimes been called 'ecologically-relevant biomarkers" (Mouneyrac & Amiard-Triquet, 2013) or even "ecological biomarkers" (Amiard-Triquet et al., 2012) since they affect populations of organisms. Advocates of biomarkers also argue that they are relatively simple to measure, require small amounts of sample (and can be non-destructive) and can provide information on the mechanisms of toxicity, particularly in an experimental setting.

However, at present the use of biomarkers in field studies, monitoring and risk assessment is not as widespread as might be expected (Hagger et al., 2008), despite numerous publications referring to their great potential (e.g., Mouneyrac and Amiard-Triquet 2013). Their limited use is due to (a) limitations of the biomarkers themselves and (b) difficulties in interpreting the implications of lower-level effects mean at the population level (ie. extrapolation problems).

14.2.1 *Limitations of Biomarkers Themselves*

There are several limitations to the use of biomarkers, particularly in the field: they are rarely stressor-specific; their correlation with biological effects is often weak; false negatives are common; and information on 'background' levels (baselines) and natural spatial and temporal variation in biomarkers is difficult to obtain in the field (Forbes et al., 2006; Lam, 2009; Hook et al., 2014).

Biomarkers are rarely specific to any one contaminant – they can be triggered by many different types of anthropogenic and natural stressors. Much work has been devoted to finding contaminant-specific biomarkers for use in chemical risk assessment, and some progress has been made (see Hook et al., 2014); often a suite of different biomarkers can collectively provide evidence for exposure and/or effects to particular stressors.

However, in many cases sub-lethal molecular responses, especially those that serve a protective function, are not correlated with effects at higher levels. For example, oxidative stress enzymes may increase in order to protect the organism from reactive oxygen species, enabling them to grow and reproduce as normal. Likewise, the reproductive capacity of an individual organism might be reduced by stress but the population as a whole is not necessarily affected since other more tolerant individuals survive and continue to reproduce. Such 'compensatory mechanisms' enable organisms to deal with stress, and are only problematic if they affect a critical proportion of the population or result in excessive energetic costs to the organisms (Hook et al., 2014; Kooijman & Metz, 1984).

Sub-organismal biomarkers, as well as organism-level endpoints, are addressed in more detail elsewhere in this book and numerous critical reviews have been published in the ecotoxicology literature, addressing in particular the lack of ecological relevance (e.g., Forbes et al., 2006; Lam, 2009). The following section is therefore restricted to a brief overview of how biomarkers and individual organism endpoints have been used to predict ecosystem-relevant effects.

14.2.2 *Extrapolation and Problems Encountered*

Some advances have been made in extrapolating from individuals to populations in ecotoxicology and ERA, mainly through different types of modelling, and mainly using laboratory data. Three main classes of population models have been used; demographic models, energy budget models, and individual-based models (Forbes et al., 2008). Several attempts have also been made to combine modelling approaches, for example, energy-budget modelling (e.g., DEBtox) with demographic modelling (e.g., Leslie matrices) (Jager et al., 2014). Information on fecundity, growth and survival of individuals at different life stages is used to model population growth rates. Using this approach, based on data from laboratory experiments, Lopes et al. (2005) successfully modelled the effects of a pesticide on population growth rates of the insect *Chironomus riparius* and Billoir et al. (2007) the effects of cadmium on populations of the crustacean *Daphnia magna*. Alonzo et al. (2008, 2016) applied these methods in radioecology, modelling a range of different organisms for which data was available (aquatic and soil invertebrates, fish, terrestrial mammals). Interestingly, although in most cases individual-level endpoints were more sensitive to those at population level, in a few cases the opposite was true (Alonzo et al., 2016).

Extrapolating further to communities or ecosystems has been more problematic, since modelling interactions between populations of species and between them and their ecosystem requires a different type of modelling and level of model complexity. However, ecologists commonly use models to interpret and predict community- or ecosystem-level processes. These types of models include food web and network models, carbon (or energy) flow models and multi-population models, and may be spatially explicit (ie. landscape models) or dynamic. Ecosystem components may be grouped together into functional groups (e.g., primary producers, primary consumers) and linked by equations representing ecological processes (e.g., primary production, consumption rates). Ecosystem models have been used to model transfer of contaminants and radionuclides in ecosystems (e.g., Scott, 2003) but have not been extensively used for toxic effects studies in ecotoxicology or radioecology. It is beyond the scope of this chapter to review the field of ecosystem modelling; the reader is directed to the numerous textbooks and journals that cover this topic. In ecotoxicology, scientists have instead sought to find ways to utilize the enormous amount of data produced from decades of standard toxicity testing on single species in the lab (see e.g., Henegar et al., 2011). The most common approach, which is now also included in chemical risk assessment guidelines (e.g., in the EU and Canada), is species sensitivity distributions (SSDs).

SSDs (van Straalen & Denneman, 1989) use common metrics from toxicity tests such as effect concentrations (EC_x, the concentration at which an x% effect is observed compared to the control) or no effect concentrations (NOEC). Effects data (often EC_{10}) for a given contaminant for a range of different species are used in the analysis. Using statistical methods, the 'HC_x' is obtained; the hazardous concentration for x% of the species (HC_5 is often used). The validity and ecological relevance of SSDs have been extensively discussed in the ecotoxicology literature (e.g., Emans et al., 1993; Forbes & Forbes, 1993; van Straalen & Denneman, 1989). Common criticisms are that all species are treated equally but in reality the most exposed organism is not necessarily that most at risk, and/or the 5% of taxa that could be at risk could be disproportionately important in the ecosystem. SSDs are also based on a rather biased set of laboratory species (which may not even co-exist in the wild), which are usually raised in optimal conditions, often in acute exposure situations, while organisms in the field usually exist in suboptimal conditions and are often exposed to chronic pollution. Lastly, no interactions between species, or between species and the environment are considered.

SSDs have been attempted a few times in radioecology (Garnier-Laplace et al., 2006, 2008, 2010). SSDs (based on EDR_{10} (10% effect dose rate) values from laboratory experiments) were used to obtain an ecosystem level (or generic) PNEDR and HDR_5. Indeed, the value of 10 μGy h^{-1} obtained in that work is now used as a generic 'no effect' ecosystem-level benchmark "assumed to ascribe sufficient protection of all ecosystems from detrimental effects on structure and function under chronic exposure" (Garnier-Laplace et al., 2008; 2010). Garnier-Laplace et al. (2013) also compared estimations of ecosystem sensitivity and risk using SSDs based on laboratory and field (Chernobyl) data and found that organisms in the field were apparently more sensitive than those in the lab.

In freshwater ecotoxicology and risk assessment, attempts have been made to utilise the enormous amount of available effects data, particularly on pesticides, to create assessment tools that can predict likely biological, ecological and community effects. PERPEST (http://www.perpest.alterra.nl/) predicts the probability of effects of a given concentration of a pesticide by using case-based reasoning to search an effects database. The output is a prediction of the probability of effects on broad taxonomic groups (e.g., macrophytes) and also on endpoints such as community metabolism. The SPEAR tool (http://www.systemecology.eu/SPEAR/about.php) uses similar effects data, together with a database of species characteristics (traits; see Sect. 14.3.2) to estimate the proportion of species at risk/not at risk. Both these tools depend on the enormous databases available, something that is currently not possible for radioecology. However, the use of biological traits in scaling up stress effects to community and ecosystem levels shows great promise (Liess & Beketov, 2011; Gibert et al., 2015).

To summarise, extrapolating from effects at lower levels to those at higher levels (especially in the field) is difficult due to modelling challenges, using lab data to extrapolate to the field and because apparent higher level endpoints (e.g., HC_5) do not take into account ecosystem interactions. This means that we miss potential effects that occur at the system-level that cannot be predicted from lower level effects (ie. non-linear responses and emergent properties; (Bradshaw et al., 2014, Forbes & Calow, 2013, Scheffer et al., 2001). Biomarkers and lower-level endpoints are perhaps best used as one of many lines of evidence and efforts need to made to improve (and use) methods that can directly detect and predict impacts at higher biological levels (Forbes et al., 2006) as well as integrating these 'top-down' and 'bottom-up' approaches (Beketov & Liess, 2012). The next two sections give an overview of ecosystem-level and ecosystem-relevant endpoints that can be (and are) used in science and risk assessment.

14.3 What Can We Measure to Detect Ecosystem-Level and Ecosystem-Relevant Effects from a Scientific Point of View?

In the disciplines of ecology and ecosystem science, higher level (ie. community- and ecosystem-level) effects of stress and disturbance are a common focus of interest and numerous methods have been used to quantify them. In particular, community ecology has long investigated how disturbance affects community structure and function and much ecological theory is based on these phenomena (patch dynamics, succession, resilience etc). The Odum brothers were among the first to start to consider anthropogenic stress to ecosystems in terms of these ecological concepts; Odum (1985) published a list of trends that might be expected in stressed ecosystems. Many of these trends are related to imbalances and inefficiencies in the system, such as less efficient nutrient and carbon cycling per unit biomass, while others are related to changes in community structure where disturbed systems favour fast-growing, small-bodied, opportunistic species.

Ecosystem-level endpoints can be divided into those that concern structural changes (Sect. 14.3.1) and those that affect ecosystem functions (Sect. 14.3.2). Structure and function may of course be tightly coupled to each other, and the relationship between biodiversity and ecosystem function has been an extremely active research field for many years (see e.g., Hooper et al., 2005; Tilman et al., 2014).

It should be remembered that ecosystem-level endpoints are not necessarily directly proportional to a given stressor (e.g., a contaminant). This is partly due to factors other than the stressor of interest affecting the ecosystem simultaneously, which may enhance, depress or mask the effects of the stressor (Heugens et al., 2001 and Sect. 14.5). In addition, interactions between species, and between species and their environment, may cause indirect (secondary) effects – for example, if the zooplankton community is negatively affected by a pesticide, the decrease in zooplankton numbers may indirectly have a positive effect on their food species, phytoplankton, through decreased grazing pressure (Fleeger et al., 2003; Friberg-Jensen et al., 2003; Hansen & Garton, 1982; Wendt-Rasch et al., 2003). Indirect effects may be very common; Fleeger et al. (2003) reviewed 150 studies where indirect effects of contaminants on aquatic food webs had been found. These included contaminant-induced changes not only on predation/grazing rates but also on competition between species and in behaviour. The authors pointed out that such indirect effects can "enhance, mask or spuriously indicate direct contaminant effects" on individual species.

In a contaminated field situation, ecosystem-level endpoints that implicitly include effects of multiple simultaneous stressors and indirect effects will therefore give a truer picture of the overall ecosystem condition than single species endpoints, since they integrate many separate factors and effects and can capture emergent properties that cannot be predicted from the parts of the whole (Sect. 14.2). In the rest of this section, I provide a brief overview and examples of some of the ecosystem-level and ecosystem-relevant endpoints that can be measured, highlighting in particular some of the newer developments, such as biological traits and ecological networks). The first two sections separate structural and functional endpoints, but there is sometimes a degree of overlap between the two (e.g., functional diversity).

14.3.1 Structural Endpoints

A common measure used is biodiversity; most simply, the number of species (species richness), but various indices can also be calculated based on the relative numbers of the different species (their evenness or dominance), for example Shannon's diversity index. Frequencies of particular types of organisms can also be quantified – for example, the number of opportunistic vs. long-lived species (r/k strategists), the number of functional groups or the frequency of occurrence of parasites. Multivariate metrics of community structure can also be used. Biomass of species or functional groups is also used as a structural endpoint. Community or foodweb structure can also be described using network metrics (described further in Sect. 14.3.3). Lastly, habitat structure and complexity can be described at a number of different spatial scales.

14.3.2 Functional Endpoints

The more traditional functional endpoints are those that measure ecosystem processes, often fluxes of carbon or energy. These these include primary production, net ecosystem production (P) and respiration (R), as well as ratios between these (P/R), and with biomass (B) – P/B and R/B (Odum, 1985)). Organic matter degradation rates (e.g., leaf litter decomposition) and grazing rates are also common, particularly in freshwater ecosystems. Other ecosystem processes include nutrient (most commonly nitrogen and phosphorus) fluxes and cycling.

Functional diversity is a measure based on the functional characteristics of the species in a community – there are a number of different ways such diversity can be quantified (Petchey & Gaston, 2002; Schleuter et al., 2010; Schmera et al., 2017). A method that has received much interest during the last few years is to evaluate 'traits' – biological characteristics of species that define them in terms of their ecological roles (Diaz et al., 2013). Although traits are related to individual species, they can inform us about the likely role of a single species, or group of functionally-similar species, in overall ecosystem function (Gibert et al., 2015). These characteristics can be physiological, morphological, behavioural or to do with life histories, e.g., body size, dispersal, mobility longevity, reproductive method, growth rates, feeding habit, nutrient requirements, pH sensitivity, salinity tolerance. Some authors distinguish between biological traits (those that have to do with the life history of the organism) and ecological traits (that describe how organism-habitat relationships).

14.3.3 Network Metrics and Ecological Network Analysis

Ecological networks are representations of nature describing species diversity, relationships between species and flows of energy, nutrients or individuals within an ecosystem (Eklöf & Allesina, 2017). They can be described at three levels – topological (based on presence/absence of interactions), quantitative (where interactions are weighted or quantified) and dynamical (also include population dynamics) (Eklöf & Allesina, 2017). Some examples are given in Table 14.2. Historically, they have been used more within theoretical ecology, but there has been increased interest in how ecological networks might be altered by anthropogenic stress (Tyliakalis & Morris, 2017), such as eutrophication-induced hypoxia (Baird et al., 2004), fishing (Tomczak et al., 2013) and land use change (Morris, 2010).

14.4 What Can/Should We Measure to Detect Ecosystem-Level and Ecosystem-Relevant Effects in Risk Assessment?

It is worth remembering that type of measurements taken should always depend on the protection goals and the kind of assessment (see Chap. 18). For example, Cormier and Suter (2008) identify 4 types of assessments: condition assessments (that detect 'impairments'); causal pathway assessments (that identify causes and their sources); predictive assessments (that identify risks and benefits of different management actions); and outcome assessments. Whatever the type of assessment, endpoints should be meaningful. All too often, risk assessments provide data that are not matched to the protection goals (Bradshaw et al., 2014; Bréchignac et al., 2016; Forbes & Calow, 2013; Rhodes et al., 2020). This may be due to the need for simple, time- and cost-effective measurements, poorly designed sampling that does not allow for (e.g.,) the separation of anthropogenic influence from natural variability (inherent limitation *sensu* Kapustka, 2008) or simply due to precedence (contrived limitation *sensu* Kapustka, 2008). Assessment endpoints should therefore ideally be ecologically relevant, susceptible (ie. exposed and sensitive), broadly applicable and integrative over space and time, practical, measurable, cost-effective, well-defined, relevant to management goals and responsive to management actions (Dolédec et al., 1999; USEPA, 2003; Elliott & Kennish, 2011).

In radiation protection, existing, planned and emergency situations (IAEA, 2014; see Chap. 13) are often referred to in order to frame the type of assessments needed. Examples of these are already-contaminated environments (e.g., the current status at Chernobyl), routine releases from nuclear power plants and accident scenarios (e.g., directly after the Fukushima incident), respectively. Not only do these three situations alter the focus of the assessment needed (e.g., a rapid screening to determine immediate human health risks *vs.* possible long-term ecological impacts which require long-term monitoring) but also the type of contamination present (e.g., a

Table 14.2 Examples of network metrics used in Ecological Network Analysis to describe the structure and properties of the network (in many cases, a foodweb-based network)

Metric name (and common abbreviation)	Description and explanation
Some basic topological network metrics, describing foodweb structure[a]	
Species richness (S)	Number of taxa in the food web
Trophic links (L)	Number of feeding interactions between taxa
Link density (L/S)	Mean number of links per species
Connectance (L/S^2)	Proportion of possible trophic links that are realised (S^2 = total number of possible links). Indicates how sensitive the system is to disturbance
Top taxa (T)	Fraction (or %) of taxa that have no predators
Intermediate taxa (I)	Fraction (or %) of taxa that have both predators and prey
Basal taxa (B)	Fraction (or %) of taxa that have no prey
Herbivores (Herb)	Fraction (or %) of taxa that only consume basal taxa
Generality (Gen)	Mean (and SD) number of prey a taxa has
Vulnerability (Vul)	Mean (and SD) number of predators a taxa has
Mean shortest path length	Mean of the shortest chain of links connecting each pair of taxa
Some more complex quantitative network metrics, based on energy or carbon pools and fluxes[b]	
Total System Throughput (TST)	The sum of all energy flows in the food web (ie. total system activity). Usually decreases when system is degraded
Finn Cycling Index (FCI)	The proportion of the total activity (TST) that is recycled in the system – a measure of the system's ability to maintain its structure and integrity
Average Path Length (APL)	Mean number of steps a unit of flux will experience within the food web, from entry to exit. It is another way of measuring recycling in the food web
Average mutual information (AMI)	Measures the organisation of the energy flows: the greater the AMI, the higher degree of specialisation in the network
System Ascendency (A)	The product of the activity (TST) and the AMI. Describes the growth and development of the system – it increases as a system matures
System Redundancy (R)	Describes the distribution of energy flow among the ecosystem pathways and indicates the system's energy reserve. An index of system resilience.
Total Production/Total Biomass ratio (ToTP/ToTB)	System turnover rate is an indicator of the average size of organisms in an ecosystem. Stress usually reduces the mean size of organisms in ecosystems.
Total primary productivity/ Total system respiration (TPP/TR)	This ratio was proposed by EP Odum and indicates the maturity/balance of an ecosystem. Ratios close to 1 indicate mature/balanced ecosystems.

[a]Sources: Dunne et al. (2004), Dunne et al. (2013) and Williams and Martinez (2000)
[b]Sources: Niquil et al. (2012) and references therein, Tomczak et al. (2013) and references therein

distinct contamination gradient with potentially high contaminant concentrations vs. a diffuse contaminant source with a potentially relatively low environmental impact in the context of natural environmental variation). The rest of this section focusses mainly on 'existing' situations and/or condition and causal assessments (Cormier & Suter, 2008).

14.4.1 Ecosystem-Based Approach to Assessment

There have been many new legislative efforts worldwide over the last decades to use a more 'ecosystem' or 'holistic' approach, relying more on ecological assessments than measurements of single species or of abiotic factors such as water chemistry and pollutant concentrations (Borja et al., 2008). These include the European Water Framework Directive (WFD) and Marine Strategy Framework Directive (MSFD) and various environmental laws and guidelines in the USA and Canada. These frameworks often require the evaluation of integrative measures of the state or condition of the ecosystem as a whole, referred to variously as ecological integrity (USEPA), ecological quality (WFD), ecological status (WFD) or environmental status (MSFD).

These integrative measures are meant to include aspects of ecosystem structure and function, as well as habitat characteristics and complexity. In most cases, the environmental status is assessed in comparison to reference conditions for the ecosystem type in question. The overall multi-metric index of environmental status is based on a number of more specific measurable metrics (indicators (WFD), indicator variables (USEPA)), often grouped into broader categories (called variously qualitative descriptors (MSFD), quality elements (WFD) or attributes (USEPA)).

Thus, in theory, similar endpoints to those described in Sect. 14.3 should also be applicable in an assessment context and indeed many of them are used. For example, MSFD Descriptors 1 (Biodiversity) and 4 (Food webs) include mention of species composition and their relative abundance, size distribution of individuals across a trophic guild, productivity of different trophic guilds and area of different categories of habitat types (EC, 2017).

Although these ambitious pieces of legislation attempt to take an interdisciplinary systemic view, they are often implemented in a reductionist manner and still rely on quite simplistic measurements that may not in practice be very robustly linked to what is trying to be assessed (Birk et al., 2012; Brown & Williams, 2016; Voulvoulis et al., 2017). For example, it is much more common to measure structural than functional endpoints (Birk et al., 2012; Hering et al., 2010; O'Brien et al., 2016). Birk et al. (2012) identified river and transitional water assessments as having the highest usage of sensitivity and trait metrics (cf. coastal waters and lakes) in the context of the WFD. They also noted that it is often assumed that structure is a proxy for function, but this is not always the case; for example, Usseglio-Polatera et al. (2000) found that trait-based analyses of benthic communities in rivers were better able to discriminate spatial environmental changes than analyses based on taxa. There have been substantial efforts to find ways to use trait-based methods in risk assessment, particularly in the context of the WFD (Dolédec et al., 1999; Statzner et al., 2001, 2005).

14.4.2 Ecosystem Services

Ecosystem services (ES) are "the ecological characteristics, functions, or processes that directly or indirectly contribute to human wellbeing: that is, the benefits that people derive from functioning ecosystems" (Costanza et al., 2017). The ES framework can be seen as way of framing and valuing ecosystems (in monetary or other terms) in a transdisciplinary way in policy and management, though there are many critics of this approach (see Schröter et al., 2014 for the main arguments).

Since ERA should measure endpoints that are of relevance to protection goals and aspects of the environment that are in some way valued by humans, ES would seem to be a natural component of ERAs (Munns, 2015), but the ES concept is only recently starting to be more common in EIA, ERA and regulatory frameworks (Geneletti, 2013; Olander & Maltby, 2014; Rosa & Sanchez, 2015). However, ES are included in guidance documents, such as those describing generic ecological assessment endpoints (GEAEs) in the USA (Munns, 2015), ERA of pesticides in the EU (Nienstedt et al., 2012) and scoping impact assessments (Landsberg et al., 2011) and have been advocated in policy impact assessments in the EU (Helming et al., 2013).

Although they have a different purpose, some functional assessment endpoints are similar to certain ES, particularly the so-called 'supporting services' (e.g., primary production, soil formation and nutrient cycling; MEA, 2005). Others are more directly linked to human well-being, though based on ecosystem properties (e.g., provision of food to humans (an ES) is reliant on, or the same as, growth and production of biomass of species used as food by humans).

The ES concept is a potentially useful way to frame or scope an environmental assessment when selecting assessment endpoints and formulating protection goals, since ES integrates social and ecological aspects, as well as multiple scales, stressors and habitats (EFSA, 2010; Forbes & Calow, 2012; Baker et al., 2013; Nienstedt et al., 2012; Landsberg et al., 2011; Maltby et al., 2018). It can also help communication with stakeholders (Baker et al., 2013; Schröter et al., 2014; Munns, 2015; Maltby et al., 2018). Using ES as an additional type of assessment endpoint can improve decision making by linking ecological risk to human well-being (Forbes & Calow, 2013; Munns, 2015; Maltby et al., 2017).

In the context of chemical ERA, the European Food Safety Authority (EFSA) published a strategy for using an ES framework to develop specific protection goals for pesticides (EFSA, 2010; Nienstedt et al., 2012). However, since then the ES approach has not been extended, despite it being applicable to a range of other chemical groups, such as oil-related pollution, 'down-the-drain' chemicals (e.g., personal care products and pharmaceuticals) and persistent organic pollutants (Maltby et al., 2017).

14.5 Linking Ecosystem Effects to a Stressor of Interest

A common challenge in both environmental research and impact assessment is identifying the causes of observed changes in the ecosystem (see Chaps. 11 and 12). In both fields, there is often a 'stressor of interest' (e.g., a particular contaminant) and undue focus may be placed on this stressor at the expense of other potentially important variables. In a natural ecosystem, a multitude of natural and anthropogenic factors act simultaneously. In theory, as wide as range of variables as possible should be quantified (though practical constraints may make this difficult), but even then disentangling the relative importance of these variables can be difficult.

Where a major pollution event from a point source has occurred, a strong gradient of a specific pollution type may occur, strong enough to cause identifiable toxic effects. In these cases, correlating the response of any given endpoint to the concentration/dose of the contaminant of interest can be possible using, for example, simple correlation or regression analysis. The endpoints used can be ecosystem-relevant or ecosystem-level, for example biodiversity indices, net primary production, or multivariate metrics that describe aspects of community structure.

However, often pollution does not reach extreme toxic levels and/or pollution may come from one or several diffuse sources and is instead spatially (and temporally) heterogeneous and often at levels that are not high enough to cause obvious direct effects. In these cases, the level of effects caused by pollutants may be in the same order or magnitude or lower than effects caused by other stressors, environmental factors and natural variation, and may interact with them. This is not to say that science or risk assessment should ignore them. Rather, we should try and determine the relative contribution of the pollutant(s) to the overall response of the ecosystem. This was argued convincingly by Van Straalen (2003) and van Straalen and van Gestel (2008) who proposed that ecotoxicology should be viewed in the more general context of stress ecology.

It should be pointed out that several of the recent EU Directives (e.g., WFD, MSFD) require the evaluation of ecological status as deviation from the natural, undisturbed condition. Much work has been put in during the last few decades or so to (a) determine scientifically robust metrics for quantifying 'status' and (b) to find scientifically robust ways to identify the level of deviation from 'natural', given the range of different sources of natural and anthropogenic variables and the frequent lack of undisturbed areas with which to compare (e.g., Hering et al., 2015; Sandin & Hering, 2004). The methods used in many of the studies mentioned in this section are useful for such evaluations, as well as in risk assessments and scientific studies.

The rest of this section gives examples of methods that have been successfully used to disentangle the relative effects of different stressors such complex situations. To date, there are few examples from the radioecology literature, but such methods are commonly used in ecology and other environmental sciences.

14.5.1 Multivariate Methods

These are a useful exploratory and analytical approaches that have been commonly used in ecology and ecotoxicology. Examples are PCA (principal component analysis), RDA (redundancy analysis) and the related PRC (principal response curves), CCA (canonical correspondence analysis) and co-inertia analysis. Van Straalen and van Gestel (2008) reviewed four case studies where such methods were used in an ecotoxicological context to discriminate the relative effects of multiple stressors (natural and anthropogenic) to both structural and functional endpoints. Several of these methods allow a quantification of the combined and relative contribution of different factors, as well as their significance, allowing conclusions such as "45% of the variance in community composition was due to natural factors, 8.6% due to metal contamination." In addition, the relative effect of different stressors on different parts of the ecosystem can be determined (e.g., which contaminants affected which taxa groups). These methods allow the identification of covarying stressors and the effects of anthropogenic stressors along environmental gradients that might otherwise confound clear contaminant patterns. Principal response curves analyse temporal changes in community composition and are always standardised to some kind of control, whether this is an 'unaffected' ecosystem or a reference point in time (van den Brink et al., 2009).

Multivariate methods have also been used to identify the relative importance of different biological traits in determining ecosystem-relevant effects in multi-stressor environments. For example, Bremner et al. (2006) used multivariate biological trait analysis (BTA) to determine that salinity, water temperature, sediment coarseness, fish diversity and fishing pressure (ie. a combination of physico-chemical, biotic and anthropogenic factors) together provided the strongest association with trait composition. They also identified the nine (out of 46) trait categories that correlated best with environmental conditions; these included traits describing reproduction, feeding mode and mobility and habitat preferences. A similar approach was used in riverine systems to determine the 'river health' or 'quality' of different parts of the river system in a monitoring and management context (Usseglio-Polatera & Beisel, 2002). Again, a number of different natural (e.g., water chemistry) and anthropogenic factors (in this case, domestic and industrial effluents and physical and environmental alterations of the river for shipping) were identified as important drivers. Sandin and Hering (2004) used PCA of water chemistry and benthic community data as a way to try and harmonise and intercalibrate assessment systems used for judging categories of 'ecological status' in different European countries the framework of the WFD.

14.5.2 Linear Models

A second group of statistical methods includes multiple regressions and GLM (Generalized Linear Models) and its various extensions, such as GLMM (Generalized Linear Mixed Models) and GAM (Generalized Additive Models). In most cases, such analyses include the selection of the most parsimonious model – ie. the one which uses the least number of variables to explain the data.

Lecomte-Pradines et al. (2014) used a combination of PCA and multiple linear regressions to examine how structural and functional aspects of nematode communities were related to radiation dose and other environmental factors in the Chernobyl Exclusion Zone. Multiple linear regression showed that nematode 'Maturity Index' (MI) was significantly positively associated with total radiation dose rate (TDR), soil organic carbon and soil pH and that 'Nematode Channel Ratio' (NCR) was negatively correlated with TDR, but with none of the other parameters. MI is an indicator of the relative occurrence of species with different life strategies (colonisers and persisters), while NCR indicates the relative abundance of bacterial- and fungal-feeders.

In an extremely complex analysis of 20 benthic community metrics and 21 abiotic metrics at more than 1000 river sites in Germany, Leps et al. (2015) used GLM to determine the relative effects of different environmental drivers at different spatial scales. These included data on physico-chemical water parameters (e.g., conductivity, nutrients, temperature, dissolved oxygen), hydromorphological quality (ie. habitat structure – river bank stabilisation, substrate type and cross-sectional form) and land use (e.g., arable, pasture, forest). The objective of the study was to aid in future management decisions, in particular in choosing appropriate restoration methods. By identifying key stressors (or groups of stressors) for particular types of rivers (small/large, upstream/downstream) and the spatial scale at which these factors were important (local/regional), they suggested that targeted measures could be selected that would have a better chance of success that generalised management solutions.

14.6 Conclusions

Radiation protection and risk assessment are based on information on effects measured at the level of the individual organism or species. However, when an environmental assessment is conducted, we are rarely concerned with the individual organism or species, but rather the ecosystem as a whole. Some people argue that we can extrapolate from effects seen using 'lower' level endpoints such as biomarkers up to 'higher' levels. Various modelling attempts have been made, but extrapolation remains a challenge and also runs the risk of missing effects that are only apparent at higher levels (emergent effects).

There is a large range of ecosystem-relevant and ecosystem-level endpoints available for use in science and risk assessment. Many are commonly used in ecology, ecotoxicology and environmental sciences, and there is an increasing recognition and use of these endpoints in other environmental legislative frameworks. It is time to use them in radioecology and radiation protection and risk assessment!

It is highly unlikely that any radiation-specific endpoints exist at any level, but especially at the ecosystem level. Rather, radiation is just one of numerous anthropogenic and natural factors that simultaneously influence ecosystem structure and function. We should recognise this and improve our study and assessment designs and analyses to determine the *relative* influence of these different stressors. This is important in order to scientifically understand ecosystem effects in radiation-contaminated areas and also in risk communication and in order to take appropriate management actions.

References

Alonzo, F., Hertel-Aas, T., Gilek, M., Gilbin, R., Oughton, D. H., & Garnier-Laplace, J. (2008). Modelling the propagation of effects of chronic exposure to ionising radiation from individuals to populations. *Journal of Environmental Radioactivity, 99*, 1464–1473.

Alonzo, F., Hertel-Aas, T., Real, A., Lance, E., Garcia-Sanchez, L., Bradshaw, C., Vives i Batlle, J., Oughton, D. H., & Garnier-Laplace, J. (2016). Population modelling to compare chronic external radiotoxicity between individual and population endpoints in four taxonomic groups. *Journal of Environmental Radioactivity, 152*, 46–59.

Amiard-Triquet, C., Amiard, J.-C., & Rainbow, P. S. (2012). *Ecological biomarkers: Indicators of Ecotoxicological effects* (464pp). CRC Press.

Baird, D., Christian, R. R., Peterson, C. H., & Johnson, G. A. (2004). Consequences of hypoxia on estuarine ecosystem function: Energy diversion from consumers to microbes. *Ecological Applications, 14*(3), 805–822.

Baker, J., Sheate, W. R., Phillips, P., & Eales, R. (2013). Ecosystem services in environmental assessment — Help or hindrance? *Environmental Impact Assessment Review, 40*, 3–13.

Beketov, M. A., & Liess, M. (2012). Ecotoxicology and macroecology – Time for integration. *Environmental Pollution, 162*, 247–254.

Billoir, E., Péry, A. R. R., & Charles, S. (2007). Integrating the lethal and sublethal effects of toxic compounds into the population dynamics of Daphnia magna: A combination of the DEBtox and matrix population models. *Ecological Modelling, 203*(3–4), 204–214.

Birk, S., Bonne, W., Borja, A., Brucet, S., Courrat, A., Poikane, S., Solimini, A., van de Bund, W., Zampoukas, N., & Hering, D. (2012). Three hundred ways to assess Europe's surface waters: An almost complete overview of biological methods to implement the Water Framework Directive. *Ecological Indicators, 18*, 31–41.

Borja, A., Bricker, S. B., Dauer, D. M., Demetriades, N. T., Ferreira, J. G., Forbes, A. T., Hutchings, P., Jia, X., Kenchington, R., Marques, J. C., & Zhu, C. (2008). Overview of integrative tools and methods in assessing ecological integrity in estuarine and coastal systems worldwide. *Marine Pollution Bulletin, 56*, 1519–1537.

Bradshaw, C., Kapustka, L., Barnthouse, L., Brown, J., Ciffroy, P., Forbes, V., Geras'kin, S., Kautsky, U., & Bréchignac, F. (2014). Using an Ecosystem Approach to complement protection schemes based on organism-level endpoints. *Journal of Environmental Radioactivity, 136*, 98–104.

Bremner, J., Rogers, S. I., & Frid, C. L. J. (2006). Matching biological traits to environmental conditions in marine benthic ecosystems. *Journal of Marine Systems, 60*, 302–316.

Bréchignac, F., Oughton, D., Mays, C., Barnthouse, L., Beasley, J. C., Bonisoli-Alquati, A., Bradshaw, C., Brown, J., Dray, S., Geras'kin, S., Glenn, T., Higley, K., Ishida, K., Kapustka, L., Kautsky, U., Kuhne, W., Lynch, M., Mappes, T., Mihok, S., Møller, A., Mothersill, C., Mousseau, T., Otaki, J., Pryakhin, E., Rhodes, O. E., Jr., Salbu, B., Strand, P., & Tsukada, H. (2016). Addressing ecological effects of radiation on populations and ecosystems to improve protection of the environment against radiation: Agreed statements from a Consensus Symposium. *Journal of Environmental Radioactivity, 158-159*, 21–29.

Brown, E. D., & Williams, B. K. (2016). Ecological integrity assessment as a metric of biodiversity: Are we measuring what we say we are? *Biodiversity and Conservation, 25*, 1011–1035.

Cormier, S. M., & Suter, G. W. (2008). A framework for fully integrating environmental assessment. *Environmental Management, 42*, 543–556.

Costanza, R., de Groot, R., Braat, L., Kubiszewski, I., Fioramonti, L., Sutton, P., Farber, S., & Grasso, M. (2017). Twenty years of ecosystem services: How far have we come and how far do we still need to go? *Ecosystem Services, 28*, 1–16.

Díaz, S., Purvis, A., Cornelissen, J. H. C., Mace, G. M., Donoghue, M. J., Ewers, R. M., Jordano, P., & Pearse, W. D. (2013). Functional traits, the phylogeny of function, and ecosystem service vulnerability. *Ecology and Evolution, 3*(9), 2958–2975.

Dolédec, S., Statzner, B., & Bournard, M. (1999). Species traits for future biomonitoring across ecoregions: Patterns along a human-impacted river. *Freshwater Biology, 42*, 737–758.

Dunne, J. A., Williams, R. J., & Martinez, N. D. (2004). Network structure and robustness of marine food webs. *Marine Ecology Progress Series, 273*, 291–302.

Dunne, J. A., Lafferty, K. D., Dobson, A. P., Hechinger, R. F., Kuris, A. M., Martinez, N. D., McLaughlin, J. P., Mouritsen, K. N., Poulin, R., Reise, K., Stouffer, D. B., Thieltges, D. W., Williams, R. J., & Zander, C. D. (2013). Parasites affect food web structure primarily through increased diversity and complexity. *PLoS Biology, 11*(6), e1001579.

Eklöf, A., & Allesina, S. (2017). Networks, ecological. In A. Hastings & L. J. Gross (Eds.), *Encyclopaedia of theoretical ecology* (pp. 470–479).

Emans, H. J. B., van der Plassche, E. J., Canton, J. H., Okkerman, P. C., & Sparenburg, P. M. (1993). Validation of some extrapolation methods used for effect assessment. *Environmental Toxicology and Chemistry, 12*, 2139–2154.

EFSA (European Food Safety Authority). (2010). Scientific Opinion on the development of specific protection goal options for environmental risk assessment of pesticides, in particular in relation to the revision of the Guidance Documents on Aquatic and Terrestrial Ecotoxicology (SANCO/3268/2001 and SANCO/10329/2002). *EFSA Journal, 8*(10), 1821, 55 pp.

EC. (2017). Commission decision (EU) 2017/848 of 17 May 2017 laying down criteria and methodological standards on good environmental status of marine waters and specifications and standardised methods for monitoring and assessment, and repealing Decision 2010/477/EU. *Official Journal of the European Union L, 125/43*, 32pp.

Elliott, M., & Kennish, M. J. (2011). Chapter 8.01 Introduction-Overview In: Kennish, M. J., & Elliott, M. (Eds), Volume 8 Human-Induced Problems (Uses and abuses); In: Wolanski E., & McLusky, D. S. (Eds) *Treatise on Estuarine & Coastal Science* (pp. 1–15) Amsterdam: Elsevier.

Fleeger, J. W., Carman, K. R., & Nisbet, R. M. (2003). Indirect effects of contaminants in aquatic ecosystems. *Science of the Total Environment, 317*, 207–233.

Forbes, V. E., & Forbes, T. L. (1993). A critique of the use of distribution-based extrapolation models in ecotoxicology. *Functional Ecology, 7*, 249–254.

Forbes, V. E., Palmqvist, A., & Bach, L. (2006). The use and misuse of biomarkers in ecotoxicology. *Environmental Toxicology and Chemistry, 25*, 272–280.

Forbes, V. E., Calow, P., & Sibly, R. M. (2008). The extrapolation problem and how population modeling can help. *Environmental Toxicology and Chemistry, 27*(10), 1987–1994.

Forbes, V. E., & Calow, P. (2012). Use of the ecosystem services concept in ecological risk assessment of chemicals. *Integrated Environmental Assessment and Management, 9*, 269–275.

Forbes, V. E., & Calow, P. (2013). Developing predictive systems models to address complexity and relevance for ecological risk assessment. *Integrated Environmental Assessment and Management, 9*(3), e75–e80.
Friberg-Jensen, U., Wendt-Rasch, L., Woin, P., & Christoffersen, K. (2003). Effects of the pyrethroid insecticide, cypermethrin, on a freshwater community studied under field conditions. I. Direct and indirect effects on abundance measures of organisms at different trophic levels. *Aquatic Toxicology, 63*, 357–371.
Garnier-Laplace, J., Della-Vedova, C., Gilbin, R., Copplestone, D., Hingston, J., & Ciffroy, P. (2006). First derivation of predicted-no-effect values for freshwater and terrestrial ecosystems exposed to radioactive substances. *Environmental Science and Technology, 40*, 6498–6505.
Garnier-Laplace, J., Copplestone, D., Gilbin, R., Alonzo, F., Ciffroy, P., Gilek, M., Agüero, A., Björk, M., Oughton, D. H., Jaworska, A., Larsson, C. M., & Hingston, J. L. (2008). Issues and practices in the use of effects data from FREDERICA in the ERICA Integrated Approach. *Journal of Environmental Radioactivity, 99*, 1474–1483.
Garnier-Laplace, J., Della-Vedova, C., Andersson, P., Copplestone, D., Cailes, C., Beresford, N. A., Howard, B. J., Howe, P., & Whitehouse, P. (2010). A multi-criteria weight of evidence approach to derive ecological benchmarks for radioactive substances. *Journal of Radiological Protection, 30*, 215–233.
Garnier-Laplace, J., Geras'kin, S., Della-Vedova, C., Beaugelin-Seiller, K., Hinton, T. G., Real, A., & Oudalova, A. (2013). Are radiosensitivity data derived from natural field conditions consistent with data from controlled exposures? A case study of Chernobyl wildlife chronically exposed to low dose rates. *Journal of Environmental Radioactivity, 121*, 12–21.
Geneletti, D. (2013). Ecosystem services in environmental impact assessment and strategic environmental assessment. *Environmental Impact Assessment Review, 40*, 1–2.
Gibert, J. P., Dell, A. I., DeLong, J. P., & Pawar, S. (2015). Scaling-up trait variation from individuals to ecosystems. *Advances in Ecological Research, 52*, 1–17.
Hagger, J. A., Jones, M. B., Lowe, D., Leonard, D. R. P., Owen, R., & Galloway, T. S. (2008). Application of biomarkers for improving risk assessments of chemicals under the water framework directive: A case study. *Marine Pollution Bulletin, 56*, 1111–1118.
Hansen, S. R., & Garton, R. R. (1982). The effects of diflubenzuron on a complex laboratory stream community. *Archives of Environmental Contamination and Toxicology, 11*, 1–10.
Helming, K., Diehl, K., Geneletti, D., & Wiggering, H. (2013). Mainstreaming ecosystem services in European policy impact assessment. *Environmental Impact Assessment Review, 40*, 82–87.
Henegar, A., Mombelli, E., Pandard, P., & Péry, A. R. R. (2011). What can be learnt from an ecotoxicity database in the framework of the REACh regulation? *Science of the Total Environment, 409*, 489–494.
Hering, D., Borja, A., Carstensen, J., Carvalho, L., Elliott, E., Feld, C. K., Heiskanen, A.-S., Johnson, R. K., Moe, J., Pont, D., Lyche Solheim, A., & van de Bund, W. (2010). The European Water Framework Directive at the age of 10: A critical review of the achievements with recommendations for the future. *Science of the Total Environment, 408*, 4007–4019.
Hering, D., Carvalho, L., Argillier, C., Beklioglu, M., Borja, A., Cardoso, A. C., Duel, H., Ferreira, T., Globevnik, L., Hanganu, J., Hellsten, S., Jeppesen, E., Kodeš, V., Lyche Solheim, A., Nõges, N., Ormerod, S., Panagopoulos, Y., Schmutz, S., Venohr, M., & Birk, S. (2015). Managing aquatic ecosystems and water resources under multiple stress – An introduction to the MARS project. *Science of the Total Environment, 503*(504), 10–21.
Heugens, E. H. W., Hendriks, A. J., Dekker, T., van Straalen, N. M., & Admiraal, W. (2001). A review of the effects of multiple stressors on aquatic organisms and analysis of uncertainty factors for use in risk assessment. *Critical Reviews in Toxicology, 31*(3), 2347–2284.
Hook, S. E., Gallagher, E. P., & Batley, G. E. (2014). The role of biomarkers in the assessment of aquatic ecosystem health. *Integrated Environmental Assessment and Management, 10*(3), 327–341.
Hooper, D. U., Chapin, F. S., Ewel, J. J., Hector, A., Inchausti, P., Lavorel, S., Lawton, J. H., Lodge, D. M., Loreau, M., Naeem, S., Schmid, B., Setälä, H., Symstad, A. J., Vandermeer, J., &

Wardle, D. A. (2005). Effects of biodiversity on ecosystem functioning: A consensus of current knowledge. *Ecological Monographs, 75*, 3–35.

IAEA. (2014). *Radiation protection and safety of radiation sources: International basic safety standards* (471pp). International Atomic Energy Agency.

ICRP. (2007). *The 2007 Recommendations of the International Commission on Radiological Protection*. ICRP Publication 103, Annals of the ICRP 37(2-4).

Jager, T., Barsi, A., Hamda, N. T., Martin, B. T., Zimmer, I. E., & Ducrot, V. (2014). Dynamic energy budgets in population ecotoxicology: Applications and outlook. *Ecological Modelling, 280*, 140–147.

Kapustka, L. (2008). Limitations of the current practices used to perform ecological risk assessment. *Integrated Environmental Assessment and Management, 4*(3), 290–298.

Kooijman, S. A. L. M., & Metz, J. A. J. (1984). On the dynamics of chemically stressed populations: The deduction of population consequences from effects on individuals. *Ecotoxicology and Environmental Safety, 8*, 254–274.

Lam, P. K. S. (2009). Use of biomarkers in environmental monitoring. *Ocean & Coastal Management, 52*, 348–354.

Landsberg, F., Ozment, S., Stickler, M., Henninger, N., Treweek, J., Venn, O., & Mock, G. (2011). *Ecosystem Services Review for Impact Assessment: Introduction and Guide to Scoping* (WRI working paper). World Resources Institute. Online at http://www.wri.org/publication/ecosystemservices-review-for-impact-assessment

Lecomte-Pradines, C. Bonzom, J.-M., Della-Vedova, C., Beaugelin-Seiller, K., Villenave, C., Gaschak, S., Coppin, F., Dubourg, N., Maksimenko, A., Adam-Guillermin, C., & Garnier-Laplace, J. (2014). Soil nematode assemblages as bioindicators of radiation impact in the Chernobyl Exclusion Zone. *Science of the Total Environment, 490*, 161–170.

Leps, M., Tonkin, J. D., Dahm, V., Haase, P., & Sundermann, A. (2015). Disentangling environmental drivers of benthic invertebrate assemblages: The role of spatial scale and riverscape heterogeneity in a multiple stressor environment. *Science of the Total Environment, 536*, 546–556.

Liess, M., & Beketov, M. (2011). Traits and stress: keys to identify community effects of low levels of toxicants in test systems. *Ecotoxicology, 20*, 1328–1340.

Lopes, C., Péry, A. R. R., Chaumot, A., & Charles, S. (2005). Ecotoxicology and population dynamics: Using DEBtox models in a Leslie modeling approach. *Ecological Modelling, 188*, 30–40.

Maltby, L., Jackson, M., Whale, G., Brown, A. R., Hamer, M., Solga, A., Kabouw, P., Woods, R., & Marshall, S. (2017). Is an ecosystem services-based approach developed for setting specific protection goals for plant protection products applicable to other chemicals? *Science of the Total Environment, 580*, 1222–1236.

Maltby, L., van den Brink, P. J., Faber, J. H., & Marshall, S. (2018). Advantages and challenges associated with implementing an ecosystem services approach to ecological risk assessment for chemicals. *Science of the Total Environment, 621*, 1342–1351.

Millennium Ecosystem Assessment. (2005). *Ecosystems and human well-being: Biodiversity synthesis* (86pp). World Resources Institute.

Morris, R. J. (2010). Anthropogenic impacts on tropical forest biodiversity: A network structure and ecosystem functioning perspective. *Philosophical Transactions of the Royal Society B, 365*, 3709–3718.

Mouneyrac, C., & Amiard-Triquet, C. (2013). Biomarkers of ecological relevance in ecotoxicology. In J. F. Férard & C. Blaise (Eds.), *Encyclopedia of aquatic ecotoxicology* (pp. 221–236). Springer.

Munns, W. R., Jr. (2015). Ecosystem services as assessment endpoints for ecological risk assessment. *Integrated Environmental Assessment and Management, 12*(3), 522–528.

Nienstedt, K. M., Brock, T. C. M., van Wensem, J., Montforts, M, Hart, A., Aagaard, A., Alix, A., Boesten, J., Bopp, S.K., Brown, C., Capri, E., Forbes, V., Köpp, H., Liess, M., Luttik, R., Maltby, L., Sousa, J. P., Streissl, F., & Hardy, A. R. (2012). Development of a framework based on an ecosystem services approach for deriving specific protection goals for environmental risk assessment of pesticides. *Science of the Total Environment, 415*, 31–38.

Niquil, N., Chaumillon, E., Johnson, G. A., Bertin, X., Grami, B., David, V., Bacher, C., Asmus, H., Baird, D., & Asmus, R. (2012). The effect of physical drivers on ecosystem indices derived from ecological network analysis: Comparison across estuarine ecosystems. *Estuarine, Coastal and Shelf Science, 108*, 132–143.

O'Brien, A., Townsend, K., Hale, R., Sharley, D., & Pettigrove, V. (2016). How is ecosystem health defined and measured? A critical review of freshwater and estuarine studies. *Ecological Indicators, 69*, 722–729.

Odum, E. P. (1985). Trends expected in stressed ecosystems. *Bioscience, 35*(7), 419–422.

Olander, L., & Maltby, L. (2014). Mainstreaming ecosystem services into decision making. *Frontiers in Ecology and Evolution, 12*(10), 539.

Petchey, O. L., & Gaston, K. J. (2002). Functional diversity (FD), species richness and community composition. *Ecology Letters, 5*, 402–411.

Rhodes, Jr, O. E., Bréchignac, F., Clare Bradshaw, et. al. (2020). Integration of ecosystem science into radioecology: A consensus perspective. *Science of the Total Environment, 740*, 140031. https://doi.org/10.1016/j.scitotenv.2020.140031

Rosa, J. C. S., & Sanchez, L. E. (2015). Is the ecosystem service concept improving impact assessment? Evidence from recent international practice. *Environmental Impact Assessment Review, 50*, 134–142.

Sandin, L., & Hering, D. (2004). Comparing macroinvertebrate indices to detect organic pollution across Europe: A contribution to the EC Water Framework Directive intercalibration. *Hydrobiologia, 516*, 55–68.

Scheffer, M., Carpenter, S., Foley, J. A., Folke, C., & Walker, B. (2001). Catastrophic shifts in ecosystems. *Nature, 413*, 591–596.

Schleuter, D., Daufresne, M., Massol, F., & Argillier, C. (2010). A user's guide to functional diversity indices. *Ecological Monographs, 80*(3), 469–484.

Schmera, D., Heino, J., Podani, J., Erös, T., & Dolédec, S. (2017). Functional diversity: A review of methodology and current knowledge in freshwater macroinvertebrate research. *Hydrobiologia, 787*, 27–44.

Schröter, M., van der Zanden, E. H., van Oudenhoven, A. P. E., Remme, R. P., Serna-Chavez, H. M., de Groot, R. S., & Opdam, P. (2014). Ecosystem services as a contested concept: A synthesis of critique and counter-arguments. *Conservation Letters, 7*, 514–523.

Scott, E. M. (2003). *Modelling radioactivity in the environment* (438pp). Elsevier Science.

Statzner, B., Bis, B., Dolédec, S., & Usseglio-Polatera, P. (2001). Perspectives for biomonitoring at large spatial scales: A unified measure for the functional composition of invertebrate communities in European running waters. *Basic and Applied Ecology, 2*, 73–85.

Statzner, B., Bady, P., Dolédec, S., & Schöll, F. (2005). Invertebrate traits for the biomonitoring of large European rivers: An initial assessment of trait patterns in least impacted river reaches. *Freshwater Biology, 50*, 2136–2161.

Tilman, D., Isbell, F., & Cowles, J. M. (2014). Biodiversity and ecosystem functioning. *Annual Review of Ecology, Evolution and Systematics, 45*, 471–493.

Tomczak, M. T., Heymans, J. J., Yletyinen, J., Niiranen, S., Otto, S. A., & Blenckner, T. (2013). Ecological network indicators of ecosystem status and change in the Baltic Sea. *PLoS One, 8*(10), e75439.

Tyliakalis, J. M., & Morris, R. J. (2017). Ecological networks across environmental gradients. *Annual Reviews of Ecology, Evolution and Systematics, 48*, 25–48.

USEPA. (2003). *Generic Ecological Assessment Endpoints (GEAEs) for Ecological Risk Assessment*. EPA/630/P-02/004F October 2003. 59 pp.

Usseglio-Polatera, P., Bournaud, M., Richoux, P., & Tachet, H. (2000). Biomonitoring through biological traits of benthic macroinvertebrates: How to use species trait databases? *Hydrobiologia, 422*(423), 153–162.

Usseglio-Polatera, P., & Beisel, J.-N. (2002). Longitudinal changes in macroinvertebrate assemblages in the Meuse river: Anthropogenic effects versus natural change. *River Research and Applications, 18*, 197–211.

van den Brink, P. J., den Besten, P. J., bij de Vaate, A., & ter Braak, C. J. F. (2009). Principal response curves technique for the analysis of multivariate biomonitoring time series. *Environmental Monitoring and Assessment, 152*, 271–281.

van Straalen, N. M., & Denneman, C. A. J. (1989). Ecotoxicological evaluation of soil quality criteria. *Ecotoxicology and Environmental Safety, 18*, 241–251.

van Straalen, N. M., & van Gestel, C. A. M. (2008). A stress ecology framework for comprehensive risk assessment of diffuse pollution. *Science of the Total Environment, 406*, 479–483.

van Straalen, N. M. (2003). Ecology becomes stress ecology. *Environmental Science and Technology, 37*(17), 324A–330A.

Voulvoulis, N., Arpon, K. D., & Giakoumis, T. (2017). The EU Water Framework Directive: From great expectations to problems with implementation. *Science of the Total Environment, 575*, 358–366.

Wendt-Rasch, L., Friberg-Jensen, U., Woin, P., & Christoffersen, K. (2003). Effects of the pyrethroid insecticide cypermethrin on a freshwater community studied under field conditions. II. Direct and indirect effects on the species composition. *Aquatic Toxicology, 63*, 373–389.

Williams, R. J., & Martinez, N. D. (2000). Simple rules yield complex food webs. *Nature, 404*, 180–183.

Chapter 15
Biomarkers and Ecological indicators for Environmental Radioactivity in Invertebrates

Katherine Raines, Jessica Goodman David Copplestone, and Jenson Lim

Abstract This chapter presents an overview of biomarkers and ecological indicators used to measure radiation stress in invertebrates. Using examples, we will show how and in what context the different biomarkers can be used and on which invertebrate species. We will also describe how the biomarker techniques can be adapted from other taxa including humans and developed for use in invertebrate species. Biomarkers generally fall into two categories – biomarkers for exposure and effect. Invertebrates have not frequently been used to investigate radiation-induced stress but they offer great potential. Here, we present biomarkers that have been used in different ecotoxicological contexts and are appropriate for measuring radiation exposure: comet assay, micronucleus assay, cytochrome p450 and oxidative stress. The majority of previous studies that have used biomarkers for environmental stressors have been conducted on either mussel for aquatic systems or earthworms in terrestrial systems. We also present an overview of how ecological indicators benefit understanding of the consequences that radiation exposure to individuals living in contaminated environments, focussing on alterations to reproduction and behaviour.

15.1 Introduction

Biomarkers are sub organism (often cellular) markers that indicate exposure and/or effects of stress. Ecological indicators, for the sake of this chapter, are measures of exposure or effect above the cellular level. Biomarker and indicator techniques have been developed for a range of taxa, including invertebrates. Invertebrates are advantageous taxa for both field and laboratory studies and can act as indicators for monitoring environmental levels of ionising radiation (IR). Invertebrates are known to

K. Raines (✉) · J. G. D. Copplestone · J. Lim
Biological and Environmental Sciences, University of Stirling, Stirling, UK
e-mail: katherine.raines@stir.ac.uk

M. D. Wood et al. (eds.), *Biomarkers of Radiation in the Environment*, NATO Science for Peace and Security Series A: Chemistry and Biology,
https://doi.org/10.1007/978-94-024-2101-9_15

respond to environmental stressors, therefore, using them to conduct assessments and experiments can aid our understanding of the consequences of exposure to IR to wildlife inhabiting contaminated areas (Prather et al., 2013).

Invertebrates make-up 90% of all known species and are present in all compartments of the environment whether terrestrial, freshwater, marine or inhabiting areas where these ecosystems interact. Invertebrates occupy key niches and provide crucial ecosystem services such as nutrient cycling, pollination and seed dispersal in terrestrial systems and nutrient cycling, water quality and decomposition processes in aquatic systems (Prather et al., 2013).

Invertebrates have several practical advantages for the monitoring the consequences of environmental radiological contamination. Many biomarkers and ecological indicators, which were originally designed for human use, have now been developed for use in other taxa, including invertebrates. One of the main benefits of conducting an assessment using invertebrates is that they are generally abundant and easier to collect then larger taxa such as birds, mammals and amphibians. This, therefore, enables good practice by conducting repeated measurements on multiple individuals and between different species occupying different ecological niches.

There is a gap in the knowledge as biomarkers which measure toxic effects at a molecular level have been shown to a rapid way of quantifying toxic levels at a molecular and individual level but so far cannot be linked to a population level (Hyne & Maher, 2003). Ecological indicators of effects used in conjunction with biomarkers give a more complete understanding as to what the consequences are to organisms from environmental stressors. The International Commission on Radiological Protection (ICRP) has summarised effect endpoints which are considered relevant to radiation exposure into four umbrella categories; mortality, morbidity, reduced reproductive success and chromosomal damage (ICRP, 2008).

Presently, key invertebrate species in both terrestrial and aquatic environments have been used as model systems for ecotoxicology with abundant literature for a range of stressors including heavy metals (Cao et al., 2017), nanoparticles (Calisi et al., 2016), pesticides (Rolke et al., 2016). However, there has been less research on the use of invertebrates exposed to environmental levels of radiation. The Organisation for Economic Co-operation and Development (OECD) has developed standardised toxicity tests involving invertebrates, for example, the Earthworm acute toxicity tests (OECD, 1984a). New papers are constantly emerging with new standardised techniques for biomarkers, however, there is still a limited range of organisms and biomarkers, however many of them which could be adapted for invertebrates (OECD, 2014a, 2016).

The majority of existing work on invertebrates and responses of biomarker and indicators to ionising radiation have predominantly been conducted in aquatic environments; both using freshwater and marine species. This chapter gives an overview of existing literature for a range of biomarkers and ecological indicators which have been used in ecotoxicology and that can practically be applied to measure exposure and/or effects of ionising radiation. This section is split into biomarkers and ecological indicators as both have valid contributions for understanding how to measure both exposure and effects of IR.

15.2 Biomarkers for Exposure

Biomarkers for exposure enable us to measure how much damage an individual has experienced from the stressors it is exposed to. However, this does not indicate what the consequences of this stress are to an organism and there are no biomarkers specific enough to be confident that ionising radiation is the sole causative agent. Biomarkers for exposure are a useful tool for long-term monitoring studies which allow comparisons over time and give indications of changing environmental conditions. Practical applications of such biomarkers include use for surveillance monitoring, detection of long-term trends and use in the design of the monitoring strategies.

15.2.1 Micronuclei

Micronuclei ("small nuclei") form when chromosomes or chromosome fragments are not integrated into the main nucleus following cell division and are usually indicative of chromosomal (cytogenetic) damage (Doherty, 2012). The micronucleus (MN) assay is a well-established test used for assessing the cytogenetic damage. Evans et al. initially established the effects of gamma rays on the induction of micronuclei in kidney bean root tips in 1959 (Evans et al., 1959). MN formation in invertebrate species has been conducted as early as 1985, where the effects of toxic chemicals on sea urchin embryos were assessed (Hose, 1985). The assay has been increasingly used in the literature and became a standard Organisation for Economic Co-operation and Development (OECD) test guideline for mammalian erythrocytes in 1997 (OECD, 1997) most recently updated in 2016 (OECD, 2016).

The MN assay is highly reliable, due to clear, simple identification methods which reduce user error. It can be used on any dividing cell, both *in vivo* and *in vitro* and the micronuclei can be detected for a long period of time following exposure (Hayashi, 2016). The method does require some adaptation to specific organisms in order to provide a viable cell culture. For example, Kirsch-Volders *et al* outline a number of different protocols used for the preparation of lymphocyte cultures (Kirsch-Volders et al., 2011). This includes the cytokinesis-block micronucleus (CBMN) assay, which has proven to be a sensitive test for radiation exposure (Fenech, 1993). This has been widely used in the literature as a reliable method in distinguishing cells that have only completed one nuclear division. The process of cellular division where a eukaryotic cell divides into two daughter cells termed cytokinesis is stopped using cytochalasin-B. Cells that have undergone one nuclear division can be identified by a binuclear appearance. Once suitable cells have been obtained from an organism, the assay follows a standard protocol.

The presence of micronuclei has been shown to strongly correlate with radiation-induced damage, which has been used to calibrate the assay in the lymphocytes of both animals and humans (Fenech, 2000; Ferlazzo et al., 2017; Vral et al., 2011).

Although, this correlation is because ionising radiation is a strong clastogenic agent and is therefore not radiation specific. The assay cannot reveal the type of chromosomal damage that has occurred and on some occasions, staining of MN-like structures may cause type I errors by forming pseudomicronuclei (Hayashi, 2016). Furthermore, the assay often exhibits poor detection at low doses and excess saturation in higher doses.

The use of MN assay on invertebrate species has been widely reported in the literature. Although, the majority of the research has been conducted on aquatic species, particularly mussels (see Table 15.1). Mussels are highly regarded for evaluating the effects of environmental pollutants (Bayne, 1976; Dixon, 2002) as they are sentinel species that can accumulate heavy metals and other pollutants. Mussels are increasingly being used within radioecological investigations (NSCEP, 2008; Cajaraville et al., 2000). There is some evidence for application in terrestrial invertebrates, for example, Sforzini et al., conducted a study on earthworms in 2012 using a combination of the comet (see below) and the MN assays (Sforzini et al., 2012).

Table 15.1 Search results from Web of science for micronucleus assay (or test) and invertebrates

Stressor	Species	Type	References
Heavy metals (Zn, Cu, Cd, Pb) and PCBs	Mussels *Mytilus galloprovincialis*	Marine	Zhang et al. (2017a)
Metal accumulation	Mussels *M.galloprovincialis*	Marine	Touahri et al. (2016)
Benzo-pyrene & tetrachloro-dibenzo-para-dioxin	Earthworms *Eisenia andrei*	Terrestrial	
Embryo toxicity and genotoxicity	Snail *Physa acuta*	Freshwater	Sánchez-Argüello et al. (2012)
Polybrominated diphenyl ethers (PBDEs)	Mussel *Dreissena polymorpha*	Freshwater	Parolini and Binelli (2012)
Cadmium	Mussel *D. polymorpha*	Freshwater	Vincent-Hubert et al. (2011)
Hexabromo-cyclododecane (HBCDD)	Benthic clam	Marine	Smolarz and Berger (2009)
Triclosan	Mussel *D. polymorpha*	Freshwater	Binelli et al. (2009)
Environmental pollution	Mussels *M.galloprovincialis*	Marine	Taleb et al. (2009)
Copper	Platyhelminth *Girardia tigrina*	Marine	Knakievicz and Ferreira (2008)
Decabromodiphenyl ether (BDE)	Mussel *D. polymorpha*	Freshwater	Riva et al. (2007)
Metamizole sodium and acetylsalicylic acid	Polyp *Hydra attenuate*	Marine	Arkhipchuk et al. (2004)
Potassium dichromate	Crayfish *Procambarus clarkia*	Freshwater	De la Sienra et al. (2003)

It would be beneficial to further develop the MN assay for a wider range of invertebrate species as it is already well established for other organisms. This would allow comparison of sensitivities to stressors across groups.

15.2.2 Comet Assay

The comet assay is a protocol used to visualise DNA breakages in individual eukaryotic cells. Since its initial development, the comet assay has been modified and applied to a wide range of organisms, including invertebrate species (Collins, 2004). Ostling and Johanson established the scientific background for the Comet Assay in 1984 (Ostling & Johanson, 1984) where they developed an electrophoresis-based method to detect relaxation of DNA coils as a result of DNA breakages. Expanding on these findings, Singh et al. combined fluorescence microscopy with gel electrophoresis as a method to visualise these lesions (Singh et al., 1988). DNA is negatively charged and will migrate towards the anode during gel electrophoresis. However, if DNA contains no breakages, the tight supercoils and relatively large structure will prevent migration. Breakages in the DNA will cause the supercoils to relax and exposed fragmented ends will permit increased migration. Further method development was conducted for single-strand DNA breaks in 1990 by Olive., et al. and the term "comet assay" was introduced (Beedanagari et al., 2014; Olive et al., 1990; Olive & Banáth, 2006). Olive et al., continued to develop an alternative version of the comet assay, specific to double-strand DNA breaks in 1991 (Olive et al., 1991). A variety of adaptations have been made to the protocol (summarised in (Collins, 2004; Olive & Banáth, 2006) to increase the method applications.

The comet assay has become a standard procedure for measuring DNA damage in a wide variety of organisms, including invertebrates. The OECD guidelines published the procedure for the *in vivo* mammalian alkaline comet assay in 2014 (OECD, 2014b, b) updated in 2016 (OECD, 2016). Very few cells are required for analysis and the assay can be conducted in any cells that can be prepared as a single suspension either *in vivo* or *in vitro*. The comet assay has been applied to a wider range of invertebrates than the MN assay (see Table 15.2) with a large proportion of studies being on the mussel in aquatic ecosystems, likely due to reasons listed previously (see Sect. 15.2.1). Earthworms are a common terrestrial invertebrate used for the assay. They are frequently used as bioindicators of soil quality and have been adopted as model organisms in toxicology assessments (Lee et al., 2008; OECD, 1984a, b).

The comet assay is appealing because results can be visualised by eye or by using computer software packages such as OpenComet (Gyori et al., 2014) and (Końca et al., 2003) which have been developed to aid analysis. Like the MN test, the Comet assay is usually calibrated using ionising radiation to induce a known amount of DNA damage, thus establishing its use as a biomarker for radiation studies. Standard curves have been generated for a range of radiation sources including alpha, gamma and X-rays (Pitozzi et al., 2006; Rössler et al., 2006). There are a

Table 15.2 Search results from Web of science for comet assay (or test) and invertebrates (2012–2018)

Type	Stressor	Species	Ecosystem type	References
Abiotic	Depleted oxygen levels	Snail *Haliotis midae*	Marine	Vosloo et al. (2013)
Chemical	Arsenic	Earthworm *Lumbricus castaneous Dendrodrilus rubidus*	Terrestrial	Button et al. (2012)
Chemical	Vitamin B12	Waterflea *Daphnia magna*	Freshwater	Kusari et al. (2017)
Chemical	Methotrexate	Waterflea *D. magna*	Freshwater	Kusari et al. (2017)
Chemical	2, 4-dinitroanisole and 2, 4, 6-trinitrotoluene	Water flea *D. carinata*	Freshwater	Prasath et al. (2016)
Chemical	Psychotropic drugs and antibiotics	Mussel *M.edulis*	Marine	Lacaze et al. (2015)
Chemical	Dechlorane Plus	Earthworm *E. fetida*	Terrestrial	Zhang et al. (2014)
Fungicide	Chlorothalonil	Mussels *Perna perna*	Marine	Guerreiro et al. (2017)
Fungicide	Carbendazim	Clam *D. faba*	Marine	Devi et al. (2013)
Heavy metal	Lead nitrate	Worm *Perinereis cultrifera*	Marine	Singh et al. (2017)
Heavy metal	Cobalt chloride	Worm Perinereis cultrifera	Marine	Singh et al. (2017)
Heavy metal	Mercury	Amphipod *Gammarus elvirae*	Freshwater	Di Donato et al. (2016)
Heavy metal	Lead	Amphipod *Gammarus elvirae,* Earthworm *E. fetida*	Freshwater, Terrestrial	Di Donato et al. (2016) and Wu et al. (2012)
Heavy metal	Copper	Polyp *Hydra magnipapillata*	Freshwater	Zeeshan et al. (2016)
Nanomaterial	Nanomaterial	Earthworm *Eisenia Andrei*	Terrestrial	Correia et al. (2017)
Heavy metal	Gypsum mine water	Amphipod *G. balcanicus*	Freshwater	Ternjej et al. (2014)
Heavy metal	Chromium	Earthworm *Eisenia fetida*	Terrestrial	Sforzini et al. (2017)
Heavy metal	Cadmium	Earthworm *Eisenia fetida*	Terrestrial	Wu et al. (2012)
Hydrocarbon	Benzo(a) pyrene	Mussel *Mytilus galloprovincialis* Shrimp *Fusitriton brasiliensis,* Earthworm *Eisenia fetida*	Marine, terrestrial	Di et al. (2017) and Rocha et al. (2012)

(continued)

Table 15.2 (continued)

Type	Stressor	Species	Ecosystem type	References
Hydrocarbons	Oil	Amphipod Gammaridea	Freshwater	Weber et al. (2013)
Hydrocarbons	Diesel Oil	Mussel *M.galloprovincialis*	Marine	Martinović et al. (2015)
Insecticide	Imidacloprid	Earthworm *E. fetida*	Terrestrial	Zhang et al. (2017b)
Insecticide	Spirotetramat	Earthworm *E. fetida*	Terrestrial	Zhang et al. (2015)
Nanomaterial	Silver oxide nanoparticles	Clam *Macoma balthica*	Marine	Dai et al. (2013)
Nanomaterial	Lipid-coated CdSe/ZnS quantum dots and CdCl2	Earthworm *E. fetida,* Ragworm *Hediste diversicolor*	Terrestrial, Marine	Saez et al. (2015)
Nanomaterial	Copper oxide nanoparticles	Clam *Macoma balthica,* Mollusc *Scrobicularia plana,* Worm *Hediste diversicolor*	Marine	Dai et al. (2013) and Buffet et al. (2013)
Nanomaterial	Carbon nanotubes	Snail *Lymnaea luteola*	Freshwater	Ali et al. (2015)
Peptide	Melittin	Waterflea *D. magna*	Freshwater	Galdiero et al. (2015)
Pesticide	Chlorpyrifos	Clam *Donax faba*	Marine	Devi et al. (2013)
Pesticide	Vinclozolin	Harlequin fly *Chironomus riparius,* Non-biting midge *Diptera, Chironomidae*	Terrestrial	Aquilino et al. (2018)
Radioactive isotope	Cobalt 60	Mussel *M.galloprovincialis*	Marine	Di et al. (2017)
Radioactive isotope	Tritiated water at an elevated temperature	Mussel *M.galloprovincialis*	Marine	Dallas et al. (2016)
Radiation	Gamma radiation	Bivalves *P. malabarica, M. casta*	Freshwater	Kumar et al. (2014)
Radiation	UV	Coral *Seriatopora hystrix,* Worm *Botryllus schlosseri*	Marine	Svanfeldt et al. (2014)

number of issues associated with the comet assay. Similarly to the MN test, the preparation of cell suspensions needs to be adapted depending on the organism. For example, Pellegri *et al* tested three different methods to extract a viable sample of haemolymph from *Daphnia* using disintegration by mincing, extraction using a pipette and the use of an amalgamator to shock the samples. They found that the rapid shock method proved to be the most effective (Pellegri et al., 2014).

The Comet assay measures gross DNA damage and is therefore not unique to radiation. Although, application of a modified assay for double-strand break detection can reduce uncertainty as it is more radiation specific (Olive, 1998; Ward, 1981). It is not possible to determine the size of each DNA fragment or the category of the lesion. Single cell suspensions are not necessarily easy to obtain from all organisms. Additionally, selection of the comets during analysis may cause bias even when carried out using computer-based software (Collins, 2004; Olive & Banáth, 2006). There are limits in the range of radiation doses that can be applied, with saturation at the upper dose limits (Collins et al., 2008). Taking into account variability between different studies (for example due to differences in methodology), Dhawan et al., suggest that the dose range for detectable damage is between 0.2 and 10 Gy (Dhawan et al., 2009).

Furthermore, DNA damage repair mechanisms alter the results in a time-dependent manner. That is, the more time between administering the radiation dose and applying the comet assay, the more DNA damage repair will have taken place (Mozdarni et al., 2007; Wang et al., 2013). Therefore, the assay should be conducted as close as possible to when the damage has been incurred (Beedanagari et al., 2014). This is not always possible in the field, so some studies have adapted the protocol to process and stabilise samples on site, examples include (Akcha et al., 2004; Hartl et al., 2010; León et al., 2007). Overall, the comet assay is a useful tool for assessing DNA damage in individual cells. However, the test should be carried out shortly after exposure to the stressor of interest, or adapted in order to process and stabilise the samples of interest. It is recommended that the test is applied as part of a panel of biomarkers in order to obtain information on the effects of the DNA damage.

15.2.3 Cytochrome p450

Cytochrome P450 (CYP) enzymes are a major class of detoxification enzyme, responsible for the metabolism of a variety of endogenous and exogenous substances. Phylogenetic analysis of the enzyme was completed, revealing the ancestry of the Cytochrome P450 family going back to the divergence of prokaryotes and eukaryotes (Nelson & Strobel, 1987). The CYP enzymes are highly conserved across species and the increasing use of sequencing technologies has resulted in the identification of further variants (Parvez et al., 2016).

CYP450 enzymes are useful biomarkers because the levels within an organism vary in response to different stressors (Hyne & Maher, 2003; Rendic & Guengerich, 2010). Studies have demonstrated that expression of the enzyme is extremely sensitive and can vary according to dietary intake, between tissue types and at different developmental stages. Research into the impact of ionising radiation on CYP activity has mainly focused on humans, as variation in enzyme availability may influence the way different drugs work (Rendic & Guengerich, 2012). Increased oxidative stress caused by ionising radiation would likely result in upregulation of the enzyme and therefore, enzyme activity may be a useful biomarker in radiation biology.

A lot of work has been conducted on vertebrate species, however, very few methods are transferable to invertebrates due to limitations such as organism size which restricts tissue specific applications, meaning that whole body measurements need to be made. In addition, invertebrates generally have lower P450 levels compared to vertebrates, making it harder to detect (Rendic & Guengerich, 2012). Due to increases in the use of insecticides, CYP enzymes in insects became a popular research area and are now, therefore, better understood (Scott et al., 1998; Scott & Wen, 2001). A number of detection methods have been identified for insects, including the use of degenerate primers, protein purification and sequencing technologies. Less is known about the activity and expression of CYP enzymes in aquatic invertebrates (Snyder, 2000). To fill the data gap for aquatic invertebrates, research has utilised previous knowledge established from insect and vertebrate studies and the increasing investment in sequencing technologies is improving our understanding in a range of aquatic invertebrates (reviewed in Rewitz et al., 2006).

Despite these limitations, there are still a high volume of studies using aquatic invertebrates which suggests that more investment in P450 work has been done since the reviews and suggestions by Snyder in 2000 and Rewitz et al., in 2006. We can see from Table 15.3, that some of the more recent papers for aquatic species are focusing on method development and analysis of new sequences.

15.2.4 Oxidative Stress

One common marker in the study of invertebrate response to environmental stimuli, as well as human neurodegenerative and cardiovascular diseases and cancer is the production or accumulation of reactive oxygen species (ROS) which can lead to a number of different effects including DNA damage (Barzilai and Yamamoto, 2004). The oxidative stress process stems from ROS generated in multiple cellular compartments and enzymes within the cell, including proteins within the plasma membrane, such as the family of NADPH oxidases (Lambeth, 2004), lipid metabolism within the peroxisomes and various cytosolic enzymes such as cyclooxygenases. The majority of cellular ROS is due to mitochondrial activity because of aerobic metabolism and ATP production in the mitochondria (Balaban et al., 2005).

It is known that exogenous sources such as ionising radiation and environmental pollutants contribute to oxidative stress production and researchers in radiation biology have exploited this phenomenon to establish “oxidative stress” as a biomarker technique. While this allows us to set “benchmarks” with levels of damage an organism has experienced from an external stimulus, it may not be specific enough to give us definitive answers as to the effects of cellular damage through environmental stimuli due to the presence of pro- and antioxidant systems (Scholze et al., 2016).

Table 15.3 Cytochrome p450 invertebrate studies found on Web of Science (2012–2018)

Type	Stressor	Species	Ecosystem type	References
Abiotic	Temperature	Sea urchin	Marine	Vergara-Amado (2017)
Chemical	Triclosan (TCS) and Triclocarban (TCC)	Rotifer *Brachionus koreanus*	Freshwater	Han et al. (2016)
Chemical	Pharmaceuticals	Mussel *M.galloprovincialis*	Marine	Gonzalez-Rey et al. (2014)
Fungicide	Azole	Isopod *Gammarus pulex*	Freshwater	Rösch et al. (2017)
Herbicide	Atrazine	Midge *Chironomus tentans*	Freshwater	Tang et al. (2017)
Heavy metal	Cadmium	Earthworm *E. fetida, Fly Chironomus dilutus,* scallop *Mizuhopecten yessoensis*	Terrestrial, freshwater, marine	Cao et al. (2017), Chen et al. (2016) and Meng et al. (2013)
Heavy metal	Copper	Earthworm *E. fetida*	Terrestrial	Cao et al. (2017)
Heavy metal	Lead	Earthworm *E. fetida*	Terrestrial	Cao et al. (2017)
Heavy metal	Zinc	Earthworm *E. fetida*	Terrestrial	Cao et al. (2017)
Hydrocarbon	Oil	Oyster *Crassostrea virginica*	Marine	Jenny et al. (2016)
Hydrocarbon	Benzo (a) pyrene	Shrimp *Litopenaeus vannamei,* Copepod *T.japonicus, Litopenaeus vannamei,* Mollusc *Chlamys farreri,* rotifer *B. koreanus*	Marine	Han et al. (2017a), Ren et al. (2014), Tian et al. (2014) and Kim et al. (2013)
Hydrocarbon	BPA-, BDE-47, and WAF	Copepod *T. japonicus, Paracylopina nana*	Marine	Han et al. (2017b)
Radiation	UV	Copepod *T. japonicus*	Marine	Puthumana et al. (2017)

In human studies, various molecules are currently used as oxidative stress biomarkers and they are measured, depending on their origin (e.g., serum, plasma, urine or tissue) or chemistry i.e. ROS-induced modification of specific molecules, using a variety of biochemical and/or molecular biology techniques. (Ho et al., 2013; Frijhoff et al., 2015). Most of these studies have been transferred to invertebrates and rely on simpler, though robust methodologies adapted from ROS studies in humans. These include chemiluminescence, fluorescence and colourimetric assays (Table 15.4).

Chemiluminescent methods are common methods for studying ROS production. The two commonly used probes in invertebrate biology are **luminol** and **lucigenin**, though other luminogenic substrates exist (Table 15.4). It is the oxidation of the luminogenic substrate by ROS that produces light which can be detected using

Table 15.4 Various methods to study ROS production for invertebrates

Type	Pollutant	Chemical	Organisms tested	References
Chemiluminescent	Heavy metals, organic contaminants, pesticides and tributyltin compounds	Luminol	Mussel *M. galloprovincialis* Clams *Venus verrucosa* & *Callista chione*	Cotou et al. (2013)
	Dibutyl phthalate (DBP)	Lucigenin	Annelid Worm *Galeolaria caespitose*	Lu et al. (2017)
Fluorescent dye	Chlorothalonil Dispersed oil Microcystin-LR Copper Antidepressants	DCF-DA	Moth *Trichoplusia ni* Mussels *Perna perna* Earthworms *E.fetida* Oyster *Crassostrea virginica* Nematode/ roundworm *Caenorhabditis elegans* *Hydra magnipapillata* Abalone *Haliotis tuberculata*	Luan et al. (2017), Guerreiro et al. (2017), Liu et al. (2017), Vignier et al. (2017), Augusti et al. (2017), Zeeshan et al. (2016) and Minguez et al. (2014)
	Chromium	DHR/DHE	Earthworms *E.hortensis* Mudflat bloodworm *Glycera dibranchiate* Earthworms *Eisenia fetida*	Tumminello and Fuller-Espie (2013), Joyner-Matos et al. (2010) and Sforzini et al. (2017)
	Chromium	TEMPO-9-AC	Earthworms *Eisenia fetida*	Sforzini et al. (2017)
	Chromium	BODIPY 581/591 C11	Earthworms *Eisenia fetida*	Sforzini et al. (2017)
	Temperature	NDA-GSH	White shrimp *L.vannamei*	De Souza et al. (2016)
	Silver nanoparticles	CDNB-GSH	Earthworm *E.fetida*	
	Temperature	NDA-γ-GC	White shrimp *L.vannamei*	De Souza et al. (2016)
	Temperature Silver nanoparticles Temperature	TBARS	White shrimp *L. vannamei* Earthworm *E. fetida* Mud crab (*Scylla serrata*)	De Souza et al., (2016), and Paital and Chainy (2014)

(continued)

Table 15.4 (continued)

Type	Pollutant	Chemical	Organisms tested	References
Colourimetric	Silver nanoparticles	NBT	Penaeid shrimp (*Penaeus vannamei*) Earthworm *E. fetida*	Munoz et al. (2000)
Decrease in UV	Silver nanoparticles	H_2O_2 decomposition	Earthworm *E. fetida*	

either a luminometer or an appropriate plate reader. This method is sensitive (sub-picomolar range), safe, rapid and simple to perform, with the light produced for a long time and so facilitating measurement (Kricka, 1991; Chen et al., 2011). This approach is complemented by the development of fluorescent probes.

Fluorescent methods are currently the most popular means of tracking intracellular ROS in various cell types. However, they are not without controversy, not just in the names/types of very similar chemical compounds, but also in their usage in different culture conditions (Gomes et al., 2005; Wardman, 2007; Yazdani, 2015). Therefore, to use these compounds with invertebrates, one has to determine the individual fluorescent probes empirically. **Dichlorofluorescin diacetate (DCFH-DA or the oxidised version, dichlorofluorescein diacetate DCF-DA)** is a useful probe for ROS production study in invertebrates (Table 15.3). It is membrane permeable and once within cells, it is cleaved by esterases to release the DCFH component which is oxidised by free radical compounds to form the fluorescent DCF. However, as there are multiple pathways in ROS production which produce free radical compounds that can lead to DCF fluorescence. Furthermore, there is a possibility of DCF leaking out of cells where they were generated and reacting with extracellular oxidants through modified versions of DCFH-DA are available which contain modification to prevent this from happening (Royall & Ischiropoulos, 1993; Ubezio & Civoli, 1994; Thornber et al., 2009). This may impact on the desire to dissect the ROS pathway in invertebrates during exposure of ionising radiation. Therefore, this probe may be more suitable for qualitative, not quantitative assessment of ROS production (Tarpey and Fridovich, 2001).

The similar chemical compound alternatives, **Dihydrorhodamine (DHR)** and **dihydroethidium (DHE)**, are also membrane permeable and oxidise by free radicals to form the fluorescent products, rhodamine and ethidium, respectively (Gomes et al., 2005). This has also been used successfully in invertebrates (Maria & Bebianno, 2011; Pragya et al., 2014) (Table 15.4). Other alternatives are **4-((9-acridinecarbonyl) amino)-2,2,6,6-tetramethylpiperidine-1-oxyl (TEMPO-9-AC)** and boron-dipyrromethene (C11-BODIPY or BODIPY C11) which do not react directly with hydroxyl and therefore detects ROS indirectly. TEMPO-9-AC measures the effects of hydroxyl radicals on other molecules which generate thiyl radicals that oxidise TEMPO-9-AC to produce the fluorescent acridine-piperidine (Borisenko et al., 2004; Cohn et al., 2008). Likewise, BODIPY C11 measures antioxidant activity during lipid oxidation by the loss (not gain, like previous dyes) of fluorescence upon interaction with peroxyl radicals (Naguib, 1998; Pap et al., 1999). Closely similar to BODIPY C11, thiobarbituric acid reactive substances (TBARS) measures lipid peroxidation during the production of reactive oxygen species. One of the key by-products of this process is the production of malondialdehyde (MDA) produced through lipid hydroperoxidases which is measured using the TBARS assay, as used in invertebrate study (Bouskill et al., 2006; Akbulut et al., 2014; Lewis et al., 2016) (Table 15.4).

Finally, 2,3-naphthalene carboxaldehyde (NDA) is another compound used to indirectly measure ROS, during the production of protein mixed disulfides (Protein-S-SG) by the combined activities of glutathione (GSH) and glutamate-cysteine ligase (GCL). NDA forms a fluorescent product through the action of GSH or GCL activity with the addition of γ-glutamylcysteine γ-GC). Hence, it is through the addition of exogenous γ-GC that one is able – by the fluorescence appearance of NDA – distinguish the enzymatic functions of GCL from GSH, which acts as a baseline readout (White et al., 2003; Geihs et al., 2010).

15.3 Ecological Indicators

Reproduction is a key life-history event for all organisms and therefore measuring alterations to reproduction are commonly used to investigate stress to organisms. It is well known that radiation has negative impacts on somatic cells; the most radiosensitive are germ cells being those with a high mitotic rate and long mitotic future (i.e. undergo many divisions) (Sazykina & Kryshev, 2006). The use of a combination of different reproductive endpoints is an informative way of evaluating impacts to both individuals and extrapolating to a population level (Lance et al., 2012). Reproductive measurements, such as those mentioned below are typically conducted in a laboratory but can also be conducted in the field or using individuals removed from the field. Invertebrates lend themselves to reproductive endpoint studies as they have the valuable advantage of relatively short life spans and generation times. Short generation times enable multiple experiments and the potential for **transgenerational** studies, which are proving important in understanding the long-term consequences of living in radiologically contaminated environments (Buisset-Goussen et al., 2014). However, there are few studies investigating the effects of radiation on reproduction in invertebrates at environmentally relevant dose rates. The majority of existing effects data comes from pest management, where invertebrates are subjected to high acute doses of radiation to induce non-lethal sterilisation for use in pest population control (Bakri et al., 2005; Dyck et al., 2005).

Commonly explored reproductive endpoints include measurements of **fecundity**; brood size and timing of reproduction are commonly measured in ecotoxicology in both aquatic and terrestrial invertebrates. Reproductive endpoints have been measured in studies using ionising radiation exposure (Buisset-Goussen et al., 2014; Gilbin et al., 2008; Parisot et al., 2015; Sarapultseva & Dubrova, 2016) and other stressors such as metals (Aránguiz-Acuña & Pérez-Portilla, 2017; Biandolino et al., 2018) and pesticides (Leitão et al., 2014; Whitehorn et al., 2012).

Fertilisation success is an important reproductive endpoint because it is sensitive and known to exhibit a measurable dose-response relationship with heavy metals (Woods et al., 2016). Fertilisation success is ecologically relevant because it has a direct impact on population dynamics (Reichelt-Brushett & Hudspith, 2016). The method of fertilisation depends on the life history of the organism and invertebrates have a range of different ways in which fertilisation occurs. Spermiotoxicity is typically assessed by adding the exposed sperm to eggs and measuring fertilisation success and hatching. There are many ways in which fertilisation success can be measured by proxy and on a large scale, an overview of some of the techniques available are listed below.

Sperm quality is affected by multiple factors and has been measured in a range of different ways in ecotoxicological studies and links very closely with fertilisation success. Fertilisation success partly depends on the quality of sperm motility and viability, which can be measured through DNA integrity, acrosomal integrity, mitochondrial activity and degree of reactive oxygen species as well as sperm density, egg quality and sperm-egg compatibility (Favret & Lynn, 2010; Vignier et al., 2017). Dividing cells are most sensitive to radiation and generally, earlier stages of spermatogenesis are more radiosensitive than the later stages (Bakri et al., 2005). IR has been known to alter sperm viability, one proposed mechanism being associated with the increase in reactive oxygen species concentrations which can account for the observed structural and functional abnormalities (Lavranos et al., 2012). The use of sub-lethal acute radiation exposure to sterilising pest insects has been used successfully implemented by on a large scale and illustrates the evidence of dose-response for IR (Bakri et al., 2005).

Sperm viability assessments using fluorescent dyes are a popular and fast method which can be used to test a range of parameters related to sperm quality. Analysis after staining has commonly been conducted using fluorescence or confocal microscopy, flow cytometry. A new technique which has been recently proposed and adapted from human studies is spectrofluorometric analysis (Gallo et al., 2018). Confocal or fluorescence microscopy is relatively easy to use to identify fluorescence signals in individual cells but it is slower and a few cells from each sample are assessed. Flow cytometry is fast and allows the evaluation of multiple parameters, however, the flow cytometer is an expensive instrument to buy and maintain and requires skilled operators. Development of spectrofluorometric analysis for use in invertebrates has been proposed as a faster analysis of sperm viability parameters after staining but without the expense of the expensive flow cytometer (Gallo et al., 2018).

Live dead staining is a common indicator for sperm viability and has been used in a range of organisms, including invertebrates for exposure to a range of stressors (Lewis & Ford, 2012) including pesticides (Chaimanee et al., 2016; Straub et al., 2016), immune challenges (Simmons, 2012). Sperm is stained SYBR®-14 dye which is a membrane permanent nucleic acid stain used to label live cells with intact membranes and fluoresces bright green. Propidium iodide (PI) dye labels cells with damaged cell membranes and fluoresces red (Favret & Lynn, 2010; Vignier et al., 2017).

Mitochondrial membrane permeability is an important parameter as the change could lead to reduced fertilisation success if the plasma membrane fails to fuse with the egg. Another parameter which could be altered by stress is the timing of the acrosome reaction, which under normal conditions in induced and binds to plasma membrane thus releasing enzymatic contents which help the sperm penetrate the egg (Favret & Lynn, 2010). Previously developed for humans, these are both sperm parameters which can be measured using flow cytometry in addition to live dead staining Mitochondrial function can be assessed using MitoTracker® Red CMXRos which stains mitochondria in live cells with accumulation of dye in each cell being dependant on membrane potential (Favret & Lynn, 2010; Lewis & Ford, 2012).

The acrosomal reaction is measured by positive staining with fluorescein isothiocyanate (FITC) conjugated *Arachis hypogaea* (peanut) lectin which binds to the acrosome in acrosome-reacted sperm and fluoresce green (Espinoza et al., 2009; Favret & Lynn, 2010; Kekäläinen et al., 2015; Li et al., 2016). Sperm chromatin state indicating DNA integrity in the sperm can be measured using fluorescence (Bonde, 2002; Li et al., 2016).

Many ecotoxicological studies have focussed on external fertilisation, whereby the sperm and the eggs are released into the environment which is typical and specialised to many marine invertebrates, named "broadcast spawners" (Lewis & Ford, 2012). This mechanism of reproduction offers advantages for ecotoxicological studies as it enables a relatively quick and easy way of establishing damage by stress to many individuals or even populations. An alternative bioassay, "Sperm Accumulated against Surface" (SAAS) would allow a quick and simple proxy measurement of fertilisation success as motile but not immotile spawn from broadcast spawners often accumulates at the surface. The use of SAAS as an indicator could provide rapid measurements to be made for a range of organisms (Falkenberg et al., 2016).

Ecotoxicological studies using broadcast spawners, therefore, dominate the literature on fertilisation success and sperm quality for exposure to a range of stressors (Lewis & Ford, 2012). The majority of studies have used Cnidaria, Echinodermata, Polychaeta, Mollusca, Crustacea, Urochordata when exposed to trace metals (Hudspith et al., 2017; Reichelt-Brushett & Hudspith, 2016), crude oil (Vignier et al., 2017), pharmaceuticals (Cunha et al., 2017; Mohd Zanuri et al., 2017).

15.3.1 *Behaviour*

Behavioural responses to stress have received less attention than developmental or reproduction, invertebrate species offer investigations into subtle alterations to behaviour which have been shown to be a sensitive indicator to sublethal stress (Faimali et al., 2017). Most examples using behaviour as an ecological indicator of stress for invertebrates have been conducted in aquatic species. A number of studies indicate that swimming behaviour may be altered by pesticides, nanoparticles, bacterial products or other chemicals and could be applied to radiation (Morgana et al., 2016). Crustacea represent the most studied model organism in marine ecotoxicology for behavioural studies, followed by cnidarian larvae and rotifers. However behavioural studies are not limited to these taxa and could be applied to most motile aquatic invertebrates (Faimali et al., 2017). Acute and chronic exposure to stress can adversely affect sensation, perception, cognition, coordination and motor function which can have consequences to feeding, predator avoidance, social relationship and reproductive success which can affect individual survival and population dynamics. Behaviour transcends single levels of biological organisation and links subcellular processes measured to a whole body response (Faimali et al., 2017). Direct mortality measurements of exposure to stressors are not sufficient particularly when environmental levels are at sublethal levels which can cause impacts but do not kill individuals or populations. Sublethal impacts could have further consequences. For example, altered behaviour has successfully been used in conjunction with biomarkers indicating exposure and effect of stress and can be an "early warning" of exposure to stress (Kunce et al., 2017; Morgana et al., 2016; Parolini et al., 2017).

Swimming is a frequently used biomarker for aquatic invertebrates. Alterations to swimming behaviour have been used as an indicator of stress for a range of aquatic organisms including invertebrates. There are four main areas where swimming in an important mechanism; population-level behaviours such as vertical or horizontal migration, predator-prey interactions and feeding (Faimali et al., 2017). Current standard experiment equipment design consists of dishes of exposed individuals monitored by one (for two – dimensional) or two (for three dimensional systems) video camera connected to a computer with software for movement analysis. Recent technological developments mean there are different methods to monitor movement; video tracking systems, manual systems and automated systems (Bownik, 2017; Faimali et al., 2017). Two-dimensional system has been criticised as this cannot encompass all swimming parameters, further criticism arises from the environmental setting. Housing vessels of inappropriate dimensions and the physical conditions of the laboratory set up have been criticised as these can influence swimming parameters (Bownik, 2017). Swimming behaviour has been standardised for some organisms in response to stressors. *Daphnia* is a common planktonic invertebrate and popular model organism used in a range of ecotoxicological studies. *Daphnia* species have specific organs that facilitate them to move in water. Standardised parameters which have been measured include; swimming time, speed, behavioural strength, hopping frequency, horizontal distribution, vertical distribution and migration, the time ratio of vertical to horizontal swimming, distance travelled, trajectory, number of turnings, resting time, sinking rate, gravitaxis, swarming and spinning (Bownik, 2017).

15.4 Considerations

Biomarkers and ecological indicators can be used in conjunction effectively, but there has to be careful thought in selection and interpretation. Much of the existing literature focussing on both general ecotoxicology with different stressors and more specifically, radiation indicates a bias towards aquatic organisms and therefore more research is encouraged for terrestrial organisms. However, there is sufficient literature to draw from when considering biomarkers and indicators and more have been adapted from use in humans and mammals to invertebrates (Sulmon et al., 2015).

Biomarkers and indicators often have very little interspecies variability as shown by the comparison between two freshwater mussel species (Farkas et al., 2017) and comparisons between earthworm species (Spurgeon et al., 2000). Biological variables such as age, sex, reproductive status and general health complicate consistency between biomarkers and indicators (Ippolito et al., 2017). Early life stages of marine organisms are used extensively in ecotoxicological studies as it is generally agreed that they are more sensitive to chemical contaminants than their adult counterparts (Hudspith et al., 2017).

Acknowledgements This work was supported by the Carnegie Trust for the Universities of Scotland [grant number RIG008296] (JL) and the Radioactivity And The Environment (RATE) programme's Transfer-Exposure-Effects (TREE) project, funded by NERC, the Environment Agency and Radioactive Waste Management Ltd. (KR, JG & DC).

References

Akbulut, S., et al. (2014). Cytoprotective effects of amifostine, ascorbic acid and N-acetylcysteine against methotrexate-induced hepatotoxicity in rats. *World Journal of Gastroenterology, 20*(29), 10158–10165.

Akcha, F., Leday, G., & Pfohl-Leszkowicz, A. (2004). Measurement of DNA adducts and strand breaks in dab (*Limanda limanda*) collected in the field: Effects of biotic (age, sex) and abiotic (sampling site and period) factors on the extent of DNA damage. *Mutation Research, 552*(552), 197–207.

Ali, S. M., Yousef, N. M., & Nafady, N. A. (2015). Application of biosynthesized silver nanoparticles for the control of land snail Eobania vermiculata and some plant pathogenic fungi. *Journal of Nanomaterials*.

Aquilino, M., Sánchez-Argüello, P., & Martínez-Guitarte, J. L. (2018). Genotoxic effects of vinclozolin on the aquatic insect Chironomus riparius (Diptera, Chironomidae). *Environmental Pollution, 232*, 563–570.

Aránguiz-Acuña, A., & Pérez-Portilla, P. (2017). Metal stress in zooplankton diapause production: Post-hatching response. *Ecotoxicology, 26*(3), 329–339.

Arkhipchuk, V. V., Goncharuk, V. V., Chernykh, V. P., Maloshtan, L. N., & Gritsenko, I. S. (2004). Use of a complex approach for assessment of metamizole sodium and acetylsalicylic acid toxicity, genotoxicity and cytotoxicity. *Journal of Applied Toxicology: An International Journal, 24*(5), 401–407.

Augusti, P. R., et al. (2017). Microcystin-LR exposure induces oxidative damage in Caenorhabditis elegans: Protective effect of lutein extracted from marigold flowers. *Food and Chemical Toxicology, 109*, 60–67. https://doi.org/10.1016/j.fct.2017.08.045

Bakri, A., et al. (2005). Fifty years of radiation biology in entomology: Lessons learned from IDIDAS. *Annals of the Entomological Society of America, 98*(1), 1–12.
Balaban, R. S., Nemoto, S., & Finkel, T. (2005). Mitochondria, oxidants, and aging. *Cell, 120*(4), 483–495.
Barzilai, A., & Yamamoto, K. (2004). DNA damage responses to oxidative stress. *DNA Repair (Amst), 3*(8–9), 1109–1115.
Bayne, B. L. (1976). *Marine mussels, their ecology and physiology*. Cambridge University Press.
Beedanagari, S., et al. (2014). Genotoxicity biomarkers. In *Biomarkers in toxicology* (pp. 729–742). Elsevier.
Biandolino, F., et al. (2018). Effects of short- and long-term exposures to copper on lethal and reproductive endpoints of the harpacticoid copepod Tigriopus fulvus. *Ecotoxicology and Environmental Safety, 147*(February 2017), 327–333.
Binelli, A., Cogni, D., Parolini, M., Riva, C., & Provini, A. (2009). In vivo experiments for the evaluation of genotoxic and cytotoxic effects of Triclosan in Zebra mussel hemocytes. *Aquatic Toxicology, 91*(3), 238–244.
Bonde, J. P. (2002). Sperm count and chromatin structure in men exposed to inorganic lead: Lowest adverse effect levels. *Occupational and Environmental Medicine, 59*(4), 234–242.
Borisenko, G. G., et al. (2004). Glutathione propagates oxidative stress triggered by myeloperoxidase in HL-60 cells. Evidence for glutathionyl radical-induced peroxidation of phospholipids and cytotoxicity. *The Journal of Biological Chemistry, 279*(22), 23453–23462.
Bouskill, N. J., et al. (2006). Differentiating copper and arsenic toxicity using biochemical biomarkers in Asellus aquaticus and Dreissena polymorpha. *Ecotoxicology and Environmental Safety, 65*(3), 342–349.
Bownik, A. (2017). Daphnia swimming behaviour as a biomarker in toxicity assessment: A review. *Science of the Total Environment, 601–602*, 194–205.
Buffet, P. E., Richard, M., Caupos, F., Vergnoux, A., Perrein-Ettajani, H., Luna-Acosta, A., Akcha, F., Amiard, J. C., Amiard-Triquet, C., Guibbolini, M., & Risso-De Faverney, C. (2013). A mesocosm study of fate and effects of CuO nanoparticles on endobenthic species (Scrobicularia plana, Hediste diversicolor). *Environmental Science & Technology, 47*(3), 1620–1628.
Buisset-Goussen, A., et al. (2014). Effects of chronic gamma irradiation: A multigenerational study using Caenorhabditis elegans. *Journal of Environmental Radioactivity, 137*, 190–197.
Button, M., Koch, I., & Reimer, K. J. (2012). Arsenic resistance and cycling in earthworms residing at a former gold mine in Canada. *Environmental Pollution, 169*, 74–80.
Cajaraville, M. P., et al. (2000). The use of biomarkers to assess the impact of pollution in coastal environments of the Iberian Peninsula: A practical approach. *Science of the Total Environment, 247*(2–3), 295–311.
Calisi, A., et al. (2016). Multibiomarker response in the earthworm *Eisenia fetida* as tool for assessing multi-walled carbon nanotube ecotoxicity. *Ecotoxicology, 25*(4), 677–687.
Cao, X., Bi, R., & Song, Y. (2017). Toxic responses of cytochrome P450 sub-enzyme activities to heavy metals exposure in soil and correlation with their bioaccumulation in Eisenia fetida. *Ecotoxicology and Environmental Safety, 144*, 158–165.
Chaimanee, V., et al. (2016). Sperm viability and gene expression in honey bee queens (*Apis mellifera*) following exposure to the neonicotinoid insecticide imidacloprid and the organophosphate acaricide coumaphos. *Journal of Insect Physiology, 89*, 1–8.
Chen, X., Tian, X., Shin, I., & Yoon, J. (2011). Fluorescent and luminescent probes for detection of reactive oxygen and nitrogen species. *Chemical Society Reviews, 40*(9), 4783–4804.
Chen, X., Li, H., Zhang, J., Ding, Y., & You, J. (2016). Does cadmium affect the toxicokinetics of permethrin in Chironomus dilutus at sublethal level? Evidence of enzymatic activity and gene expression. *Environmental Pollution, 218*, 1005–1013.
Cohn, C. A., Simon, S. R., & Schoonen, M. A. (2008). Comparison of fluorescencebased techniques for the quantification of particle-induced hydroxyl radicals. *Particle and Fibre Toxicology, 5*, 2.
Collins, A. R., et al. (2008). The comet assay: Topical issues. *Mutagenesis, 23*(3), 143–151.
Collins, A. R. (2004). The comet assay for DNA damage and repair: Principles, applications, and limitations. *Molecular Biotechnology, 26*(3), 249–261.

Correia, B., Lourenco, J., Marques, S., Nogueira, V., Gavina, A., da Graça Rasteiro, M., Antunes, F., Mendo, S., & Pereira, R. (2017). Oxidative stress and genotoxicity of an organic and an inorganic nanomaterial to Eisenia andrei: SDS/DDAB nano-vesicles and titanium silicon oxide. *Ecotoxicology and Environmental Safety, 140*, 198–205.

Cotou, E., et al. (2013). Comparative study of biochemical and immunological biomarkers in three marine bivalves exposed at a polluted site. *Environmental Science and Pollution Research International, 20*(3), 812–822. https://doi.org/10.1007/s11356-012-1150-3

Cunha, S. C., Pena, A., & Fernandes, J. O. (2017). Mussels as bioindicators of diclofenac contamination in coastal environments. *Environmental Pollution, 225*, 354–360.

da Silva Rocha, A. J., Santos, T. C. A., Gomes, V., Bícego, M. C., de Albergaria Barbosa, A. C. R., Rocha, M. J. D. A. C., Hasue, F. M., & Van Ngan, P. (2012). Assessment of trophic transfer of benzo (a) pyrene genotoxicity from the post-larval pink shrimp F. brasiliensis to the juvenile Florida pompano T. carolinus. *Environmental Toxicology and Pharmacology, 34*(3), 969–976.

Dai, L., Syberg, K., Banta, G. T., Selck, H., & Forbes, V. E. (2013). Effects, uptake, and depuration kinetics of silver oxide and copper oxide nanoparticles in a marine deposit feeder, Macoma balthica. *ACS Sustainable Chemistry & Engineering, 1*(7), 760–767.

Dallas, L. J., Bean, T. P., Turner, A., Lyons, B. P., & Jha, A. N. (2016). Exposure to tritiated water at an elevated temperature: Genotoxic and transcriptomic effects in marine mussels (M.galloprovincialis). *Journal of Environmental Radioactivity, 164*, 325–336.

de Souza, D. M., et al. (2016). Antioxidant enzyme activities and immunological system analysis of Litopenaeus vannamei reared in biofloc technology (BFT) at different water temperatures. *Aquaculture, 451*, 436–443.

Devi, Y., & Mishra, A. (2013). Histopathological alterations in gill and liver Anotomy of fresh water, air breathing fish Channa Punctatus after pesticide Hilban®(Chlorpyrifos) treatment. *Advances in Bioresearch, 4*(2).

Dhawan, A., et al. (2009). *The comet assay in toxicology*. Royal Society of Chemistry.

Di, Y., Aminot, Y., Schroeder, D. C., Readman, J. W., & Jha, A. N. (2017). Integrated biological responses and tissue-specific expression of p53 and ras genes in marine mussels following exposure to benzo (α) pyrene and C60 fullerenes, either alone or in combination. *Mutagenesis, 32*(1), 77–90.

Dixon, D. R. J. (2002). Marine invertebrate eco-genotoxicology: A methodological overview. *Mutagenesis, 17*(6), 495–507.

Di Donato, G., De Matthaeis, E., Ronci, L., & Setini, A. (2016). Genotoxicity biomarkers in the amphipod Gammarus elvirae exposed in vivo to mercury and lead, and basal levels of DNA damage in two cell types. *Chemistry and Ecology, 32*(9), 843–857.

Doherty, A. T. (2012). The in vitro micronucleus assay. *Methods in Molecular Biology, 817*, 121–141.

Dyck, V.A., Hendrichs, J. & Robinson, A.S., 2005. Sterile insect technique: Principles and practice in area-wide integrated pest management,

Espinoza, J. A., et al. (2009). Integrity of mitochondrial membrane potential reflects human sperm quality. *Andrologia, 41*(1), 51–54.

Evans, H. J., Neary, G. J., & Williamson, F. S. (1959). The relative biological efficiency of single doses of fast neutrons and gamma-rays on Vicia Faba roots and the effect of oxygen. *International Journal of Radiation Biology and Related Studies in Physics, Chemistry and Medicine, 1*(3), 216–229.

Faimali, M., et al. (2017). Old model organisms and new behavioral end-points: Swimming alteration as an ecotoxicological response. *Marine Environmental Research, 128*, 36–45.

Falkenberg, L. J., Havenhand, J. N., & Styan, C. A. (2016). Sperm accumulated against surface: A novel alternative bioassay for environmental monitoring. *Marine Environmental Research, 114*, 51–57.

Farkas, A., et al. (2017). Interspecies comparison of selected pollution biomarkers in dreissenid spp. inhabiting pristine and moderately polluted sites. *Science of the Total Environment, 599–600*, 760–770.

Favret, K. P., & Lynn, J. W. (2010). Flow-cytometric analyses of viability biomarkers in pesticide-exposed sperm of three aquatic invertebrates. *Archives of Environmental Contamination and Toxicology, 58*(4), 973–984.

Fenech, M. (1993). The cytokinesis-block micronucleus technique and its application to genotoxicity studies in human populations. In *Environmental health perspectives* (pp. 101–107). National Institute of Environmental Health Science.

Fenech, M. (2000). The in vitro micronucleus technique. *Mutation Research – Fundamental and Molecular Mechanisms of Mutagenesis, 455*(1–2), 81–95.

Ferlazzo, M. L., Bourguignon, M., & Foray, N. (2017). Functional assays for individual radiosensitivity: A critical review. *Seminars in Radiation Oncology, 27*(4), 310–315.

Frijhoff, J., et al. (2015). Clinical relevance of biomarkers of oxidative stress. *Antioxidants & Redox Signaling, 23*(14), 1144–1170.

Gallo, A., Boni, R., & Tosti, E. (2018). Sperm viability assessment in marine invertebrates by fluorescent staining and spectrofluorimetry: A promising tool for assessing marine pollution impact. *Ecotoxicology and Environmental Safety, 147*, 407–412.

Galdiero, E., Maselli, V., Falanga, A., Gesuele, R., Galdiero, S., Fulgione, D., & Guida, M. (2015). Integrated analysis of the ecotoxicological and genotoxic effects of the antimicrobial peptide melittin on Daphnia magna and Pseudokirchneriella subcapitata. *Environmental Pollution, 203*, 145–152.

Geihs, M. A., et al. (2010). Effect of melatonin in the antioxidant defense system in the locomotor muscles of the estuarine crab Neohelice granulata (Decapoda, Brachyura). *General and Comparative Endocrinology, 166*(1), 72–82.

Gilbin, R., Alonzo, F., & Garnier-Laplace, J. (2008). Effects of chronic external gamma irradiation on growth and reproductive success of *Daphnia magna*. *Journal of Environmental Radioactivity, 99*(1), 134–145.

Gomes, A., Fernandes, E., & Lima, J. L. (2005). Fluorescence probes used for detection of reactive oxygen species. *Journal of Biochemical and Biophysical Methods, 65*(2–3), 45–80. https://doi.org/10.1016/j.jbbm.2005.10.003

Gonzalez-Rey, M., Mattos, J. J., Piazza, C. E., Bainy, A. C. D., & Bebianno, M. J. (2014). Effects of active pharmaceutical ingredients mixtures in mussel Mytilus galloprovincialis. *Aquatic Toxicology, 153*, 12–26.

Guerreiro, A. S., et al. (2017). Antifouling biocides: Impairment of bivalve immune system by chlorothalonil. *Aquatic Toxicology, 189*, 194–199. https://doi.org/10.1016/j.aquatox.2017.06.012

Gyori, B. M., et al. (2014). OpenComet: An automated tool for comet assay image analysis. *Redox Biology, 2*(1), 457–465.

Han, C., Lim, Y. H., & Hong, Y. C. (2016). Ten-year trends in urinary concentrations of triclosan and benzophenone-3 in the general US population from 2003 to 2012. *Environmental Pollution, 208*, 803–810.

Han, J., Kim, D. H., Seo, J. S., Kim, I. C., Nelson, D. R., Puthumana, J., & Lee, J. S. (2017a). Assessing the identity and expression level of the cytochrome P450 20A1 (CYP20A1) gene in the BPA-, BDE-47, and WAF-exposed copepods Tigriopus japonicas and Paracyclopina nana. *Comparative Biochemistry and Physiology Part C: Toxicology & Pharmacology, 193*, 42–49.

Han, S., Wang, B., Wang, M., Liu, Q., Zhao, W., & Wang, L. (2017b). Effects of ammonia and nitrite accumulation on the survival and growth performance of white shrimp Litopenaeus vannamei. *Invertebrate Survival Journal, 14*(1), 221–232.

Hartl, M. G. J., Grigson, S. J. W., & Sinet, E. (2010). Maintenance of bivalve hemocytes for the purpose of delayed DNA strand break assessment using the comet assay. *Environmental and Molecular Mutagenesis, 51*(1), 64–68.

Hayashi, M. (2016). The micronucleus test—Most widely used in vivo genotoxicity test—. *Genes and Environment, 38*(1), 18.

Ho, E., Karimi Galougahi, K., Liu, C. C., Bhindi, R., & Figtree, G. A. (2013). Biological markers of oxidative stress: Applications to cardiovascular research and practice. *Redox Biology, 1*, 483–491.

Hose, J. E. (1985). Potential uses of sea urchin embryos for identifying toxic chemicals: Description of a bioassay incorporating cytologic, cytogenetic and embryologic endpoints. *Journal of Applied Toxicology, 5*(4), 245–254.

Hudspith, M., Reichelt-Brushett, A., & Harrison, P. L. (2017). Factors affecting the toxicity of trace metals to fertilization success in broadcast spawning marine invertebrates: A review. *Aquatic Toxicology, 184*, 1–13.

Hyne, R. V., & Maher, W. A. (2003). Invertebrate biomarkers: Links to toxicosis that predict population decline. *Ecotoxicology and Environmental Safety, 54*(3), 366–374.

ICRP. (2008). Environmental protection – The concept and use of reference animals and plants. ICRP Publication 108. *Annals of the ICRP, 38*(4–6).

Ippolito, A., et al. (2017). Natural variability of enzymatic biomarkers in freshwater invertebrates. *Environmental Science and Pollution Research, 24*(1), 732–742.

Jenny, M. J., Walton, W. C., Payton, S. L., Powers, J. M., Findlay, R. H., O'Shields, B., Diggins, K., Pinkerton, M., Porter, D., Crane, D. M., & Tapley, J. (2016). Transcriptomic evaluation of the American oyster, Crassostrea virginica, deployed during the Deepwater horizon oil spill: Evidence of an active hydrocarbon response pathway. *Marine Environmental Research, 120*, 166–181.

Joyner-Matos, J., et al. (2010). Hydrogen sulfide induces oxidative damage to RNA and DNA in a sulfide-tolerant marine invertebrate. *Physiological and Biochemical Zoology, 83*(2), 356–365. https://doi.org/10.1086/597529

Kekäläinen, J., et al. (2015). Lectin staining and flow cytometry reveals female-induced sperm acrosome reaction and surface carbohydrate reorganization. *Scientific Reports, 5*(15321), 1–11.

Kim, R. O., Kim, B. M., Jeong, C. B., Nelson, D. R., Lee, J. S., & Rhee, J. S. (2013). Expression pattern of entire cytochrome P450 genes and response of defensomes in the benzo [a] pyrene-exposed monogonont rotifer Brachionus koreanus. *Environmental Science & Technology, 47*(23), 13804–13812.

Kirsch-Volders, M., et al. (2011). In vitro genotoxicity testing using the micronucleus assay in cell lines, human lymphocytes and 3D human skin models. *Mutagenesis, 26*(1), 177–184.

Knakievicz, T., & Ferreira, H. B. (2008). Evaluation of copper effects upon Girardia tigrina freshwater planarians based on a set of biomarkers. *Chemosphere, 71*(3), 419–428.

Końca, K., et al. (2003). A cross-platform public domain PC image-analysis program for the comet assay. *Mutation Research – Genetic Toxicology and Environmental Mutagenesis, 534*(1–2), 15–20.

Kricka, L. J. (1991). Chemiluminescent and bioluminescent techniques. *Clinical Chemistry, 37*(9), 1472–1481.

Kumar, M. P., Shyama, S. K., Sonaye, B. S., Naik, U. R., Kadam, S. B., Bipin, P. D., D'costa, A., & Chaubey, R. C. (2014). Evaluation of γ-radiation-induced DNA damage in two species of bivalves and their relative sensitivity using comet assay. *Aquatic Toxicology, 150*, 1–8.

Kunce, W., Stoks, R., & Johansson, F. (2017). Single and mixture impacts of two pyrethroids on damselfly predatory behavior and physiological biomarkers. *Aquatic Toxicology, 190*, 70–77.

Kusari, F., O'Doherty, A. M., Hodges, N. J., & Wojewodzic, M. W. (2017). Bi-directional effects of vitamin B12 and methotrexate on Daphnia magna fitness and genomic methylation. *Scientific Reports, 7*(1), 1–9.

Lacaze, E., Pédelucq, J., Fortier, M., Brousseau, P., Auffret, M., Budzinski, H., & Fournier, M. (2015). Genotoxic and immunotoxic potential effects of selected psychotropic drugs and antibiotics on blue mussel (Mytilus edulis) hemocytes. *Environmental Pollution, 202*, 177–186.

Lambeth, J. D. (2004). NOX enzymes and the biology of reactive oxygen. *Nature Reviews. Immunology, 2004*(3), 181–189.

Lance, E., et al. (2012). Modelling population-level consequences of chronic external gamma irradiation in aquatic invertebrates under laboratory conditions. *Science of the Total Environment, 429*, 206–214.

Lavranos, G., et al. (2012). Investigating ROS sources in male infertility: A common end for numerous pathways. *Reproductive Toxicology, 34*(3), 298–307.

Lee, B. T., et al. (2008). Progress in earthworm ecotoxicology. In *Advanced environmental monitoring* (pp. 248–258).

Leitão, S., et al. (2014). Effects of azoxystrobin, chlorothalonil, and ethoprophos on the reproduction of three terrestrial invertebrates using a natural Mediterranean soil. *Applied Soil Ecology, 76*, 124–131.

León, G., et al. (2007). Genotoxic effects in wild rodents (*Rattus rattus* and *Mus musculus*) in an open coal mining area. *Mutation Research/Genetic Toxicology and Environmental Mutagenesis, 630*(1–2), 42–49.

Lewis, C., & Ford, A. T. (2012). Infertility in male aquatic invertebrates: A review. *Aquatic Toxicology, 120–121*, 79–89.

Lewis, C., et al. (2016). Ocean acidification increases copper toxicity differentially in two key marine invertebrates with distinct acid-base responses. *Scientific Reports, 6*, 21554.

Li, N., et al. (2016). Quality decline and oxidative damage in sperm of freshwater crab Sinopotamon henanense exposed to lead. *Ecotoxicology and Environmental Safety, 130*, 193–198.

Liu, T., et al. (2017). Oxidative stress and gene expression of earthworm (Eisenia fetida) to clothianidin. *Ecotoxicology and Environmental Safety, 142*, 489–496. https://doi.org/10.1016/j.ecoenv.2017.04.012

Lu, Y., et al. (2017). Exposure of spermatozoa to dibutyl phthalate induces abnormal embryonic development in a marine invertebrate Galeolaria caespitosa (Polychaeta: Serpulidae). *Aquatic Toxicology, 191*, 189–200. https://doi.org/10.1016/j.aquatox.2017.08.008

Luan, S., et al. (2017). Emamectin benzoate induces ROS-mediated DNA damage and apoptosis in Trichoplusia Tn5B1-4 cells. *Chemico-Biological Interactions, 273*, 90–98. https://doi.org/10.1016/j.cbi.2017.06.004

Maria, V. L., & Bebianno, M. J. (2011). Antioxidant and lipid peroxidation responses in Mytilus galloprovincialis exposed to mixtures of benzo(a)pyrene and copper. *Comparative Biochemistry and Physiology Part C: Toxicology and Pharmacology, 154*(1), 56–63.

Martinović, R., Kolarević, S., Kračun-Kolarević, M., Kostić, J., Marković, S., Gačić, Z., Kljajić, Z., & Vuković-Gačić, B. (2015). Genotoxic potential and heart rate disorders in the Mediterranean mussel Mytilus galloprovincialis exposed to Superdispersant-25 and dispersed diesel oil. *Marine Environmental Research, 108*, 83–90.

Meng, X. L., Liu, M., Jiang, K. Y., Wang, B. J., Tian, X., Sun, S. J., Luo, Z. Y., Qiu, C. W., & Wang, L. (2013). De novo characterization of Japanese scallop Mizuhopecten yessoensis transcriptome and analysis of its gene expression following cadmium exposure. *PLoS One, 8*(5), e64485.

Minguez, L., et al. (2014). Assessment of cytotoxic and immunomodulatory properties of four antidepressants on primary cultures of abalone hemocytes (Haliotis tuberculata). *Aquatic Toxicology, 153*, 3–11. https://doi.org/10.1016/j.aquatox.2013.10.020

Muñoz, M., et al. (2000). Measurement of reactive oxygen intermediate production in haemocytes of the penaeid shrimp, Penaeus vannamei. *Aquaculture., 191*(1–3), 89–107.

Mohd Zanuri, N. B., Bentley, M. G., & Caldwell, G. S. (2017). Assessing the impact of diclofenac, ibuprofen and sildenafil citrate (Viagra®) on the fertilisation biology of broadcast spawning marine invertebrates. *Marine Environmental Research, 127*, 126–136.

Morgana, S., et al. (2016). Swimming speed alteration in the early developmental stages of *Paracentrotus lividus* sea urchin as ecotoxicological endpoint. *Marine Environmental Research, 115*, 11–19.

Mozdarni, H., Nasirian, B., & Haeri, S. A. (2007). In vivo γ-rays induced initial DNA damage and the effect of famotidine in mouse leukocytes as assayed by the alkaline comet assay. *Journal of Radiation Research, 48*(2), 129–134.

Naguib, Y. M. (1998). A fluorometric method for measurement of peroxyl radical scavenging activities of lipophilic antioxidants. *Analytical Biochemistry, 265*(2), 290–298.

National Service Center for Environmental Publications (NSCEP). (2008). *An introduction to freshwater mussels as biological indicators including accounts of interior basin, Cumberlandian, and Atlantic slope species*.

Nelson, D. R., & Strobel, H. W. (1987). Evolution of cytochrome P-450 proteins. *Molecular Biology and Evolution, 4*(6), 572–593.

OECD. (1984a). OECD 207 -Earthworm, Acute Toxicity Tests. *OECD Guideline for Testing of Chemicals, 207*(April), 9.

OECD. (1984b). *OECD guidelines for testing of chemicals: Earthworm acute toxicity test* (OECD Guideline No. 207).

OECD. (2014a). OECD TG487 – In-vitro mammalian cell micronucleus test. *Organisation for Economic Co-operation and Development publishing*, (September), pp. 1–26.

OECD. (1997). *Test no. 473: In vitro mammalian chromosome aberration test*. OECD Publishing.

OECD. (2016). *Test no. 489: In vivo mammalian alkaline comet assay*. OECD Publishing.

OECD. (2014b). *Test no. 489: In vivo mammalian alkaline comet assay, OECD guidelines for the testing of chemicals*. OECD Publishing.

Olive, P. L., et al. (1991). DNA double-strand breaks measured in individual cells subjected to gel electrophoresis. *Cancer Research, 51*(17), 4671–4676.

Olive, P. L. (1998). The role of DNA single- and double-strand breaks in cell killing by ionizing radiation. *Radiation Research, 150*(5), S42–S51.

Olive, P. L., & Banáth, J. P. (2006). The comet assay: A method to measure DNA damage in individual cells. *Nature Protocols, 1*(1), 23–29.

Olive, P. L., Banáth, J. P., & Durand, R. E. (1990). Heterogeneity in radiation-induced DNA damage and repair in tumor and normal cells measured using the comet assay. *Radiation Research, 122*(1), 86–94.

Ostling, O., & Johanson, K. J. (1984). Microelectrophoretic study of radiation-induced DNA damages in individual mammalian cells. *Biochemical and Biophysical Research Communications, 123*(1), 291–298.

Paital, B., & Chainy, G. B. N. (2014). Effects of temperature on complexes I and II mediated respiration, ROS generation and oxidative stress status in isolated gill mitochondria of the mud crab Scylla serrata. *Journal of Thermal Biology, 41*, 104–111. https://doi.org/10.1016/j.jtherbio.2014.02.013

Pap, E. H., et al. (1999). Ratio-fluorescence microscopy of lipid oxidation in living cells using C11-BODIPY(581/591). *FEBS Letters, 453*(3), 278–282.

Parisot, F., et al. (2015). DNA alterations and effects on growth and reproduction in *Daphnia magna* during chronic exposure to gamma radiation over three successive generations. *Aquatic Toxicology, 163*, 27–36.

Parolini, M., & Binelli, A. (2012). Cyto-genotoxic effects induced by three brominated diphenyl ether congeners on the freshwater mussel Dreissena polymorpha. *Ecotoxicology and Environmental Safety, 79*, 247–255.

Parolini, M., et al. (2017). Benzoylecgonine exposure induced oxidative stress and altered swimming behavior and reproduction in *Daphnia magna*. *Environmental Pollution, 232*, 236–244.

Parvez, M., et al. (2016). Molecular evolutionary dynamics of cytochrome P450 monooxygenases across kingdoms: Special focus on mycobacterial P450s. *Scientific Reports, 6*.

Pellegri, V., Gorbi, G., & Buschini, A. (2014). Comet assay on *Daphnia magna* in eco-genotoxicity testing. *Aquatic Toxicology, 155*, 261–268.

Pitozzi, V., et al. (2006). Calibration of the comet assay for the measurement of DNA damage in mammalian cells. *Free Radical Research, 40*(11), 1149–1154.

Pragya, P., et al. (2014). Over-expression of superoxide dismutase ameliorates Cr(VI) induced adverse effects via modulating cellular immune system of Drosophila melanogaster. *PLoS One, 9*(2), e88181.

Prasath, A., Panneerselvan, L., Provatas, A., Naidu, R., & Megharaj, M. (2016). Genotoxicity assessment of acute exposure of 2, 4-dinitroanisole, its metabolites and 2, 4, 6-trinitrotoluene to Daphnia carinata. *Ecotoxicology, 25*(10), 1873–1879.

Prather, C. M., Pelini, S. L., Laws, A., Rivest, E., Woltz, M., Bloch, C. P., Del Toro, I., Ho, C. K., Kominoski, J., Newbold, T. S., & Parsons, S. (2013). Invertebrates, ecosystem services and climate change. *Biological Reviews, 88*(2), 327–348.

Puthumana, J., Lee, M. C., Park, J. C., Kim, H. S., Hwang, D. S., Han, J., & Lee, J. S. (2017). Ultraviolet B radiation induces impaired lifecycle traits and modulates expression of cyto-

chrome P450 (CYP) genes in the copepod Tigriopus japonicus. *Aquatic Toxicology, 184*, 116–122.
Reichelt-Brushett, A., & Hudspith, M. (2016). The effects of metals of emerging concern on the fertilization success of gametes of the tropical scleractinian coral *Platygyra daedalea. Chemosphere, 150*, 398–406.
Ren, X., Pan, L., & Wang, L. (2014). Metabolic enzyme activities, metabolismrelated genes expression and bioaccumulation in juvenile white shrimp Litopenaeus vannamei exposed to benzo [a] pyrene. *Ecotoxicology and Environmental Safety, 104*, 79–86.
Rendic, S., & Guengerich, F. P. (2010). Update information on drug metabolism systems—2009, part II: Summary of information on the effects of diseases and environmental factors on human cytochrome P450 (CYP) enzymes and transporters. *Current Drug Metabolism, 11*(1), 4–84.
Rendic, S., & Guengerich, P. F. (2012). Summary of information on the effects of ionizing and non-ionizing radiation on cytochrome P450 and other drug metabolizing enzymes and transporters. *Current Drug Metabolism, 13*(6), 787–814.
Rewitz, K. F., et al. (2006). Marine invertebrate cytochrome P450: Emerging insights from vertebrate and insect analogies. *Comparative Biochemistry and Physiology – C Toxicology and Pharmacology, 143*(4), 363–381.
Riva, C., Binelli, A., Cogni, D., & Provini, A. (2007). Evaluation of DNA damage induced by decabromodiphenyl ether (BDE-209) in hemocytes of Dreissena polymorpha using the comet and micronucleus assays. *Environmental and Molecular Mutagenesis, 48*(9), 735–743.
Rolke, D., et al. (2016). Large-scale monitoring of effects of clothianidin-dressed oilseed rape seeds on pollinating insects in Northern Germany: Effects on honey bees (*Apis mellifera*). *Ecotoxicology*, 0–1.
Rösch, A., Gottardi, M., Vignet, C., Cedergreen, N., & Hollender, J. (2017). Mechanistic understanding of the synergistic potential of azole fungicides in the aquatic invertebrate Gammarus pulex. *Environmental Science & Technology, 51*(21), 12784–12795.
Rössler, U. et al., 2006. The sensitivity the alkaline comet assay in detecting DNA lesions induced by X rays, gamma rays and alpha particles. Radiation Protection Dosimetry, 122(1–4), pp.154–159.
Royall, J. A., & Ischiropoulos, H. (1993). Evaluation of 2′,7′-dichlorofluorescin and dihydrorhodamine 123 as fluorescent probes for intracellular H2O2 in cultured endothelial cells. *Archives of Biochemistry and Biophysics, 302*(2), 348–355.
Saez, G., Aye, M., De Meo, M., Aimé, A., Bestel, I., Barthélémy, P., & Di Giorgio, C. (2015). Genotoxic and oxidative responses in coelomocytes of Eisenia fetida and Hediste diversicolor exposed to lipid-coated CdSe/ZnS quantum dots and CdCl2. *Environmental Toxicology, 30*(8), 918–926.
Sánchez-Argüello, P., Aparicio, N., & Fernández, C. (2012). Linking embryo toxicity with genotoxic responses in the freshwater snail Physa acuta: Single exposure to benzo (a) pyrene, fluoxetine, bisphenol a, vinclozolin and exposure to binary mixtures with benzo (a) pyrene. *Ecotoxicology and Environmental Safety, 80*, 152–160.
Sarapultseva, E. I., & Dubrova, Y. E. (2016). The long-term effects of acute exposure to ionising radiation on survival and fertility in *Daphnia magna. Environmental Research, 150*, 138–143.
Sazykina, T., & Kryshev, I. I. (2006). Radiation effects in wild terrestrial vertebrates – the EPIC collection. *Journal of Environmental Radioactivity, 88*(1), 11–48.
Scholze, A., Jankowski, J., Pedraza-Chaverri, J., & Evenepoel, P. (2016). Oxidative stress in chronic kidney disease. *Oxidative Medicine and Cellular Longevity, 2016*, 8375186.
Scott, J. G., Liu, N., & Wen, Z. (1998). Insect cytochromes P450: Diversity, insecticide resistance and tolerance to plant toxins. *Comparative Biochemistry and Physiology – C Pharmacology Toxicology and Endocrinology, 121*(1–3), 147–155.
Scott, J. G., & Wen, Z. (2001). Cytochromes P450 of insects: The tip of the iceberg. *Pest Management Science*, 958–967.
Sforzini, S., et al. (2012). Genotoxicity assessment in *Eisenia andrei* coelomocytes: A study of the induction of DNA damage and micronuclei in earthworms exposed to B[a]P- and TCDD-spiked soils. *Mutation Research/Genetic Toxicology and Environmental Mutagenesis, 746*(1), 35–41.

Sforzini, S., et al. (2017). Mode of action of Cr(VI) in immunocytes of earthworms: Implications for animal health. *Ecotoxicology and Environmental Safety, 138*, 298–308. https://doi.org/10.1016/j.ecoenv.2017.01.013

Simmons, L. W. (2012). Resource allocation trade-off between sperm quality and immunity in the field cricket, *Teleogryllus oceanicus*. *Behavioral Ecology, 23*(1), 168–173.

Singh, N. P., et al. (1988). A simple technique for quantitation of low levels of DNA damage in individual cells. *Experimental Cell Research, 175*(1), 184–191.

Singh, N., Bhagat, J., & Ingole, B. S. (2017). Genotoxicity of two heavy metal compounds: Lead nitrate and cobalt chloride in polychaete Perinereis cultrifera. *Environmental Monitoring and Assessment, 189*(7), 1–11.

Smolarz, K., & Berger, A. (2009). Long-term toxicity of hexabromocyclododecane (HBCDD) to the benthic clam Macoma balthica (L.) from the Baltic Sea. *Aquatic Toxicology, 95*(3), 239–247.

Snyder, M. J. (2000). Cytochrome P450 enzymes in aquatic invertebrates: Recent advances and future directions. *Aquatic Toxicology, 48*(4), 529–547.

Spurgeon, D. J., et al. (2000). Relative sensitivity of life-cycle and biomarker responses in four earthworm species exposed to zinc. *Environmental Toxicology and Chemistry, 19*(7), 1800–1808.

Straub, L., et al. (2016). Neonicotinoid insecticides can serve as inadvertent insect contraceptives. *Proceedings of the Royal Society B: Biological Sciences, 283*(1835).

Sulmon, C., et al. (2015). Abiotic stressors and stress responses: What commonalities appear between species across biological organization levels? *Environmental Pollution, 202*, 66–77.

Svanfeldt, K., Lundqvist, L., Rabinowitz, C., Sköld, H. N., & Rinkevich, B. (2014). Repair of UV-induced DNA damage in shallow water colonial marine species. *Journal of Experimental Marine Biology and Ecology, 452*, 40–46.

Taleb, Z. M., Benali, I., Gherras, H., Ykhlef-Allal, A., Bachir-Bouiadjra, B., Amiard, J. C., & Boutiba, Z. (2009). Biomonitoring of environmental pollution on the Algerian west coast using caged mussels Mytilus galloprovincialis. *Oceanologia, 51*(1), 63–84.

Tang, G., Yao, J., Li, D., He, Y., Zhu, Y. C., Zhang, X., & Zhu, K. Y. (2017). Cytochrome P450 genes from the aquatic midge Chironomus tentans: Atrazine-induced up-regulation of CtCYP6EX3 enhanced the toxicity of chlorpyrifos. *Chemosphere, 186*, 68–77.

Tarpey, M. M., & Fridovich, I. (2001). Methods of detection of vascular reactive species: Nitric oxide, superoxide, hydrogen peroxide, and peroxynitrite. *Circulation Research, 89*(3), 224–236.

Ternjej, I., Mihaljević, Z., Ivković, M., Previšić, A., Stanković, I., Maldini, K., Želježić, D., & Kopjar, N. (2014). The impact of gypsum mine water: A case study on morphology and DNA integrity in the freshwater invertebrate, Gammarus balcanicus. *Environmental Pollution, 189*, 229–238.

Thornber, K., et al. (2009). Reactive oxygen species and lipoxygenases regulate the oncogenicity of NPM-ALK-positive anaplastic large cell lymphomas. *Oncogene, 28*(29), 2690–2696.

Tian, S., Pan, L., & Zhang, H. (2014). Identification of a CYP3A-like gene and CYPs mRNA expression modulation following exposure to benzo [a] pyrene in the bivalve mollusk Chlamys farreri. *Marine Environmental Research, 94*, 7–15.

Touahri, H. G., Boutiba, Z., Benguedda, W., & Shaposhnikov, S. (2016). Active biomonitoring of mussels Mytilus galloprovincialis with integrated use of micronucleus assay and physiological indices to assess harbor pollution. *Marine Pollution Bulletin, 110*(1), 52–64.

Tumminello, R. A., & Fuller-Espie, S. L. (2013). Heat stress induces ROS production and histone phosphorylation in celomocytes of Eisenia hortensis. *Invertebrate Survival Journal, 10*, 50–57.

Ubezio, P., & Civoli, F. (1994). Flow cytometric detection of hydrogen peroxide production induced by doxorubicin in cancer cells. *Free Radical Biology & Medicine, 16*(4), 509–516.

Vergara-Amado, J., Silva, A. X., Manzi, C., Nespolo, R. F., & Cárdenas, L. (2017). Differential expression of stress candidate genes for thermal tolerance in the sea urchin Loxechinus albus. *Journal of Thermal Biology, 68*, 104–109.

Vignier, J., Volety, A. K., Rolton, A., Le Goïc, N., Chu, F. L., Robert, R., & Soudant, P. (2017). Sensitivity of eastern oyster (Crassostrea virginica) spermatozoa and oocytes to dispersed oil: Cellular responses and impacts on fertilization and embryogenesis. *Environmental Pollution, 225*, 270–282.

Vincent-Hubert, F., Arini, A., & Gourlay-Francé, C. (2011). Early genotoxic effects in gill cells and haemocytes of Dreissena polymorpha exposed to cadmium, B [a] P and a combination of B [a] P and Cd mutation research/genetic. *Toxicology and Environmental Mutagenesis, 723*(1), 26–35.

Vosloo, A., Laas, A., & Vosloo, D. (2013). Differential responses of juvenile and adult south African abalone (Haliotis midae Linnaeus) to low and high oxygen levels. *Comparative Biochemistry and Physiology Part A: Molecular & Integrative Physiology, 164*(1), 192–199.

Vral, A., Fenech, M., & Thierens, H. (2011). The micronucleus assay as a biological dosimeter of in vivo ionising radiation exposure. *Mutagenesis, 26*(1), 11–17.

Wardman, P. (2007). Fluorescent and luminescent probes for measurement of oxidative and nitrosative species in cells and tissues: Progress, pitfalls, and prospects. *Free Radical Biology & Medicine, 43*(7), 995–1022.

Wang, Y., et al. (2013). Evaluation of the comet assay for assessing the dose-response relationship of DNA damage induced by ionizing radiation. *International Journal of Molecular Sciences, 14*(11), 22449–22461.

Ward, J. F. (1981). Symposium on radical processes in radiobiology and carcinogensis; some biochemical consequences of the spatial distribution ionizing radiation-produced free radicals. *Radiation Research, 86*, 185–195.

Weber, L., Carvalho, L., Sá, N., Silva, V., Beraldini, N., Souza, V., & Conceição, M. (2013). Genotoxic effects of the water-soluble fraction of heavy oil in the brackish/freshwater amphipod Quadrivisio aff. lutzi (Gammaridea) as assessed using the comet assay. *Ecotoxicology, 22*(4), 642–655.

White, C. C., et al. (2003). Fluorescence-based microtiter plate assay for glutamate-cysteine ligase activity. *Analytical Biochemistry, 318*(2), 175–180.

Whitehorn, P. R., et al. (2012). Neonicotinoid pesticide reduces bumble bee colony growth and queen production. *Science (New York, N.Y.), 336*(6079), 351–352.

Woods, R. M., et al. (2016). Environmental factors limiting fertilisation and larval success in corals. *Coral Reefs, 35*(4), 1433–1440.

Wu, B., Liu, Z., Xu, Y., Li, D., & Li, M. (2012). Combined toxicity of cadmium and lead on the earthworm Eisenia fetida (Annelida, Oligochaeta). *Ecotoxicology and Environmental Safety, 81*, 122–126.

Yazdani, M. (2015). Concerns in the application of fluorescent probes DCDHF-DA, DHR 123 and DHE to measure reactive oxygen species in vitro. *Toxicol in Vitrology, 30* (1 Pt B), 578–582.

Zeeshan, M., et al. (2016). ROS dependent copper toxicity in hydra-biochemical and molecular study. *Comparative and Biochemistry Physiology Part C: Toxicology Pharmacology, 185-186*, 1–12. https://doi.org/10.1016/j.cbpc.2016.02.008. Epub 2016 Mar 3.

Zhang, L., Ji, F., Li, M., Cui, Y., & Wu, B. (2014). Short-term effects of Dechlorane Plus on the earthworm Eisenia fetida determined by a systems biology approach. *Journal of Hazardous Materials, 273*, 239–246.

Zhang, Q., Zhang, G., Yin, P., Lv, Y., Yuan, S., Chen, J., Wei, B., & Wang, C. (2015). Toxicological effects of soil contaminated with spirotetramat to the earthworm Eisenia fetida. *Chemosphere, 139*, 138–145.

Zhang, Y. F., Chen, S. Y., Qu, M. J., Adeleye, A. O., & Di, Y. N. (2017a). Utilization of isolated marine mussel cells as an in vitro model to assess xenobiotics induced genotoxicity. *Toxicology In Vitro, 44*, 219–229.

Zhang, Y., Zhang, L., Feng, L., Mao, L., & Jiang, H. (2017b). Oxidative stress of imidaclothiz on earthworm Eisenia fetida. *Comparative Biochemistry and Physiology Part C: Toxicology & Pharmacology, 191*, 1–6.

Part V
Biomarker Use and Responses

Chapter 16
Biomarkers of Radiation and Risk Assessment by Ionizing Radiation, Countermeasures for Radiation Protection of Environment, Workers and Public

Luan QAFMOLLA

Abstract Institute of Applied Nuclear Physics (IANP) is main responsible institution in country for managing of methodologies and techniques used for environmental/epidemiological samples collected for levels of radiation monitoring and radionuclide identification activities. In Albania, was well established the legal framework of legislation and regulations for using of radioactive materials: Law No. 8025, date 9.11.1995 "*On Ionizing Radiation Protection*", amended No. 9973, July 28th 2008. Also, the process of establishment of more laws, regulations, code of practices in radiology, nuclear medicine etc., started based on IAEA documents, elements of Joint Convention, as well as part of Interregional Projects & EU Directives, about all issues related to the policy issues for the application of biomarkers in the field of human health. Both types, of environmental and human body samples, are as indicators of biological markers, signaling events in biological systems and those are classifying into three types: [a] those of exposure, [b] biological effects, and [c] susceptibility. Once exposure has occurred, a continuum of biological events may to be detected. These events may serve as markers of the initial exposure, internal dose, biologically effective dose and some other parameters for evaluation of exposures. Even, before exposure occurs, there may be biological differences, between humans that cause some individuals to be more susceptible to environmentally induced disease. Biomarkers, therefore, are robust tools that can be used to clarify the relationship between ionizing exposure radiation & environment health impairment. The Institute studied in collaboration with Working Hygiene and Epidemiological Research Department in Institute of Public Health (IPH) in Albania, the effects of radiation caused by Chernobyl & Fukushima NPP accidents, carried out: radiation level monitoring, radionuclide identification activities, analyses of environmental human body laboratory samples to the individual workers and public exposed to radiation. Also, was assessed the ambient monitoring using chemical or physical analyses of food, air, water, soil etc., coupled with measurement for

L. QAFMOLLA (✉)
Private Consultant, former employee of Institute of Applied Nuclear Physics, Tirana, Albania

M. D. Wood et al. (eds.), *Biomarkers of Radiation in the Environment*, NATO Science for Peace and Security Series A: Chemistry and Biology,
https://doi.org/10.1007/978-94-024-2101-9_16

estimation of actual human intake at these areas, and by biomarkers of exposure was study the effects in body fluids such as: blood, urine, saliva, or some limited samples for reproductive and developmental systems, follicular fluids, cells, and semen. It is known, the thyroid cancers attributed to ^{131}I radioisotope exposure, as well as by other radionuclide's which have contaminated the environment, and for that reason it was important to evaluate patterns of excess absolute and relative risks by external/internal irradiation over time.

Keywords Radiation exposures monitoring · Biomarkers for human health · Biological effect · Environment contamination

16.1 Introduction

16.1.1 Challenges for Establishment of Albanian Legislation and Regulations According EU

In the Republic of Albania is well established the legal framework with laws and regulations involving practices and applications of using radioactive materials or devices in medicine, researches, agriculture, industries, environment protection/control and education for the safety, security and radiation protection from ionizing exposure radiation. Based at the main Law No. 8025, date 9.11.1995 "*On Ionizing Radiation Protection*", amended No.9973, July 28th 2008; some other important regulations are approved by Albanian institutions:

- Regulation on "Safe management radioactive waste in Republic of Albania", Decision No. 08, date 07 January 2010 of Council of Ministers.
- Regulation on "Categorization of radioactive sources in the Republic of Albania", Decision No. 09, date 07 January 2010 of Council of Ministers
- Regulations on "Licensing and inspection of activities with sources of ionizing radiation" Decision No.10, Date 07 January 2010 of Council of Ministers.
- Regulation on "Safe transport of radioactive materials", Decision No. 488, date 23 June 2010 of Council of Ministers.
- Regulation on "Safe handling with ionizing radiation sources", Decision No. 543, Date 7 July 2010 of Council of Ministers.
- Regulation on "Physical protection of the radioactive materials in the Republic of Albania", Decision No 344, date 29 April 2011 of Council of Ministers.
- Regulation on "Protection of the employees professionally exposed to ionizing radiation sources" Decision No 590 date 18 August 2011 of Council of Ministers.
- Regulation on "The permitted levels of the radon concentration on buildings and water, guide levels of radionuclide's building materials, as well as permitted levels of radionuclide's in food and cosmetic products", Decision No 591 date 18 August 2011 of Council of Ministers.

- Regulation on "Public protection from the discharges in the environment, determination of sampling, regions and frequency of measurement" Decision No 313, date 09 May 2012 of Council of Ministers.
- Regulation on "Public safety to exposures caused from ionizing radiation sources" Decision No 481, date 25 July 2012 of Council of Ministers.
- Regulation on "Safety on Medical exposure with ionizing radiations" No 229 date 20 March 2013

The process of establishment of more laws, regulations, code of conduct, code of practices in radiology, nuclear medicine etc., started based on IAEA documents, elements of Joint Convention, as well as part of Interregional Projects & EU Directives, about all issues related to the policy issues for the application of radioactive materials including biomarkers in the fields of environment and human health monitoring. So, for instance at the Regulation No. 481, date 25.7.2012 "*Public protection from exposures by ionizing radiation devices*", was improved our understanding to the diseases, providing new knowledge of disease mechanisms and processes providing a means for improved health management through the earlier diagnosis of disease and the delivery of more efficacious and safer therapies. In fact our above mentioned laws and regulations needed to be in accordance with the directives of the European Parliament Commission, as well as with the European Economic and Social Committee and the Committee of the regions documents, especially: "Safe, Innovative and Accessible Medicines" which as the main focus have on the developing technologies and tools for gathering information on various classes of biomolecules, biomarkers and understanding relationships among them, including the related regulatory mechanisms.

16.1.2 Ionizing Radiation Exposures and Determining Risk Health

Ionizing radiation is a known carcinogen but the magnitude of health risk at low doses and dose rates, for instance below 100 mSv or 0.1 mSv min^{-1}, remains controversial due to a lack of direct human evidence. Epidemiological studies of radiation exposed populations can provide evidence of risk. Much more information is needed for interpreting ionizing radiation measurements for determining health risks, because several factors should be considered before making qualitative or quantitative evaluation of exposures by ionizing radiation. For example, the concentration, duration and the time of exposure, and physicochemical nature of the radioactive agents are all relevant to the selection of an appropriate marker of the ionizing radiation exposure. Specialists have proposed that there are two key factors governing interpretation of radiation measurements: [1] measurements have no meaning until interpreted and [2] measurements only have meaning in terms of how they are interpreted. Thus, recorded or reported radiation measurements have no inherent meaning by themselves, they are just numbers.

Radiation safety specialists have the advantage for interpreting radiation measurements based on knowledge of comparative readings from background and other sources. Most people without this specialized knowledge do not know that we live in a sea of radiation, which surrounds us all the time. Furthermore, screaming dosimeter instrument may sound alarming but radiation risks depend on many other factors, such as the type of radiation and the duration of exposure (Jonson, 2014).

16.1.3 General Considerations on Biomarkers for Use in Epidemiological Studies

Biomarkers allow new ways of understanding disease processes and the ways in which medicines work to counteract disease. Within the practice of evidence-based medicine this knowledge can be used to improve disease diagnosis, to improve the safety and efficacy of existing medicines and to develop new medicines and targeted therapies. A biomarker has been defined as "*any measurement reflecting an interaction between a biological system and an environmental agent*, which may be chemical, physical or biological agent". Biomarkers can be used for multiple purposes in epidemiological investigation included: [a] estimation or validation of received dose, thus improving the validity of a correlation between exposure and biological responses; [b] investigation of individual susceptibility; and [c] early detection of a radiation induced health effect (Committee on Biological Markers of the National Research Council, 1987).

Biomarkers are robust tools that can be used to address many different issues confronting environmental health scientists. Biomarkers that indicate the occurrence of an internal dose, a biologically effective dose, or the presence of an incipient disease can be useful in hazard identification, for example, as the qualitative step that causally associates an environmental agent with an adverse effect (Fowle, 1984). Biomarkers can also be used to determine dose-response relationships and to estimate risk, especially at the low doses relevant to most environmental samples. Another major role of biomarkers is clarification of the extent of exposure in human populations. Methods of direct or indirect measurement of total exposure through analysis of body fluids (e.g. IAEA, 1969) are far more likely to be of value in epidemiological studies than are most of the modeling and ambient monitoring approaches now in use. Biomarkers of exposure also hold the promise of demonstrating, which individuals in a potentially affected population (e.g., residents or workers in the neighborhood of a hazardous radioactive wastes facility) have inordinate levels of exposure. Developments in the field of biomarkers are also likely to lead to a more accurate determination of the proportion of highly susceptible people within the population and of the results of human or public exposure (Ashford, 1986). For present purposes, the effects on, or responses of, an organism to an exposure are considered in the context of the relationship of exposure to health impairment or the probability of health impairment. An effect is defined as: an actual health impairment or (by general consensus) recognized disease; an early precursor of a disease process that indicates a potential for impairment of health; or an event peripheral to any disease process but correlated with it and thus predictive of development of impaired health. A biological marker of an effect or response, then, can be any change that is qualitatively or quantitatively predictive of health impairment or potential impairment resulting from exposure.

16.1.4 Collection and Use of Biological Samples in Epidemiological Studies

Differences among species and individual variations in physiological characteristics such as sex, age, and health status can significantly affect the absorption and distribution of the chemical and its metabolites. Individual response to environmental temperature, such as the ingestion of large quantities of water, also may affect absorbed dose. Blood flow, capillary permeability, transport into an organ or tissue, the number of receptor sites, and route of administration (which determines the path of the parent compound or its metabolites in the body) can all influence internal or biologically effective dose. Exposures to environmental agents are classically evaluated by mathematical modeling based upon assumptions concerning emission sources, environmental fate, and the location of individuals in space and time. Exposures, are also evaluated by ambient monitoring using chemical or physical analyses of food, air, water, or soil, coupled with measurement or estimation of actual human intake of these media, and by biological markers of exposure including measurements in body fluids such as blood, urine, saliva, cerebral spinal fluid for reproductive and developmental systems, follicular fluids, amniotic fluids and cells, and semen. Examination of other biological samples, such as hair, feces, or teeth, may prove useful. The use of such biological markers is a more preferable means for accurately estimating exposure than are the more indirect approaches of modeling or environment monitoring (Ashford, 1986; Fowle, 1984).

16.2 Material and Methods of Sampling Analysis in IANP

In IANP exist some different methods and techniques for evaluation and determination of the alpha, beta, gamma nuclides level contents in natural samples or aquatic discharges by research Labs in the country. We are describing our simple methods used, in order to realize the cooperation between above-mentioned institutions, promoting environmental safety and security in natural resources management. The methods used for measurements of the background level, in order to determine the activity/concentration of the component elements (nuclides) by alpha or gamma spectrometry, beta-gamma total measurements, as well as by radiochemistry separation analysis to the specific nuclides, are consolidated in IANP already (Suomela, 1993b; Suomela et al., 1993; LMRI-CEA, 2006).

16.2.1 Epidemiological Liquid and Aqueous Samples Analyses

The introduction of the European Commission's Water Framework Directive (WFD; 2000/60/EC) established a new era in environmental risk assessment. In addition to incorporating the compliance of chemical quality standards, the key objective of the WFD is the general protection of the aquatic environment in its entirety. This new approach emphasizes the need for an integrated environmental risk assessment and offers the potential for the incorporation of biological effects measures, including the use of biomarkers in this process. A variety of biological samples can be used for biomarker measurements in epidemiological studies, given appropriate ethics approvals and informed consent and do depend on the nature of the internal/external radiation exposure. These include aqueous or fluid biological samples like: urine, blood, saliva and semen or solid biological samples like: faeces, hair, hair follicle cells, and nail clippings. The fluid or aqueous samples are submitted the measurements of gamma nuclides procedures in order to determine their radioactivity (concentration) in epidemiological Lab of IPH. To provide a suitable and repeatedly geometry of the detection (measurements) are used the Marinelli beakers with 500 ml volume for aqueous environmental samples and Laboratory tubes for epidemiological samples. The "Anal-Spec" (system) equipment with NaI (TI) detector (Ø = 2 × 2 inch), as well as the Lab tubes or Marinelli beakers with volumes 5, 10, 50–500 ml volume, putted into the lead shielding place. This apparatus performs the determination of natural/artificial radionuclide, gamma radiation measurement. The system was provided with digital suite (set) for the gamma spectrometry. The device has the multi-channel analyzer, its spectroscopic amplificatory, the high voltage system, memory with a scintillation integral detector. The device has in its structure a standard NaI (TI) detector (Ø = 1–2 inch), and another G-M detector. "Anal-Spec" apparatus was connected with PC system storing in its memory over 74 specters at 1024 channels that it has. The spectra are processed by its TMCA software.

Technical Parameters "Anal-Spec" system are:

- Radionuclide identification and spectrum analysis;
- Multi-channel analizator its spectroscopic amplificatory, the high voltage system;
- NaJ (TI) detector (Ø = 1–2 inch), as well NaI (TI) detector with tungsten protection support etc.

Some other specifications "Anal-Spec" are as below:

- Some other detectors may to be used: NaI, BGO, CdWo, CdZnTe, Plastic;
- Selected High – Voltage (HV): 50–1275 volt diapason; Type/Shape: digital filter

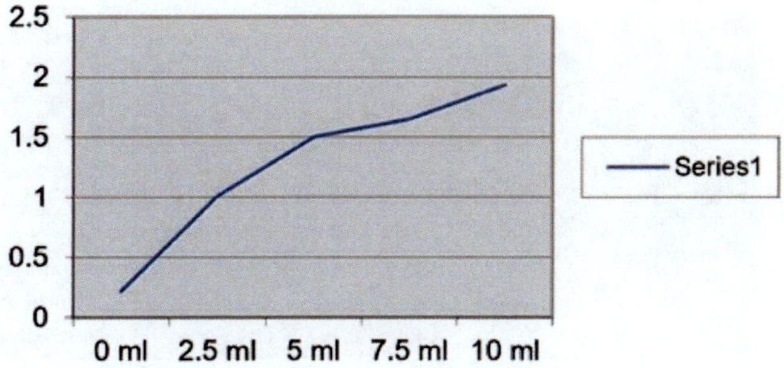

Fig. 16.1 Calibration curve established for the results comparison of the aqueous samples

- The energy ranges: NaI (Tl) detector 20 KeV–2,5 MeV; Geiger-Muller 60 KeV–1,6 MeV.
- Sensitivity: 137Cs > 10 imp/sec for NaI (1 × 2 inch) detector.

Calibration curve: In beginning of detection of each gamma nuclides was established the calibration curve (Fig. 16.1), using the ^{137}Cs standard solution (LMRI-CEA, 2006) with data as below: *Standard solution:* ^{137}Cs, *Type:* ELE-1, *Vials number:* 228/27703, *Produced by:* LMRI-CEA France on 15 April 2006, *Specific activity*: $A_S = 93, 76$ Bq/g (2%) and Cc = 0,999 gr/cm^3 (20^0).

16.2.2 Environmental Samples Analysis

In cases of a radiological/nuclear a release of short-lived fission and activation products to the environment can be expected. Some of these radionuclides, mainly lanthanides and actinides, will form with HDEHP di(2-etyl-hexyl) phosphoric acid and consequently interfere with the subsequent counting of the Cerenkov radiation from yttrium-90 (^{90}Y). According to the standard procedure, interfering nuclides such as uranium, thorium, radium and their decay products, as well as isotopes cesium, potassium and strontium are separate from the samples by an extraction with HDEHP. The determination of strontium-90 (^{90}Sr) in equilibrium with yttrium-90, is accomplished by monitoring Cerenkov radiation of high energetic beta particles (2,27 MeV) from yttrium-90 in a liquid scintillator counter. Yttrium-90 is the decay product of strontium-90. The chemical yield of yttrium-90 is determined by adding a known amount of inactive yttrium carrier. The amount of yttrium recovered is determined by acidimetric titration of the sample in the scintillation vial with “Titriplex III”. In IANP are in use two methods of determination of strontium −90 in food and environmental samples in the absence and presence of short-lived activation and fission products. By using the above-mentioned procedures and a low-level liquid scintillator, a lower limit of detection of 10 MBq/sample can be reached for beta nuclides.

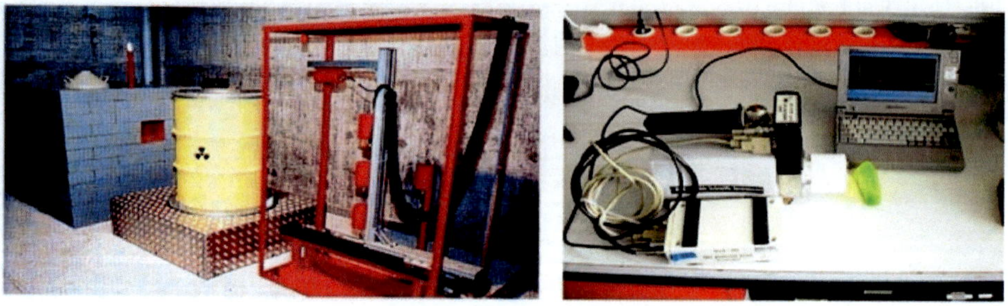

Fig. 16.2 Facilities and equipment used for sample analysis

16.2.3 Transuranic Environmental Samples Analysis

The radioactivity of the uranium, thorium, radium isotopes released in the environment from discharges of radiochemistry division at the IANP, or by contamination of NPP accidents was performed by electrodeposition of coprecipitation method with an ammoniac alkaline earth phosphate precipitate, adding a known amount of uran-232 marker solution to a sample 500 ml. The purification of uranium by the ion exchange procedure and electrodeposition on stainless steel disc, when measured by alpha spectrometry, normally gives alpha peaks with a frequency wave measured of about 50 keV. This means that the peaks from the different uranium isotopes are well separated and easily identified and qualified. The radioactivity of the uranium isotopes deposited on the stainless steel disc is measures by counting the alpha particles of uranium isotopes in an alpha spectrometer. The minimum detectable activity for a counting time of 1000 min, is about 10 MBq for each of the uranium isotopes present in the sample (Fig. 16.2).

16.3 Results

In this study are represented some results of analysis performed by our Laboratory teams for environmental samples, in order to probate the validity of our measurements and methodology of measures, and based at the data received by some of aqueous samples was established respectively the table results for the standard samples with known concentration of the ^{137}Cs nuclide, as well as for the environmental samples collected very close with resources that supply the Shkodra Lake (Farkas, 1980; HASL, 1983). As well, we have determined the lower limit of detection of concentration of the epidemiological or environmental samples for alpha, beta and gamma particles (Suomela, 1993a, 1993b). The calibration curve for the standard samples was established, while for the environmental aqueous samples made the mean (average) of the measurement for each of the resource samples and later all the measurement are compared with the average measure of the standard sample. Mean value ≈0,23 Bq is background (Table 16.1).

Table 16.1 The measurements of the ^{137}Cs standard radioactive solutions

No	Marinelli Beaker	Volume of standard solution	Volume (500 ml)	Measurements (Bq)	Mean value (Bq)
1	Blank	0 ml	500	0,02; 0,75; 0,02; 0,27;	≈0,23
2	Vessel no. 1	2,5 ml	497,2 + 2,5	2,03; 9,35; 0,78; 5,77,	≈1,01
3	Vessel no. 2	5 ml	495 + 5	0,43; 2, 07; 1,45; 1,15	≈1,50
4	Vessel no. 3	7,5 ml	492,5 + 7,5	4,13; 0,12; 0,02; 0,47;	≈1,65
5	Vessel no. 4	10 ml	490 + 10	0,67; 2,68; 0,87; 3,67;	≈1,93

Table 16.2 The measurements of the aqueous samples resources supplier of Shkodra Lake

Aqueous sample	Resources	Measurements (Bq)	Mean value (Bq)
Vessel no. 1	Syri i Vrakes	0,00; 0,00; 0,00; 1,87; 0,08	0,760 (zero)
Vessel no. 2	Syri i Sumajve	3,60; 6,51; 4,07; 0,00; 0,05	3,34
Vessel no. 3	Syri i Gjonit	0,00; 0,00; 1,13; 1,52; 2,48	1,49
Vessel no. 4	Syri i Virit	0,00; 0,00; 0,00; 0,00; 0,00	0 (zero)
Vessel no. 5	Syri i Vakes	0,05; 0,03; 0,02; 0,00; 0,00	0,033 (zero)
Vessel no. 6	Syri i Zi	2,82; 2,55; 5,05; 1,95, 8,25	3,80
Vessel no. 7	Vija e Mullirit	0,02; 0,025; 0,13; 0,32; 1,72	0,27 (zero)

The results received in Table 16.2, by samples of resources: Syrit te Zi = 3,80 Bq; Syri Sumajve = 3,34 Bq and Syri i Gjonit = 1,49 Bq; shown that concentration of nuclides (gamma total) in the supplying resources of the Shkodra Lake is ≅5–15 times above the background.

The measurements were carried out by rapid method for determination of activity (concentration) of gamma total nuclides by environmental samples, but underlining: by above mentioned method, was impossible to determine the effects and role of specific nuclides. So, for that reason confirm the necessity to realize the measurements of the aqueous samples resources using other systems and contemporary methods: [a] Instrumental method of the gamma spectrometry analysis; [b] Instrumental method of the liquid scintillates or X-ray fluorescence method; [c] Alpha spectrometry for determination of the uranium, thorium, plutonium and americium nuclides.

Table 16.3 shows the data which given: the collective effective dose takes by patients (calculated and measured by TLD-100 chips; how is used the technetium ^{99m}Tc nuclide, marked by phosphon for bones scintigraphy, and how was evolved the patients numbers during 2010–2015 based at the examinations performed in NML in University Hospital "Mother Theresa" in Tirana.

Article 2 of the national regulation for treatment/management of liquid radioactive waste discharged to the environment (Albanian Government, 1996) are

Table 16.3 The collective effective dose using the ^{99m}Tc nuclide marked with phosphon

Years	No. Exam (s) bones scintigraphy	Examination number %	Collective effective dose SE = man Sv
2010	247	19,5	3,22
2013	355	27,9	3,47
2015	478	37,6	3,89

Table 16.4 Concentration limits & total radioactivity for nuclides at the discharged liquid radioactive waste

No	Radiotoxicity group	Total activity (Bq/day)	Concentration (Bq/l)
1	Very higher	4×10^4	4×10^3
2	Higher	4×10^5	4×10^4
3	Medium	4×10^6	4×10^5
4	Lower	4×10^7	4×10^6

foreseen the level of radioactivity for specific nuclides, which shown in Table 16.4 are in accordance with WAC recommended by IAEA, EC.

16.3.1 Interpretation of Measurements and Content Validity for Risk Assessment

Measurements are one of the principal building blocks of quantitative risk assessment. If measurements are invalid, it is likely that the risk assessments constructed from those measurements will also be invalid. Measurement validity characterizes the extent to which a biomarker of a phenomenon has *content validity*: for instance, pertains to the underlying phenomenon; or *construct validity*: for instance, correlates with other relevant characteristics of the underlying phenomenon; and *criterion validity*: for instance, to predicts some component of the underlying phenomenon. These three components of measurement validity are best assessed in terms of the extent or degree to which they apply to the underlying phenomenon. While, the content validity is the extent to which a marker "incorporates the domain of the phenomenon under study". For instance, a biomarker of internal dose will have content validity if it reflects the dose contributed by all routes of exposure. A biomarker of effect will have content validity, if it encompasses the essential characteristics of the disease it represents. In other words, the marker must pertain to the appropriate target organ, or its relationship to the natural history of the disease in question must be unambiguous. To properly assess content validity, one must consider the extent to which the marker pertains to the phenomenon (exposure, effect) of interest or, the extent to which the marker represents a relevant feature of that phenomenon. However, it is possible to strengthen determinations of content validity if judgments are made by a group of experts. The focus of such judgments should be the degree to which the marker represents the underlying phenomenon.

16.4 Conclusions

1. Adaptability of the national legal framework and its integration for using of a number of scientific, economic, regulatory and governance challenges need to be addressed if biomarker applications must be incorporated into clinical practices for health innovation.
2. Moreover, much more research studies in these areas are still needed to understand the mechanism of the relationship between exposure by ionizing radiation and health effect. Biomarkers can be used to gain insight into these mechanisms, as well as to describe the empirical associations between exposures by ionizing radiation and results.
3. The framework presented in my research paper may serve as a basis for evaluating the validity of biomarkers for research and for quantitative risk assessments. At present, there are few valid biological markers that can be used to conduct quantitative risk assessments. Before a marker is useful in risk assessment, it should be shown to have content, criterion, and construct validity, and it should be shown to be reliable. More other studies should be performed by Albanian institutions, in order to establish background levels, the range of normal, confounding factors, and optimize collection and analytical techniques.
4. If studies are to be useful in risk assessment, they must be generalized but, more importantly, they must be internally valid. If separate studies are conducted for use in risk assessments, efforts should be made to use similar markers and to pay attention to confounding factors.
5. The results received in our study, by rapid method of the measurements carried out for environmental epidemiological liquid or aqueous samples, and all other measured samples, reflect our first attempt for applying of such method in alpha, betta, gamma, and nuclides determination, taking in consideration factors of risk assessment.
6. The guidance levels that we have in use were taken from IAEA Basic Safety Standards (BSS) for different procedures of safety and security of public and environment control and protection and are adapted in accordance with issues of National Regulations and Code of Practice that exist in Albania already.

References

Albanian Government. (1996). *Rregullore per trajtimin e mbetjeve radioaktive ne Shqiperi*, Gazeta Shqiptare Nr. 45, fq 12,13.

Ashford, N. A. (1986). Policy considerations for human monitoring in the workplace. *Journal Occupational Medicine, 28*, 563–568.

Committee on Biological Markers of the National Research Council. (1987). Biological markers in environmental health research. *Environmental Health Perspectives, 74*, 3–9.

Farkas A. (1980). *Micro-chemical Acts – Radio-cesium determination, Poland.*

Fowle, J. R., III. (1984). *Workshop proceedings: Approaches to improving the assessment of human genetic risk-human biomonitoring Report No. EPA-600/9–84-016*; Office of Health and Environmental Assessment, U.S. Environmental Protection Agency, Washington, DC.

HASL – Cs – 01. (1983). *Determination of ^{137}Cs in environmental samples.*

IAEA. (1969). Technical Report 85. Quick methods for radionuclides analysis No. 85.

Jonson R. (2014). *Interpretation of radiation measurements, health physics society midyear meeting* Baton Rouge, LA, Radiation Safety Counseling Institute, February 12, pp. 9–10.

LMRI – CEA. (2006). Cedex- France, Certificate of Standard Solution of 137Cs, April.

Suomela, J. (1993a). *Method for determination of U-isotopes in water* (SSI-report 93-14). Swedish Radiation Protection Institute.

Suomela J. (1993b). *Method for determination of radium-226 in water by liquid scintillation counting*. SSI-report 93-12, Sweden.

Suomela, J., Wallberg, L., & Melin, J. (1993). *Methods for determination of strontium-90 in food and environmental samples by Cerenkov counting* (SSI-rapport 93-11, ISSN 0282-4434). Swedish Radiation Protection Institute.

Chapter 17
Application of the International System of Radiation Protection in Fit-for-Purpose Assessment of Impacts

Tamara L. Yankovich

Abstract The fundamental safety objective is *"to protect people and the environment from harmful effects of ionizing radiation"* and this needs to be done *"without unduly limiting the operation of facilities or the conduct of activities that give rise to radiation risks"* (IAEA (INTERNATIONAL ATOMIC ENERGY AGENCY), Fundamental Safety Principles, Safety Fundamentals No. SF-1, 2006). In doing so, it is necessary to apply the system of radiation protection in a *'justifiable'* and balanced (or *'optimized'*) manner, such that the amount of effort is commensurate with the risk (or the potential for effect), i.e., applying a *"graded approach"*. In addition, account needs to be taken of the type of exposure situation being considered (planned, emergency or existing), and how to apply the graded approach in this context.

Evaluation of the potential for effect(s) is achieved through a preliminary evaluation of exposure, where exposure is then compared to relevant, relatively conservative criteria for the purposes of screening to determine whether or not there is the possibility of significant adverse effects. Selection of appropriate screening-level criteria is dependent upon the exposure situation and the prevailing circumstances (e.g., technical, social, economic).

In cases where a preliminary evaluation indicates that there might be significant adverse effects, a detailed evaluation is then carried out to determine whether or not there are significant adverse effects. Depending on the outcome of the detailed evaluation, which includes a comparison of exposure with appropriate criteria, a decision may be made to proceed with planning and implementation of the facility or activity under consideration. At different phases in the lifetime of a facility or activity, the types of criteria that are applied to take decisions, to verify that implementation is occurring in accordance with approved plans, and to evaluate impacts will change.

The current paper provides an overview of the application of the IAEA's safety standards and supporting programmes in the application of criteria in assessing

T. L. Yankovich (✉)
International Atomic Energy Agency (IAEA), Vienna International Centre, Vienna, Austria
e-mail: t.yankovich@iaea.org

M. D. Wood et al. (eds.), *Biomarkers of Radiation in the Environment*, NATO Science for Peace and Security Series A: Chemistry and Biology,
https://doi.org/10.1007/978-94-024-2101-9_17

impacts, in the context of the international system of radiation protection, the different exposure situations, and the application of a graded approach, to ensure and demonstrate protection of the public and the environment.

17.1 Introduction

Human activities involving the use of radiation and radioactive material can result in radiation exposure to people (workers, patients, the public) and flora and fauna in the environment. This exposure needs to be regulated and monitored in accordance with international safety standards and national legislation and regulations. The fundamental safety objective of the International Atomic Energy Agency (IAEA) is *"to protect people and the environment from harmful effects of ionizing radiation"* (IAEA, 2006). This objective is applicable to all circumstances giving rise to radiological risks[1] (IAEA, 2018a) and needs to be carried out *"without unduly limiting the operation of facilities or the conduct of activities that give rise to radiation risks"* (IAEA, 2006). Therefore, in the planning and implementation of activities during which there might be exposure to ionizing radiation, a graded approach should be applied, such that the amount of effort and regulatory oversight are commensurate with the risk (or the potential for effects). In doing so, it is necessary to apply robust, technically defensible approaches and technologies to:

- Characterize and monitor exposure of people, flora and fauna to ionizing radiation (IAEA, 2005) and other stressors (including chemical, physical and/or biological stressors);
- Establish robust and relevant criteria to evaluate risk;
- Compare the level of exposure to relevant criteria reflecting the risk;
- Evaluate the relative risks and impacts associated with relevant stressors, so that they can be prioritized, and where possible, mitigative measures can be put in place to prevent or minimize risks and impacts; and
- In complex cases, evaluate the effects themselves, to determine their significance and whether alternative approaches should be applied, or additional measures taken to minimize risks and impacts to people and the environment.

---

[1] From the perspective of the IAEA safety standards, risk is defined as a *"multiattribute quantity expressing hazard, danger or chance of harmful or injurious consequences associated with exposures or potential exposures. It relates to quantities such as the probability that specific deleterious consequences may arise and the magnitude and character of such consequences"* (IAEA, 2019). Radiation risks are *"Detrimental health effects of exposure to radiation (including the likelihood of such effects occurring), and any other safety related risks (including those to the environment) that might arise as a direct consequence of: (a) Exposure to radiation; (b) The presence of radioactive material (including radioactive waste) or its release to the environment; (c) A loss of control over a nuclear reactor core, nuclear chain reaction, radioactive source or any other source of radiation"* (IAEA, 2019).

Characterization, evaluation and prioritization of risks and impacts from ionizing radiation and radioactive material can be undertaken in the context of the international system of radiation protection, and tested through international model validation and data compilation programmes, such as those described in the sections that follow.

17.1.1 Objectives

The objectives of the current paper are to provide an overview of:

- The international system of radiation protection for the different situations in which exposure can occur, in the context of relevant IAEA safety standards that can be applied in support of protection of people and the environment, applying a graded approach; and
- International model validation and data compilation programmes of the IAEA as a mechanism for evaluating and prioritizing potential impacts.

17.2 Principles of Radiation Protection

Appropriate characterization, evaluation and prioritization of radiation risks and impacts in the context of those from other stressors can be achieved through application of the three general principles of radiation protection (IAEA, 2014; ICRP, 2007), which are:

1. Justification;
2. Optimization of protection and safety; and
3. Limitation of exposure.

Justification is defined as *"the process of determining ... whether a practice is, overall, beneficial; i.e. whether the expected benefits to individuals and to society from introducing or continuing the practice outweigh the harm (including radiation detriment) resulting from the practice"* (IAEA, 2014, 2019).

Optimization of protection and safety is defined as *"the process of determining what level of protection and safety would result in the magnitude of individual doses, the number of individuals (workers and members of the public) subject to exposure and the likelihood of exposure being "as low as reasonably achievable, economic and social factors being taken into account" (ALARA)"* (IAEA, 2014, 2019).

Limitation of exposure involves the setting of criteria and ensuring that measures are taken to prevent exposure from exceeding set criteria.

In accordance with these principles, the international system of radiation protection can be applied to ensure there is net benefit to individuals and to society, as well as to present and future generations, and that a pragmatic and balanced outcome is achieved, taking account of all relevant factors in optimization of protection and safety (IAEA, 2014). In doing so, there is a need to assess exposure to ionizing radiation and other stressors, and the corresponding risks and effects of each stressor to facilitate their prioritization in decision-making.

17.3 Assessment of Exposure over Lifetime of Facilities and Activities

Evaluation of the potential for effect(s) involves an assessment of exposure, where exposure is then compared to relevant criteria to determine whether or not there is the possibility of significant adverse effects (IAEA, 2018a, b). Over the course of the lifetime of a facility or activity, monitoring needs to be undertaken to assess the actual exposure relative to the anticipated exposure (as defined and approved during planning), as well as relative to selected criteria indicating the effectiveness of the implementation of plans, the possible loss of control of radioactive releases and/or the potential for effects to people and/or the environment. The phases in the lifetime of a facility or activity (Fig. 17.1) can include:

1. Preliminary evaluation;
2. Detailed evaluation;
3. Planning;
4. Implementation and verification; and
5. Post-facility or activity management.

17.3.1 Preliminary Evaluation

As a first step in the lifetime of a facility or activity, a relatively conservative, screening-level assessment (i.e., a 'preliminary evaluation') is undertaken to estimate exposure and quickly '*screen out*' non-issues, so that focus can then be placed on stressors that *might* cause significant detrimental effects (e.g. IAEA, 2001; Brown et al., 2008; US-DOE, 2004; Yankovich et al., 2014; IAEA GSG-15, in preparation) (Fig. 17.1). This involves conducting a desk-based study to estimate exposure, making use of available information and data, and if needed, information and measurements collected as part of targeted field studies. In doing so, conservative assumptions are made to estimate exposure, which is compared to a relatively conservative screening criterion (as described in Sect. 17.4). If the estimated exposure exceeds the criterion, it can be concluded that there might be significant adverse effects related to the facility or activity under consideration and a more detailed evaluation is needed to determine whether or not this is the case. In cases where there is a perception of risk by interested parties (also called 'stakeholders') that

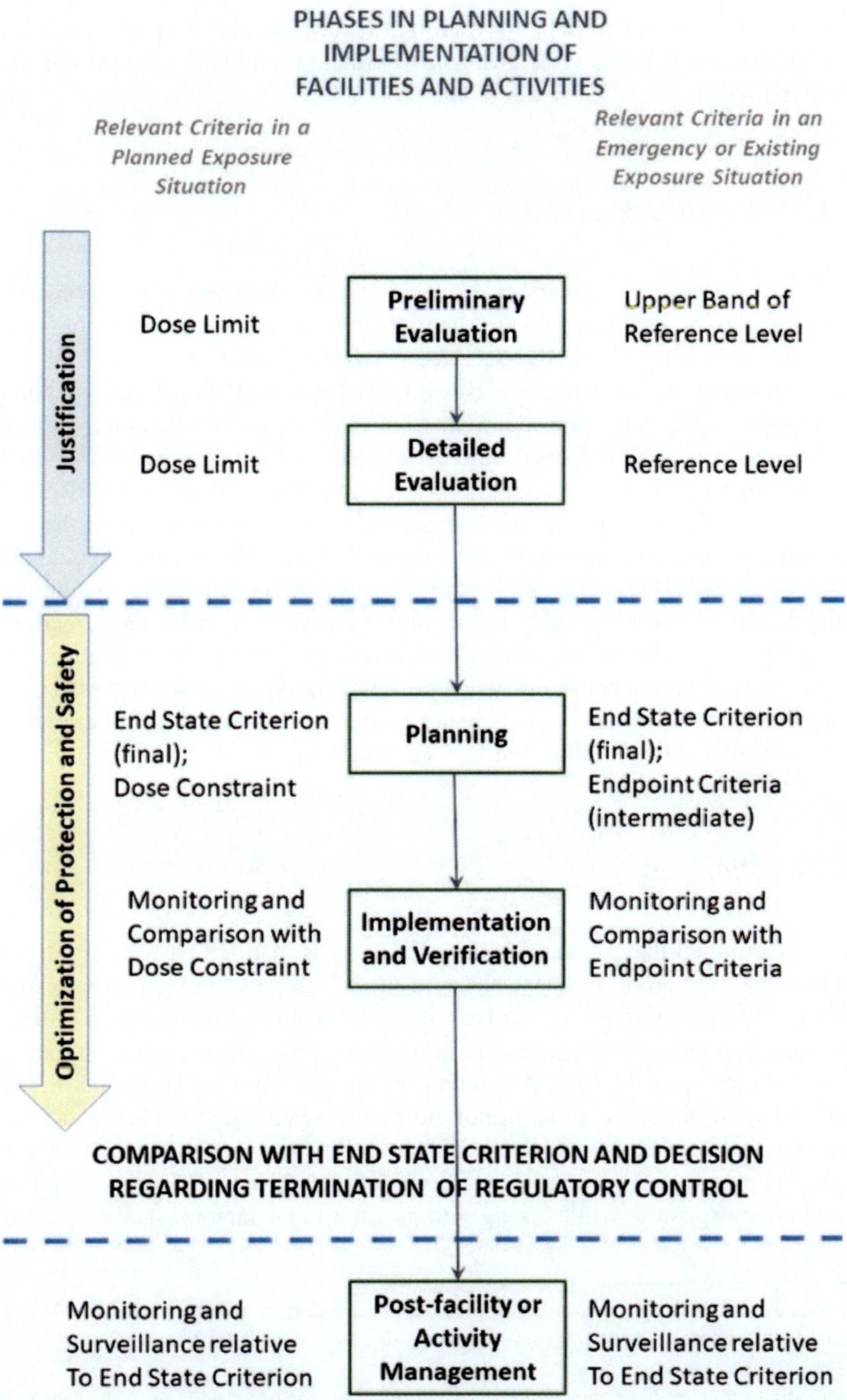

Fig. 17.1 Relevant criteria at different stages of the lifetime of a facility or activity in planned, emergency and existing exposure situations

does not reflect the actual risk, efforts might be needed to clarify understanding of the situation, and in some cases, further assessment might beneeded to address perceived risk.

17.3.2 Detailed Evaluation

In cases where a preliminary evaluation indicates that there might be significant adverse effects, a 'detailed evaluation' will then be undertaken to determine whether or not there are significant adverse effects associated with the facility or activity under consideration, and whether it is justified to proceed (Fig. 17.1). Depending on the outcome of the detailed evaluation, which includes an evaluation to determine whether or not there is net benefit (indicating that the facility or activity is justified) and a comparison of exposure with appropriate criteria (as described in Sect. 17.4), there may be a decision to proceed with planning and implementation of the facility or activity under consideration. In cases where the possibility of detrimental effects cannot be excluded based on an exposure assessment, or where the situation is too complex and/or the uncertainty is too high to determine what the effects could potentially be, an effects assessment may be undertaken to complement the exposure assessment and to determine whether or not significantly adverse effects could occur. The outcomes of such assessment can then be considered in an iterative manner to re-evaluate the possible significance of effects.

17.3.3 Planning

If it is determined that a given facility or activity is justified based on preliminary and detailed evaluation, the next phase in the lifetime of the facility or activity is planning. The planning phase involves the identification and evaluation of options, for example, related to siting, design, protective actions[2] or remedial actions[3] to be taken and others, as relevant, depending on the exposure situation and prevailing circumstances. In particular, an important part of planning is the estimation of the exposure that would be expected if the facility or activity were undertaken. Appropriate options are then selected through optimization of protection and safety to minimize exposure, while taking into account other factors, such as feasibility,

[2]A protective action is *"An action for the purposes of avoiding or reducing doses that might otherwise be received in an emergency exposure situation or an existing exposure situation"* (IAEA, 2019).

[3]A remedial action is *"The removal of a source or the reduction of its magnitude (in terms of activity or amount) for the purposes of preventing or reducing exposures that might otherwise occur in an emergency or in an existing exposure situation"* (IAEA, 2019). A remedial action is a type of protective action.

socioeconomics, expectations of interested parties and others, as relevant. Once developed, the plan is then submitted for approval by the regulatory body or other relevant authority.

17.3.4 Implementation and Verification

If the plan to embark on the facility or activity is approved, implementation of the approved plan is undertaken. In support of implementation, a monitoring programme needs to be established to ensure that implementation is carried out in accordance with the plan throughout the entire lifetime of the facility or activity. This includes verification that exposure does not exceed that approved by the regulatory body or other relevant authority.

Near the end of the lifetime of a facility or activity, the conditions on the site or in the area where activities are being undertaken, including the actual exposure and other criteria, need to be compared to relevant criteria to verify the approved end state conditions have been achieved (as described in Sect. 17.4) and regulatory control can be terminated for all or part of the site or area under consideration.

17.3.5 Post-Facility or Activity Management

The final phase of the lifetime of a facility or activity is post-facility or activity management. During this phase, any post-facility or activity control measures, which have been defined in the decommissioning plan or remediation plan (approved as part of the planning phase), need to be put into place. In doing so, decisions need to be made regarding whether the site or area is considered suitable for restricted or unrestricted use, and monitoring and surveillance are continued, as necessary, depending on the conditions. This includes monitoring of exposure relative to relevant criteria, as described in the sections that follow.

17.4 Assessment of Risk and Impacts in Different Exposure Situations

An integral part of evaluating whether the anticipated risks and impacts might be significant is the comparison of exposure to relevant criteria. Such criteria will differ in different phases of the lifetime of a facility or activity (as depicted in Fig. 17.1) and in different exposure situations (as described in the sections that follow).

The IAEA General Safety Requirements, Part 3 (IAEA, 2014) identifies three exposure situations (planned, emergency and existing exposure situations), based on the recommendations of the International Commission on Radiological Protection (ICRP, 2007), in which different criteria are applied to evaluate risk. Such criteria can serve as useful benchmarks in planning studies of exposure and effects for application in the setting of pragmatic regulatory requirements for nuclear facilities and activities, taking account of the situation and prevailing circumstances.

17.4.1 Definition of Exposure Situations

A planned exposure situation arises from the planned operation of a source or from a planned activity that results in an exposure due to a source (IAEA, 2014, 2019). In such situations, planning is undertaken prior to embarking on the activity during which exposure might occur, thereby restricting (controlling) exposures and their likelihood of occurring from the beginning. The primary means of controlling exposure in planned exposure situations is through proper design of facilities, processes and equipment, good operating procedures, and adequate training.

17.4.1.1 Planned Exposure Situations

In a planned exposure situation, release of radioactive substances to the environment is planned and controlled (IAEA, 2018c). Therefore, due to the controlled nature of such situations, a firm public dose limit of 1 mSv/a is set for exposure to the representative person[4] from all exposure pathways, above which exposure is not considered acceptable (IAEA, 2014).

17.4.1.2 Emergency Exposure Situations

An emergency exposure situation arises as a result of an accident, a malicious act or any other unanticipated event, and requires prompt action in order to avoid or reduce adverse consequences (IAEA, 2014, 2015, 2018d). In emergency exposure situations, short-term focus is placed on human, as opposed to environmental, protection; once the emergency is under control and focus can start to be placed on the longer-term situation, environmental protection is considered as part of optimization of protection and safety (Copplestone et al., 2018).

[4] The representative person is *"An individual receiving a dose that is representative of the doses to the more highly exposed individuals in the population"* (IAEA, 2019).

17.4.1.3 Existing exposure Situations

By comparison, an existing exposure situation is a situation of exposure that already exists when a decision on the need for control needs to be taken (IAEA, 2014, 2019). Such situations can include areas with natural background radiation, sites or areas with residual radioactive material from past practices that were never subject to regulatory control or that were subject to regulatory control but not in accordance current standards, and sites or areas with residual radioactive material from a nuclear or radiological emergency after an emergency has been declared to be ended (IAEA, 2014).

In both emergency and existing exposure situations, there has been a loss of control of radioactive material with a need to regain control of the situation, and as a result, relevant radiological criteria are set on a case-by-case basis, taking account of the prevailing circumstances (e.g., technical, social, economic) and applying the principles of justification, optimization of protection and safety and a graded approach (as described above). In such cases, agreed long-term criteria might be less stringent than the 'firm' dose limit that is set in planned exposure situations, and instead, are selected within a range. This more flexible 'target' criterion, known as a 'reference level', is set in emergency and existing exposure situations.

By definition, a reference level is *"the level of dose, risk or activity concentration above which it is not appropriate to plan to allow exposures to occur and below which optimization of protection and safety would continue to be implemented"* (IAEA, 2019). In the case of emergency exposure situations, the reference level can be set within the range of 20–100 mSv/a (ICRP, 2007; IAEA, 2014), depending upon the nature of the exposure, feasibility of reducing or preventing exposure, experience, expectations of interested parties (i.e., stakeholders), availability of resources and other prevailing circumstances (IAEA, 2014). By comparison, in existing exposure situations, the reference level can be set within the range of 1–20 mSv/a, again, depending upon the prevailing circumstances. In addition to the reference level, for practicality and efficiency, corresponding derived criteria that are easily measurable might also be established (IAEA GSG-15, *in prep.*). For example, this could include a 'derived reference level', which is defined as a *"numerical value expressed in an operational or measurable quantity, corresponding to the reference level set in dose"* (ICRP, 2014).

17.4.2 Establishment of Criteria

Depending on the exposure situation and the associated prevailing circumstances, different criteria could be considered relevant at different phases of the lifetime of a facility or activity (Fig. 17.1). Relevant criteria for different phases in each exposure situation are described in the sections that follow.

17.4.2.1 Criteria in a Planned Exposure Situation

Justification The dose limit in a planned exposure situation is the maximum dose that is considered acceptable. Therefore, applying the principle of justification (as described in Sect. 17.2 above), facilities or activities for which the annual effective dose to members of the public is anticipated to exceed the dose limit (i.e., 1 mSv/a) are not justified and above this limit, it is not appropriate to plan to allow exposures to occur; below the dose limit, optimization of protection and safety is implemented.

Optimization of Protection and Safety Through the process of optimization of protection and safety, the optimum criterion can be set during planning of a facility or activity, such that the exposure and corresponding dose and risk are as low as reasonably achievable (ALARA). In the case of a planned exposure situation, the dose limit is used as a starting point to set a more conservative, 'optimum' dose constraint, which serves as a boundary in defining the range of options for a facility or activity (Fig. 17.1). The dose constraint can be achieved through design of facilities, processes and equipment, implementation of good operating procedures, and adequate training.

During the operation of a facility or when undertaking a planned activity in a planned exposure situation, monitoring is undertaken to measure exposure and a dose assessment is conducted, based on exposure. The actual dose is then be compared to the dose constraint to ensure and demonstrate compliance, while verifying that there is no loss of control of radioactive material (for example, as indicated by an increasing trend in radioactive discharge or in activity concentrations measured in environmental media, such as air, water or soil).

17.4.2.2 Criteria in an Emergency Exposure Situation

Justification The reference level in an emergency exposure situation represents the maximum dose that is considered acceptable, applying the principle of justification, similar to how this principle is applied in a planned exposure situation (see Sect. 17.4.2.1). In an emergency exposure situation, the reference level is typically set within the range of 20 and 100 mSv acute or annual effective dose, taking account of dose contributions from all exposure pathways (IAEA, 2014, 2015). Above the reference level, it is not appropriate to plan to allow exposures to occur, and below this level, optimization of protection and safety is implemented.

The reference level in an emergency exposure situation is used in conjunction with the goals of emergency response[5] and the specific time-frame in which specific

[5] *"In a nuclear or radiological emergency, the goals of emergency response are: (a) To regain control of the situation and to mitigate consequences; (b) To save lives; (c) To avoid or to minimize severe deterministic effects; (d) To render first aid, to provide critical medical treatment and to manage the treatment of radiation injuries; (e) To reduce the risk of stochastic effects; (f) To keep*

goals are to be achieved (IAEA, 2015). For example, in the phase of an emergency when urgent actions need to be taken in response to the emergency (i.e., in the 'urgent response phase'), an effective dose of 100 mSv (acute or annual) might be justified as a starting point for implementing and optimizing a protection strategy (IAEA, 2015). Therefore, in such a case, the reference level would be set at 100 mSv acute or annual effective dose, representing the upper level above which exposure would not be appropriate and below which optimization of protection and safety would be undertaken. By comparison, in later phases of an emergency, for example, during the transition[6] from an emergency exposure situation to an existing exposure situation, an effective dose of 20 mSv/a might be justified as the starting point for implementing and optimizing a protection strategy (i.e., reference level = 20 mSv/a), as 20 mSv/a is the upper value in the range within which the reference level is selected in an existing exposure situation (ICRP, 2007; IAEA, 2014, 2015).

Optimization of Protection and Safety Once the reference level has been set, it is then used as a starting point for optimization of protection and safety to minimize radiological doses to the extent possible. In the case of an emergency exposure situation, as part of emergency preparedness, the anticipated exposure in the event of an emergency related to the facility or activity being planned is estimated, and the facility or activity and corresponding operating procedures are designed to minimize the projected dose (i.e., *"the dose that would be expected to be received if planned protective actions were not taken"*; IAEA, 2019), in the event of an emergency. In addition, the projected doses are estimated for all relevant scenarios of possible emergencies and taken into consideration in the development of emergency preparedness and response plans.

As part of emergency preparedness and planning, the anticipated dose reduction following implementation of protective actions is estimated to determine the anticipated residual dose (i.e., *"the dose expected to be incurred after protective actions have been terminated (or after a decision has been taken not to take protective actions)"*; IAEA, 2019). The estimated reduction in dose following protective actions would then be compared to the anticipated residual dose for a representative range of emergency scenarios and corresponding protective actions, in support of emergency preparedness and response planning and decision-making (e.g., as part of the facility design phase and siting, such that potential radiological impacts are minimized prior to embarking of the facility or activity). Such estimations can then be used to identify appropriate protective actions for relevant emergency scenarios

---

the public informed and to maintain public trust; (g) To mitigate, to the extent practicable, non-radiological consequences; (h) To protect, to the extent practicable, property and the environment; (i) To prepare, to the extent practicable, for the resumption of normal social and economic activity" (IAEA, 2015).

[6] The transition phase of an emergency is *"the period of time after the emergency response phase13 when (a) the situation is under control, (b) detailed characterization of the radiological situation has been carried out and (c) activities are planned and implemented to enable the emergency to be declared terminated"* (IAEA, 2018d).

and to refine the design of facilities, activities and processes, as well as the emergency preparedness and response plans, such that exposure is minimized.

In the event of an emergency, protective actions would be taken in accordance with the emergency preparedness plan, as part of emergency response, for the purposes of avoiding or reducing doses that might otherwise be received in an emergency exposure situation (IAEA, 2019). During the emergency, measurements would be taken to determine radiological exposure. Based on the resultant measurement data, additional protective actions and other response actions would be undertaken, as necessary, based on the measured data.

Following implementation of protective actions, monitoring and/or modelling would be conducted to determine the actual dose being incurred. The projected dose (prior to taking protective actions), which is estimated as part of emergency preparedness, would then be compared to the residual dose (after taking protective actions) determined during emergency response to verify the effectiveness of protective actions and adjust emergency response plans, as necessary, to minimize the dose incurred. Such information needs to be documented and used as input to longer-term planning, for example, in an existing exposure situation (see Sect. 17.4.2.3).

Monitoring and modelling of the situation to determine the actual dose is also undertaken for consideration in other decision-making, for example, to determine when control of the source has been regained, to identify appropriate protective actions and other response actions that need to be taken, and to determine when the emergency has been terminated and transitioning to a planned or existing exposure situation can be undertaken (IAEA, 2018d).

17.4.2.3 Criteria in an Existing Exposure Situation

Justification As in an emergency exposure situation (see Sect. 17.4.2.2), the reference level represents the maximum dose that is considered acceptable in an existing exposure situation; however, in the case of an existing exposure situation, the reference level is typically expressed as an annual effective dose to the representative person, falling within the range of 1–20 mSv, or other corresponding quantity (IAEA, 2014). Above the reference level, it is not appropriate to plan to allow exposures to occur, and below this level, optimization of protection and safety is implemented.

The process of remediation[7] is undertaken to address an existing exposure situation, following the five phases described in Sect. 17.3 above (as depicted in Fig. 17.1) (IAEA GSG-15, *in preparation*). During the preliminary evaluation (see Sect. 17.3.1), for the purposes of screening, it is necessary to set a relatively conservative

[7]Remediation is defined as *"Any measures that may be carried out to reduce the radiation exposure due to existing contamination of land areas through actions applied to the contamination itself (the source) or to the exposure pathways to humans"* (IAEA, 2019).

screening criterion that has been approved by the regulatory body or other relevant authority and that falls within the possible range of reference levels for an existing exposure situation (1–20 mSv/a) (e.g., the upper bound of the reference level range). The projected dose (prior to remediation) is then compared to the approved screening criterion and through this comparison, a determination is made regarding whether or not remediation *might be* justified.

If the projected dose prior to remediation exceeds the screening criterion, then remediation might be justified and detailed evaluation (see Sect. 17.3.2) is undertaken to determine whether or not remediation *is*, in fact, justified. In doing so, the reference level is established, taking account of the prevailing circumstances and in consultation with interested parties (i.e., stakeholders), and approved by the regulatory body or other relevant authority. The projected dose prior to remediation is then compared to the approved reference level, and if the reference level is exceeded, remediation is deemed justified.

Optimization of Protection and Safety Early in the planning phase of remediation (see Sect. 17.3.3), it is necessary to decide on the desired end state that is to be achieved through remediation of a contaminated site or area. The end state is *"the final status of a site or area at the end of activities for decommissioning and/or remediation, including approval of the radiological and physical conditions of the site and remaining structures"* (IAEA, 2019; IAEA GSG-15, in preparation). A corresponding end state criterion then needs to be established, for example, based on intended current and future use of the land and its resources, availability of resources, and other factors (IAEA GSG-15, in preparation). This end state criterion is *"a set of conditions that need to be met to verify that remediation has been completed and the defined end state has been achieved"* (IAEA GSG-15, in preparation). Achievement of the end state will ultimately lead to the release of part or all of a site or area from regulatory control or other restrictions, and subsequent post-remediation management (see Sect. 17.3.5; IAEA GSG-15, in preparation). By definition, the end state criterion should be more stringent than the reference level (as the reference level is the level above which it is not appropriate to allow exposure to occur and below which optimization of protection and safety should continue).

In addition to the need to define the desired end state, during remediation planning, it is also necessary to define relevant end points (equivalent to 'milestones') that are to be achieved at different stages during the implementation of remediation, in accordance with the approved remediation plan. The end state criterion for remediation, and the corresponding intermediate end point criteria, should be defined in the environmental impact assessment (EIA) for the remediation (IAEA GSG-15, in preparation). In addition, the EIA should include an estimate of the projected dose prior to remediation and the residual dose expected to be incurred after completion of individual remedial actions, as well as after the overall remediation is complete. The expected dose reduction should be documented in the EIA and later used to verify that the remediation is being implemented according to plan, and ultimately, has been completed.

A given end point can be used to indicate completion of an individual remedial action or a group of related remedial actions. For each end point, an end point criterion will need to be established, which is *"typically the level of contamination beyond which further decontamination or remediation is considered unnecessary"* (IAEA, 2019). A given end point is *"often calculated on the basis of a level of dose or risk that is considered acceptable"* (IAEA, 2019; IAEA GSG-15, in preparation).

As remediation is a systematic, step-wise process (covering Preliminary evaluation, Detailed evaluation, Planning, Implementation and verification, and Post-remediation management, as depicted in Fig. 17.1), end point criteria can be used to indicate when one step in the remediation process is complete and the next step should begin (IAEA GSG-15, in preparation). In addition, they can serve as benchmarks against which to compare the progress and effectiveness of remediation relative to the approved remediation plan during implementation and verification monitoring (see Sect. 17.3.4). Such comparisons should be documented, for example, as a record of site or area conditions and to answer future questions regarding how decisions were made when updating the remediation plan, establishing controls and/or setting restrictions for or terminating regulatory control of contaminated lands.

In cases where the remediation is not being implemented as planned, based on a comparison of the residual dose and the end point criterion for a given remedial action or set of remedial actions, the remediation plan will need to be reviewed and updated, as appropriate, and an assessment will need to be conducted to verify that any changes to the remediation plan will lead to the planned end state of remediation, as specified in the EIA or related regulatory conditions for the remediation (IAEA GSG-15, in preparation).

Upon completion of remediation, surveys are undertaken, and a dose assessment is conducted (based on monitoring data and surveys), as part of the verification that the end state criterion has been met. This verification includes a comparison of the residual dose (following remediation) against the dose that has been defined in the EIA and approved as part of the conditions necessary to achieve the end state criterion. Once it is verified that the end state criterion has been met, it is possible to initiate the post-remediation management phase (see Sect. 17.3.5).

17.5 Practical Application of the International System of Radiation Protection

An international system of radiation protection has been established to ensure and demonstrate the protection and safety of people and the environment from harmful effects of ionizing radiation in different exposure situations and over the lifetimes of nuclear facilities and activities. Application of this system necessitates the assessment of exposure and the corresponding dose, and the definition of appropriate, fit-for-purpose criteria, taking account of the situation and the prevailing circumstances. For example, such fit-for-purpose criteria can include the end state criterion, which is used to demonstrate that the final end state for a site or area (e.g., based on intended land-use) has been achieved at the end of the lifetime of a given facility or activity, and end point criteria, which indicate the completion of specific actions or sets of actions during the lifetime of a facility or activity. From a practical point of view, such criteria need to be scientifically-based, defensible and measurable.

In recent years, significant efforts have been undertaken to develop measurable indicators of exposure, effects, susceptibility and disease through application of biomarkers (e.g., Audette-Stuart and Yankovich, 2012; Festarini et al., 2016; Hagger et al., 2006; Hagger et al., 2009; Hagger et al., 2011; Mitchell et al., 2007; Owen et al., 2008; Van der Oost et al., 2005). A biomarker is any measurement reflecting an interaction between a biological system and a potential hazard, which may be chemical, physical or biological (WHO, 1993). Biomarkers can be used to detect changes in biological responses (e.g., molecular, cellular, physiological or behavioural) and can reflect normal biological processes, adaptive responses, and positive or negative effects following exposure to stressors (e.g., radiological, chemical, physical and/or biological stressors; individual or multiple stressors).

Although significant progress is being made in the development of biomarker techniques, uncertainty remains regarding their applicability in a regulatory context, for example, due to unanswered questions related to their sensitivity, practicality and reproducibility (Hagger et al., 2006). In addition, interpretation of effects can be difficult due to poor detectability of the biomarker 'signal' relative to the baseline 'noise' due to responses to other stressors, and there can be significant complexity in biological responses in cases of exposure to multiple stressors. For example, responses can change for different combinations of stressors and at different levels of exposure. As a result, although promising, further work is needed in order to apply biomarkers to demonstrate protection of people and the environment from the perspective of regulatory compliance.

Through the development and improvement of analytical tools, for example, to better detect exposure and effects, and to manage the large volumes of data that can be generated from environmental studies and for varying levels of biological organization, capabilities to interpret data and discern trends could improve. This could then facilitate the application of biomarkers as tools in the regulation of nuclear facilities and activities. International networking through programmes to compile data and compare different approaches for environmental impact assessment, can serve as an important mechanism in further developing analytical tools, which could lead to the application of biomarkers in a regulatory context. In addition, such initiatives could strengthen harmonization of the system of radiation protection (Fig. 17.2), for example, through:

- Identification of needs, for example, in the case of IAEA, needs of Member States for assessment tools and parameter values (IAEA, 2004, 2010) to estimate exposure and impacts of nuclear facilities and activities, and capacity-building on environmental impact assessment;

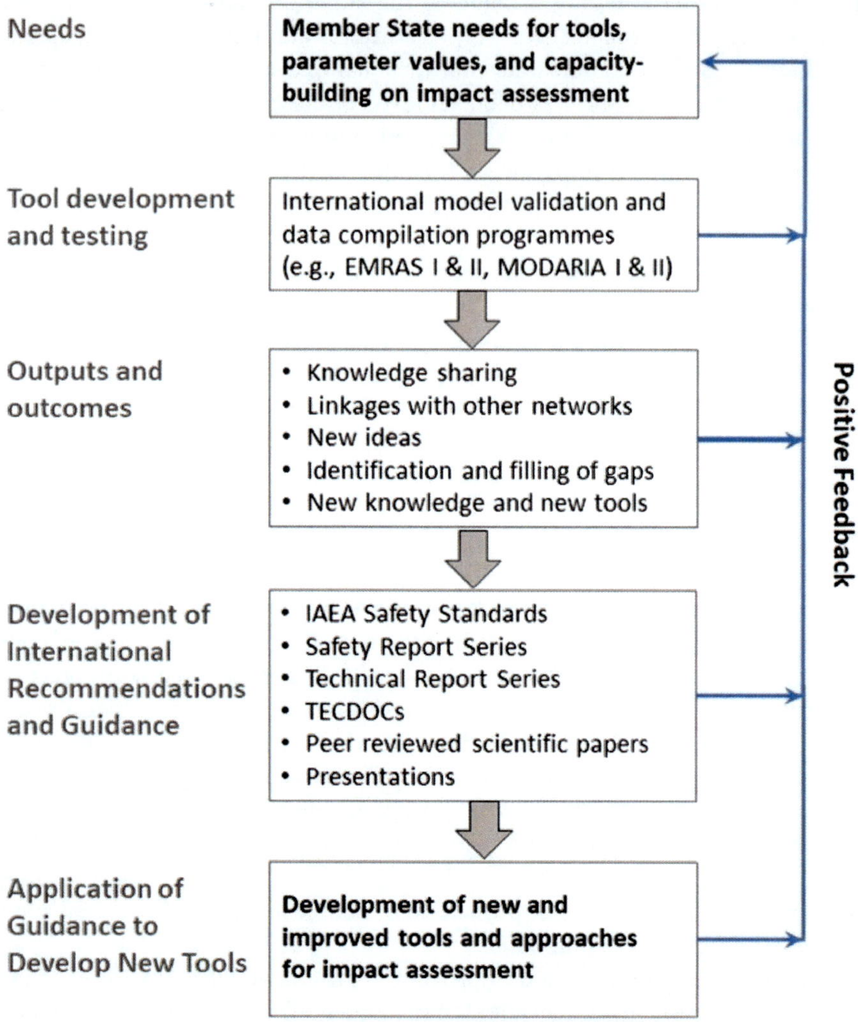

Fig. 17.2 Iterative approach to ensuring and demonstrating protection and safety

- Tool development and testing, for example, through international inter-comparison and data compilation programmes; examples of such programmes include the IAEA's Environmental Modelling for Radiation Safety (EMRAS I and II) and Modelling and Data for Radiological Impact Assessment (MODARIA I and II) programmes (e.g., IAEA, 2012);
- Evaluation of outputs and outcomes generated using different tools and approaches, so that consistency, accuracy and predictability can be assessed, and approaches can be harmonized, based on inputs from participants in the network;
- Documentation of key considerations and leveraging with other initiatives and programmes (e.g., BRITE) to facilitate compatibility, consistency and consensus-building in approach.

Within its mandate of coordination, establishment of mechanisms for information exchange, and the development of safety standards and other guidance, the IAEA plays an important role in contributing to the achievement of the above steps through interactions with international experts, organizations, and initiatives, such as BRITE. The outputs and outcomes of such activities can then feed back in to form the technical basis that informs higher-level IAEA safety standards and supporting guidance to meet the needs of Member States. In doing so, such activities and programmes can serve as mechanisms to leverage efforts to ensure and demonstrate protection and safety of people and the environment, and to further strengthen international networking in a harmonized, coherent and iterative manner (Fig. 17.2) to the benefit of the international community.

References

Brown, J. E., Alfonso, B., Avila, R., Beresford, N. A., Copplestone, D., Proehl, G., & Ulanovsky, A. (2008). The ERICA tool. *JER, 99*(9), 1371–1383.

Copplestone, D., Hirth, G. A., Johansen, M. P., Lazo, E. N., Takala, J., Sakai, K., & Yankovich, T. L. (2018). Implementation of the integrated approach in different types of exposure scenarios. *Annals of the ICRP, 47*(3–4), 304–312.

Festarini, A., Shultz, C., Stuart, M., Kim, S. B., & Ferreri, C. (2016). Cellular responses to tritium exposure in rainbow trout: HTO- and OBT-spiked feed exposure experiments. *CNL Nuclear Review, 5*(1), 115–172.

Hagger, J. A., Jones, M. B., Leonard, P., Owen, R., & Galloway, T. S. (2006). Biomarkers and integrated environmental risk assessment: Are there more questions than answers? *Integrated Environmental Assessment and Management, 2*(4), 312–329.

Hagger, J. A., Galloway, T. S., Langston, W. J., & Jones, M. B. (2009). Application of biomarkers to assess the condition of European marine sites. *Environmental Pollution, 157*(7), 2003–2010.

Hagger, J. A., Galloway, T. S., & Copplestone, D. (2011). The use of biomarkers as simple, rapid cost-effective techniques to aid in an integrated approach to environmental management and risk assessment with particular emphasis on radionuclides. *Treatise on Estuarine and Coastal Science, 4*, 339–353.

IAEA (INTERNATIONAL ATOMIC ENERGY AGENCY). (2001). *Generic models for use in assessing the impact of discharges of radioactive substances to the environment* (IAEA Safety Report Series No. 19). IAEA.

IAEA (INTERNATIONAL ATOMIC ENERGY AGENCY). (2004). *Sediment distribution coefficients and concentration factors for Biota in the marine environment* (IAEA Technical Reports Series No. 422). IAEA.

IAEA (INTERNATIONAL ATOMIC ENERGY AGENCY). (2005). *Environmental and source monitoring for purposes of radiation protection* (IAEA Safety Series No. RS-G-1.8). IAEA.

IAEA (INTERNATIONAL ATOMIC ENERGY AGENCY). (2006). *Fundamental safety principles* (IAEA Safety Fundamentals No. SF-1).

IAEA (INTERNATIONAL ATOMIC ENERGY AGENCY). (2010). *Handbook of parameter values for the prediction of radionuclide transfer in terrestrial and freshwater environments* (Technical Reports Series No. 472). IAEA.

IAEA (INTERNATIONAL ATOMIC ENERGY AGENCY). (2012). *Environmental Modelling for Radiation Safety (EMRAS) — A Summary Report of the Results of the EMRAS Programme (2003–2007) (IAEA-TECDOC-1678).* IAEA.

IAEA (INTERNATIONAL ATOMIC ENERGY AGENCY). (2014). *Radiation protection and safety of radiation sources: International Basic Safety Standards* (IAEA Safety Series No. general safety requirements (GSR) Part 3). IAEA.

IAEA (INTERNATIONAL ATOMIC ENERGY AGENCY). (2015). *Preparedness and response for a nuclear or radiological emergency* (IAEA Safety Series No. GSR Part 7). IAEA.

IAEA (INTERNATIONAL ATOMIC ENERGY AGENCY). (2018a). *Radiation protection of the public and the environment* (IAEA Safety Series No. GSG-8). IAEA.

IAEA (INTERNATIONAL ATOMIC ENERGY AGENCY). (2018b). *Prospective radiological environmental impact assessment for facilities and activities* (IAEA Safety Series No. GSG-10). IAEA.

IAEA (INTERNATIONAL ATOMIC ENERGY AGENCY). (2018c). *Regulatory control of radioactive discharges to the environment* (IAEA Safety Series No. GSG-8). IAEA.

IAEA (INTERNATIONAL ATOMIC ENERGY AGENCY). (2018d). *Arrangements for the termination of a nuclear or radiological emergency* (IAEA Standards Series No. GSG-11). IAEA.

IAEA (INTERNATIONAL ATOMIC ENERGY AGENCY). (2019). *IAEA Safety Glossary: Terminology used in nuclear safety and radiation protection: 2018 edition*. IAEA.

IAEA (INTERNATIONAL ATOMIC ENERGY AGENCY). (in preparation). *Remediation strategy and process for areas affected by past activities or events* (IAEA Safety Series No. GSG-15). IAEA.

ICRP (INTERNATIONAL COMMISSION ON RADIOLOGICAL PROTECTION), The 2007 recommendations of the international Commission on radiological protection, ICRP Publication No. 103, annals of the ICRP 37 no. 2–4 (2007).

ICRP (INTERNATIONAL COMMISSION ON RADIOLOGICAL PROTECTION), Radiological protection against radon exposure, ICRP Publication No. 126, annals of the ICRP 43 no. 3 (2014).

Mitchel, R. E. J., Audette-Stuart, M., & Yankovich, T. (2007). Effects of ionizing radiation combined with other stressors, on non-human Biota. In C. Mothersill, I. Mosse, & C. Seymour (Eds.), *Multiple stressors: A challenge for the future* (pp. 31–38).

Owen, R., Galloway, T. S., Hagger, J. A., Jones, M. B., & Depledge, M. H. (2008). Biomarkers and environmental risk assessment: Guiding principles from the human health field. *Marine Pollution Bulletin, 56*, 613–619.

Stuart, M., & Yankovich, T. (2012). Bystander effects in bullfrog tadpoles. *Radioprotection, 46*(6), S497–S502.

US-DOE (UNITED STATES DEPARTMENT OF ENERGY), RESRAD-BIOTA: A tool for implementing a graded approach to Biota dose evaluation, Interagency steering committee on radiation standards ISCORS technical report 2004–02, DOE/EH-0676 (2004).

Van der Oost, R., Porte, C., & van den Brink, N. W. (2005). Biomarkers in environmental assessment. In *Ecotoxicological testing of marine and freshwater ecosystems: Emerging techniques, trends and strategies* (pp. 87–152). Taylor & Francis.

WHO (WORLD HEALTH ORGANIZATION). (1993). *Biomarkers and risk assessment: Concepts and principles, international programmes on chemical safety (IPCS)* (Environmental health criteria 155). WHO.

Yankovich, T. L., Proehl, G., Telleria, D., & Berkovskyy, V. (2014). Generic screening models for assessing exposures to the public and ICRP reference animals and plants. In *Proceedings of the 3rd international conference on radioecology and environmental radioactivity*. Barcelona, 7–12 September 2014.

Chapter 18
Practicalities of Mainstreaming Biomarker Use – A Canadian Perspective

Steve Mihok and Malcolm McKee

Abstract This paper discusses the use of biomarkers within the environmental protection framework that has evolved since the Canadian Nuclear Safety Commission (CNSC) was given a broad mandate for the protection of the environment under the *Nuclear Safety and Control Act* (*NSCA*) in 2000. Unique insights have been obtained through environmental assessments for major nuclear projects conducted under the *Canadian Environmental Assessment Act*, and through the Environmental Effects Monitoring (EEM) requirements at uranium mines and mills (*Metal Mining Effluent Regulations* under the *Fisheries Act* (*FA*)). Altogether, the Canadian nuclear sector now has 17 years of experience in applying biological evidence in decision-making. Key examples are discussed where improved effluent controls were implemented at uranium mines for three substances (U, Mo, Se) based on risk assessments and supporting biological evidence. In the case of U, potential for localized harm from the chemical toxicity rather than radiological toxicity of U was identified at three older mines through environmental risk assessment. Evidence of potential harm in the field was also obtained from a community ecology bioindicator (benthic invertebrate biodiversity). This led to the improvement of effluent controls for U that were straightforward to implement under the *NSCA*. In the case of Mo, the weight of evidence for potential health effects on moose and other riparian wildlife from ecological risk assessments (supported by field evidence from Sweden) prompted improved controls based on precaution and pollution prevention. A technological solution was readily available for reducing Mo in effluent and was therefore implemented. In the case of Se, population level effects in fish and individual level effects in waterfowl in the USA led to selenium risks being evaluated downstream of uranium mines. Biomarkers (larval teratogenic deformities) played a pivotal role in attributing observed effects (harm) to the probable cause (selenium accumulation in the environment). However, as technological solutions were not straightforward, effort was required to build a consensus on achievable effluent control targets in a multi-stakeholder and multi-jurisdictional context. Through site-specific research

S. Mihok · M. McKee (✉)
Canadian Nuclear Safety Commission, Ottawa, ON, Canada
e-mail: Malcolm.mckee@canada.ca

M. D. Wood et al. (eds.), *Biomarkers of Radiation in the Environment*, NATO Science for Peace and Security Series A: Chemistry and Biology,
https://doi.org/10.1007/978-94-024-2101-9_18

and the latest scientific literature, criteria for selenium risk evaluation and water treatment system improvements were agreed upon and implemented. Within the EEM program, similar issues have arisen in managing a robust and defensible regulatory framework for controls on multiple hazardous substances across many mining sectors. However, an initial review of biomarkers resulted in the selection of only ecologically-relevant parameters (fish health and population indicators, benthic invertebrate biodiversity) as triggers for regulatory action. Altogether, these and other parallel experiences are discussed in terms of the desirable attributes of biological effects monitoring in a Canadian regulatory context.

18.1 Introduction

Biomarkers have been used for many years in ecotoxicology to understand how chemicals can impact on human and ecosystem health (Forbes et al., 2006; Snape et al., 2004). They have also been used as indicators of radiation exposure (Chaudhry, 2008; Hinton et al., 2004), and continue to be a valuable tool for studying radiation effects in people (Hall et al., 2017) and non-human biota (Gagnaire et al., 2017). Their potential for practical biological monitoring has been thoroughly assessed in Canada for hazardous substances (Bahamonde et al., 2016). This was done recently in terms of the lessons learned from a government-mandated Environmental Effects Monitoring (EEM) program under Environment and Climate Change Canada (ECCC). This program has applied to uranium mines as well as other metal mines since 2002. In the nuclear sector as a whole, the Canadian Nuclear Safety Commission (CNSC) has used biological information to fulfill its environmental protection mandate since 2000. The CNSC's mandate includes both nuclear and hazardous substances. However, science-based decision-making at the CNSC has not yet incorporated biomarkers in any routine context. Similarly, an approach based on bioindicators of ecosystem health is still many years away from implementation (Bréchignac et al., 2016). Altogether, for biomarkers to enter the mainstream of nuclear regulation, there are many steps that still need to be taken. In particular biomarkers still need to be more directly linked to ecologically-significant endpoints, and practical and transparent systems need to be developed for their implementation in a multi-stakeholder context.

To help researchers better understand regulatory needs for research, we first provide an overview of the CNSC's environmental protection framework. The CNSC is unique in that it was the first nuclear regulator to have a broad mandate inclusive of environmental protection. The *Nuclear Safety and Control Act* (*NSCA*) expanded the CNSC's mandate from the protection of persons to include the protection of the environment in 2000. The CNSC began exercising this mandate well before a practical consensus for radiological environmental protection was reached internationally (Stark et al., 2017). In 2000, the International Commission on Radiological Protection (ICRP) was still in the process of developing a position on a system for environmental radiological protection (ICRP, 2003).

The lessons learned from implementing an environmental protection mandate in the context of a public hearing process have provided a unique Canadian perspective on the use of scientific evidence in regulation. These lessons can help focus future efforts on how the science of biomarkers can contribute to basic regulatory functions in an environment with diverse stakeholders (Oughton, 2016). Typical regulatory functions are clear and entail risk assessment and management, licensing, compliance, enforcement and adaptive management. The interests of stakeholders are more difficult to summarize and vary internationally. To illustrate how biomarkers have, or could have, contributed to certain CNSC activities, we have prepared a retrospective overview of some issues of regulatory interest since 2000.

18.2 Environmental Protection in the Canadian Nuclear Sector

The CNSC is the responsible federal body in terms of licensing nuclear facilities in Canada. It's framework for environmental protection is publicly available in a consolidated report (CNSC 2017b). This recent report reflects 17 years of implementing broad responsibilities under the *NSCA*. In exercising this mandate, the CNSC has interpreted legal requirements in a manner that is consistent with other Canadian legislation and international guidance, and has done so with considerable stakeholder participation. For example, the CNSC respects basic environmental tenets held by many stakeholders such as the precautionary principle, the "polluter pays" principle, and the concepts of pollution prevention, sustainable development and adaptive management. The key legal requirements that the CNSC is obligated to enforce are: to prevent "*unreasonable risk*", and to ensure that "*reasonable precaution*" is taken to control releases. Other major Canadian legislation affecting these generic requirements and providing guiding principles is in the *Canadian Environmental Protection Act* (*CEPA* 1999), the *Canadian Environmental Assessment Act* (*CEAA*), the *Species at Risk Act* (*SARA*), the *Fisheries Act* (*FA*), the *Metal Mining Effluent Regulations* (*MMER*) and the *Migratory Birds Convention Act* (*MBCA*).

The CNSC's licensing approach is risk informed and incorporates science-based decision making. Key decisions routinely involve prospective and ongoing ecological and human health risk assessments that are re-evaluated and revised throughout the entire life cycle of major facilities, e.g. nuclear power reactors, uranium mines and mills, fuel conversion and fabrication facilities, waste management facilities, etc. The CNSC recognizes that uncertainty exists in science and that performance must be assessed against indicators and targets that are based on sound science. For nuclear substances, uncertainty is primarily accommodated by the well-known radiation protection goal to keep all releases as low as reasonably achievable (ALARA), social and economic factors being taken into account. For hazardous substances, this is done through the application of the best available technology and techniques economically achievable (BATEA). This means that most anthropogenic contaminants released by nuclear facilities are being controlled to levels far below those that might cause measurable impacts. Emerging issues are few and are typically related to hazardous substances; environmental protection issues resulting from releases of nuclear substances or their cumulative radiation effects are minimal (Mihok & Thompson, 2012).

This primary emphasis on pollution prevention has been particularly effective in terms of simplifying the need for biological monitoring in a surveillance context for nuclear substances. Similarly, formal investigations triggered by unexpected monitoring results for hazardous substances are few (McKee & Phaneuf, 2014); these are discussed in later sections. Campaign monitoring of biota is nevertheless routinely done to support and verify risk assessment assumptions for both types of substances. It is also occasionally done in an operational context or for emerging issues, particularly to test whether modelling is sufficiently conservative. Biological monitoring typically targets only a few practical bioindicators, and is therefore limited in scope.

In recent years, regulatory transparency at the CNSC has included the evolution of a documented system for environmental protection through cooperation with the Canadian Standards Association (CSA). This has resulted in the development of guidance/standards specific to the Canadian nuclear sector (CNSC 2017b). These documents have provided licensees with more certainty in regulatory expectations for protection of the environment. In recent years, this logic was formalized through inclusion of CSA standards in the License Conditions Handbooks of major licenses. Development of CNSC requirements and guidance has also involved a broader early system of interaction with other stakeholders; specifically, the publication of Discussion Papers with subsequent disposition of public comments.

The technical details of how the CNSC has approached environmental risk assessment are documented in the CSA standard N288.6 "*Environmental risk assessment at Class I nuclear facilities and uranium mines and mills*". This document addresses the design, implementation, and management of an environmental risk assessment that aligns with internal and external stakeholder expectations, while incorporating current best practices. CSA N288.6 incorporates the ecotoxicological methods developed by various jurisdictions for the effects of hazardous substances on non-human biota. For nuclear substances, the standard has adopted the international consensus that radiation effects on non-human biota should be addressed in terms of effects on populations of non-human biota rather than individuals (UNSCEAR, 2011). Individuals are nevertheless being assessed for impacts on threatened or endangered species, a *SARA* requirement. The quantitative information available for setting radiological benchmarks comes from laboratory studies of survival, growth or reproduction endpoints in individuals. Many significant data gaps however remain for ecological inference (Garnier-Laplace et al., 2015a).

Biomarkers are not mentioned in CSA N288.6 but are briefly discussed in the context of biological effects monitoring in CSA N288.4 "*Environmental monitoring programs at Class I nuclear facilities and uranium mines and mills*". The value of biomarkers is recognized in terms of tracking responses at low levels of biological organization, e.g. to provide early warning of potential effects at higher levels.

However, CSA N288.4 does not go beyond a simple discussion of how biomarkers might meet the ultimate objective of providing ecologically-relevant information for population level effects. Regulators are clearly interested in early indicators of potential harm, but must also temper actions based on sound evidence tying observed effects to controllable causes. This is particularly difficult in a multi-contaminant context, e.g. at uranium mines and mills. Most facilities release several nuclear and hazardous substances in parallel; hence even the root causes of changes in a well-studied biological indicator (benthic invertebrate biodiversity) are open to interpretation (Kilgour et al., 2018). Typically, these types of scientific nuances are very difficult to manage both from a regulatory perspective and in terms of public perception of "unreasonable risk". Hence, a strict pollution prevention approach remains the foundation of regulatory philosophy.

The above practical and scientific issues are not unique to Canada (Colin et al., 2016; Sanchez & Porcher, 2009). Similarly, despite a sophisticated system of environmental regulation and decades of experience in applying pollution prevention, emerging issues still exist. These have required new scientific insights and ongoing risk management activities to develop or optimize controls. Emerging issues have also required the building of scientific consensus on specific environmental protection objectives in a multi-stakeholder and multi-jurisdictional environment. The remainder of this paper will focus on where and how biomarkers have or have not contributed to environmental regulation at the CNSC. This is done by providing examples of how biological evidence as a whole has influenced key environmental protection issues in the nuclear sector in Canada since the *NSCA* came into force in 2000.

18.3 Emerging Issues for Hazardous Substances in Canada

18.3.1 Releases of Uranium

In 2000, in parallel with implementing new responsibilities under the *NSCA*, the CNSC was also working with other government departments under *CEPA* 1999 on the "impacts of releases of radionuclides from nuclear facilities on non-human biota" (Environment Canada and Health Canada 2004). The outcome of this assessment under *CEPA* 1999 was the allocation of certain risk management activities to the CNSC. Follow-up activities were related to uranium as a chemical hazard at a few older sites (McKee & Phaneuf, 2014). Modelling methods and critical toxicity thresholds for detailed ecological risk assessment were still in development at that time, especially for nuclear substances. Thresholds were based on interpretation of ecologically-relevant effects on survival, growth or reproduction in organisms studied in the laboratory. Environmental monitoring data supporting the assessment (under the former Atomic Energy Control Board), while of some value, were of limited use as they were not collected with this objective in mind. Hence, the assessment was based on best use of measured media concentrations, mainly from compliance monitoring, supplemented by fish tissue concentrations. These data contributed to the calculation of hazard quotients relative to Estimated No-Effects Values (ENEVs) for representative biota using an ecotoxicological approach. ENEV's were chosen from Critical Toxicity Values in the literature and weighted by a safety or uncertainty factor. There was little use of evidence from biomarker studies. However, studies documenting certain effects (e.g. DNA strand breaks, chromosome aberrations) did influence the choice of a high RBE weighting factor of 40 for alpha radiation, relative to other choices (Chambers et al., 2006). Biological effects monitoring data from the field were limited at that time, with the exception of biodiversity data from periodic benthic invertebrate community sampling at operating uranium mines and mills.

The *CEPA* 1999 assessment concluded that there was potential for localized harm to organisms resulting from past and current releases of uranium and uranium compounds from three older facilities as a result of uranium chemical toxicity (Environment Canada and Health Canada, 2004). There was no evidence for harm from radiation effects from radionuclides, including uranium itself. Potential uranium toxicity issues were not identified for modern facilities. These conclusions led to a requirement for improvements in effluent quality through the modernization and optimization of water treatment systems using improved water management practices and standard chemical treatment technology to reduce uranium. This posed a regulatory challenge as at no time had any mine exceeded the Saskatchewan provincial regulatory effluent limit of 2.5 mg/L. Regulatory justification under the *NSCA* was to require treatment based on reasonable precaution to protect the environment, and to control releases through BATEA. Optimization was determined to be capable of achieving releases of about 0.1 mg/L (SENES, 2006). In subsequent years, further field (Robertson & Liber, 2007) and laboratory studies were undertaken to characterize the role of uranium in observed effects as opposed to other factors (Goulet et al., 2015).

Improvements in effluent treatment saw releases decline from past highs of metric tonnes per year at one facility (CNSC and Environment Canada 2009) to current releases at the same facility of a few hundred kg per year (CNSC 2017c). Releases at modern facilities now all range in the tens of kg per year. Reporting of uranium-specific risk management activities ended in 2010 (CNSC and Environment Canada 2012) though public reporting continues within the annual regulatory oversight reporting by CNSC staff to the Commission. Uranium has been found to be particularly toxic at very low concentrations (Canadian Environmental Quality Guideline of 15 μg/L) to sensitive invertebrates (CCME, 2011). Attribution of effects to the chemical toxicity of uranium itself versus other factors (Vanhoudt et al., 2012; Mathews et al., 2009) continues to be an active area of research for benthic invertebrates (Massarin et al., 2010), in the context of both population effects (Biron et al., 2011) and community effects (Kilgour et al., 2018). Many studies have now also been completed on other organisms through coordinated research in Europe (STAR, COMET) and elsewhere, e.g. epigenetic and transcriptional studies of zebra fish exposed to depleted uranium (Gombeau et al., 2017). The ongoing interests of the research community in biomarkers unique to uranium chemical toxicity are one clear application for "mainstreaming" biomarkers (Guéguen et al., 2016). Practical indicators would be a welcome addition to the regulator's toolbox, which currently relies on labour-intensive surveys of benthic invertebrates for biological inference.

18.3.2 Releases of Molybdenum

When the *NSCA* came into force in 2000, the CNSC's new mandate included both hazardous and nuclear substances. At that time, monitoring at uranium mines and mills demonstrated that molybdenum was accumulating in sediments far downstream of effluent release points. Levels were in excess of those predicted in facility-specific environmental risk assessments. The emergence of molybdenum as a contaminant of concern was a function of accumulation over time, but also of a change in the extraction process for uranium a few years earlier. This change resulted in molybdenum being diverted for operational reasons to the effluent circuit rather than to the refined product (yellowcake) or tailings. At this time, molybdenum releases were not being regulated through prescribed effluent limits by any jurisdiction. In particular, the federal *Metal Mining and Effluent Regulations* (*MMER*, administered by ECCC) had only set limits for arsenic, copper, cyanide, lead, nickel, zinc, Ra-226, and total suspended solids and pH. This was on the assumption that a short list of limits would be sufficient to capture toxicity issues for other metals/metalloids in mixed effluents. As of 2017, proposed amendments to the *MMER* still do not include effluent limits for molybdenum (Government of Canada, 2017), though it's periodic characterization within effluent continues to be required.

Environmental risk assessments conducted under the environmental mandate of the CNSC in the early 2000s indicated potential for harm to riparian wildlife from a continuation of uncontrolled releases of molybdenum. Conclusions were based on modelling and limited data, with almost no information on exposure of wildlife, let alone health or other effects. Altogether, these analyses highlighted only the potential for harm in species that rely on aquatic macrophytes in their diet (muskrats, moose, waterfowl, etc.). Aquatic life such as fish were not predicted to be at risk; as the primary issue was accumulation of molybdenum in sediments and macrophytes over time rather than the water concentration itself. At that time, evidence of actual harm from excess molybdenum in the environment was available from independent studies in southwest Sweden where mysterious deaths of moose began in 1982 (Frank et al., 1994). This phenomenon was attributed to the mobilization of molybdenum in the environment as a result of liming of soils to ameliorate the effects of acid rain. By 2004, the weight of evidence from Sweden strongly suggested that a well-known, but complex, physiological interaction was responsible for unusual moose mortality. This involved interactions between molybdenum and other micronutrients, as in domestic ruminants (Frank, 2004). This conclusion was based on physiological biomarkers and a demonstration of mechanisms that could account for the observed phenomenon. Specifically, a combination of clinical chemistry, trace element analysis, and biochemistry correlated with pathological findings corroborated molybdenosis and molybdenum-induced disturbances of copper metabolism as the probable etiological factor.

Evidence from Sweden showing potential for harm to valued wildlife was reinforced by studies conducted elsewhere, where causes of moose mortality were not clear (Alaska, Minnesota). Combined with the conclusions of several site-specific risk assessments, this led to special field investigations at two uranium mines in Canada to measure molybdenum in the riparian food chain. Results helped to delineate the spatial extent of accumulation of molybdenum downstream under different circumstances. More importantly, this work verified exposure to elevated molybdenum in major consumers of macrophytes such as muskrats, which were more amenable to study than moose. This work was not intended to demonstrate evidence of harm, just exposure, and did not involve any biomarker or other biomedical monitoring of wildlife health.

Altogether, the weight of scientific evidence for potential harm resulted in molybdenum effluent controls being required by the CNSC. Effluent treatment for molybdenum was implemented at operating uranium mines and mills by the mid 2000s, as a result of either a CNSC licence condition or the licensees' commitment to continuous improvement and proactive risk management. Currently control of molybdenum releases is a CNSC expectation for proposed new operations based on BATEA. In the absence of federal or provincial legislated effluent limits for molybdenum, the CNSC requires licensees to develop facility-specific controls and action levels within their environmental protection program codes of practice. As traditional chemical precipitation technology was adequate, it was possible to retrofit existing treatment plants, and cost-effective to incorporate a molybdenum treatment circuit during the design of new facilities. Scientific and other practical issues were first discussed at the Commission hearing in 2004 for the new uranium mine proposed for Cigar Lake, Saskatchewan. The debate is documented in the transcripts of the hearing and the accompanying environmental impact study under *CEAA*. Altogether, the need for effluent treatment technology for molybdenum removal was not extensively debated. However, there were differences of opinion on interpretation of the scientific evidence and environmental targets (Ecometrix Inc., 2008; Government of Saskatchewan, 2017).

In terms of the environmental protection framework at the CNSC, a risk management decision on molybdenum was taken as an operational, precautionary measure without relying on any further field evidence from biomarkers or bioindicators. Action was also taken without confirmed field evidence of harm, and did not involve the development of new regulations or prescribed limits. It was justified in a pollution prevention context. With time this decision to implement effluent controls has proven to be a practical solution for protecting downstream environments from significant accumulation of molybdenum. Molybdenum now mostly reports to tailings and is well-managed. Regulatory attention on molybdenum has also resulted in proponents employing more novel risk management approaches, e.g. AREVA Resources is considering the inclusion of a molybdenum recovery and packaging circuit to manage molybdenum as a by-product rather than a waste-product (Areva

Resources Canada Inc., 2011). Altogether, as a result of many years of improved effluent treatment (CNSC 2017c), recent releases (2012–2016) from modern facilities have been on the order of tens of kg per year. This compares favourably to the performance of older facilities where releases between 1998 and 2003 (pre-treatment period) were in the tens of thousands of kg per year (up to 43,000) and are now substantially lower at one to two thousand kg per year.

Molybdenum concentrations continue to be monitored regularly in downstream environments to evaluate the efficacy of the installed treatment technology with respect to recovery in the receiving environment. To date, surface water concentrations have declined below criteria for the protection of aquatic life (Heijerick et al., 2012; Government of Saskatchewan, 2017). Follow-up studies on the health of riparian wildlife have yet to be undertaken. A field assessment is nevertheless nearing completion on whether aquatic macrophtyes in key habitats have decreased to levels below a riparian wildlife risk-based modelling toxicity threshold. Biomarkers could therefore still play a role in continuing adaptive management, e.g. if updated risk assessments suggest wildlife continue to be at risk without any active remediation. Therefore, innovation in the development of physiological biomarkers for diagnosing wildlife molybdenosis would be a welcome addition to the regulatory toolbox. Similarly, field-friendly, non-lethal methods for tracing exposure of wildlife to contaminants unique to uranium mining remains a fruitful area for research (Froehlich et al., 2016). Public stakeholders, in particular indigenous peoples in northern Canada, are specifically interested in unequivocal evidence for a limited spatial footprint of activities related to uranium mining. Wildlife make a significant contribution to their traditional lifestyles, but have yet to be extensively sampled in the nuclear sector in Canada.

18.3.3 Releases of Selenium

In 2000, the CNSC was aware of population level effects in fish and individual level effects in waterfowl from releases of selenium related to industrial activities in the USA (Skorupa, 1998). Selenium was then being considered as a potential cause of fish deformities and population changes in waters receiving tailings from past historical uranium mining activities. In 2006, the growing evidence for concerns at currently operating facilities was formally assessed (Wismer & McKee, 2007) in support of a Commission public hearing/licensing decision. Altogether, initial field observations and the weight of scientific evidence resulted in direction from the Commission to licensees to improve controls for selenium in the uranium mining sector. This was done under the *NSCA* and well-implemented by 2008/9. Action on this topic involved considerable planning and dialogue with other regulators and industry experts (Ecometrix Inc., 2008). In parallel, independent field research was initiated by various academic groups with the mine operators providing access to their sites. Field studies showed that selenium was the probable cause of teratogenic deformities in the offspring of fish downstream of uranium mines (Muscatello et al.,

2008; Muscatello et al., 2006). Biomarkers (larval teratogenic deformities) played a pivotal role in attributing observed effects (harm) to probable causes (Se biomagnification from lower trophic levels via sediment biofilms to benthic macroinvertebrates and benthic foraging fish) (Wiramenden et al., 2010; Tse et al., 2012).

Selenium toxicity in fish has continued to be an intense topic of research and regulatory interest (Janz, 2012). A consensus on water quality criteria was only recently reached in the USA (US EPA, 2016). It includes multiple criteria and is the first time a biological criterion (egg/ovary Se tissue concentration in fish) has taken precedence over a water quality criterion. The academic community continues to examine a broader criterion based on all organisms (DeForest et al., 2017). In Canada, technological solutions to control selenium releases were implemented in a BATEA context through retro-fitting existing facilities by industry a decade ago. However, adaptive management has not been as straightforward as for molybdenum and continues to be refined. The Saskatchewan government has established selenium limits within its provincial uranium mine licenses. However, these limits are at a concentration that the federal regulator (the CNSC) still considers to be inadequately protective as teratogenicity has been confirmed for releases at these levels.

As an essential micro-nutrient, selenium is a food-chain contaminant rather than a water-borne contaminant. It has a complex biogeochemistry with lentic aquatic ecosystems significantly more sensitive than lotic ecosystems. This site-specific sensitivity poses a regulatory challenge as shown in the recent ten-year review of the efficacy of the Canadian *MMER*'s. This global review failed to identify a specific end-of-pipe selenium concentration that could be used for all mines nationally. Instead, the current proposal is to use a specific effluent concentration value as a trigger for the monitoring and formal reporting of fish tissue (egg/ovary) concentrations. This will provide a national dataset for evaluation during future reviews of the regulations (Government of Canada, 2017). There is also a national initiative to develop a framework for a risk management plan for selenium for five industrial sectors under *CEPA* 1999 (Environment Canada and Health Canada, 2015). Proposed regulations for coal mining are also being considered (Environment and Climate Change Canada, 2017). However, standardized federal effluent limits under *CEPA* 1999 are unlikely in the near future due to the challenge posed by the environmental specificity of this contaminant. The CNSC therefore continues to regulate selenium by requiring control and treatment based on facility-specific risk assessments and monitoring programs.

In the uranium mining sector (CNSC 2017c), new controls on selenium releases have been reasonably effective, though this substance still remains a challenge. Recent (2016) annual releases from the five active sites have averaged 13 kg with the highest release being ~30 kg. This compares favourably to releases at the main site of concern which had past releases as high as 120 kg per year. With targeted treatment for selenium, releases have been reduced to a five year average of 17 kg (SD = 3) per year.

In summary, the latest scientific literature and site-specific biomarker research were both critical and very informative for evaluating selenium risks in Canada. Various lines of evidence indicated the need for water treatment system improvements to prevent further harm to fish and other organisms. A consensus on action was achieved among diverse stakeholders under the *NSCA*. However, action has proceeded along different tracks under the auspices of different regulatory authorities with as yet no concrete action for other industrial sectors. Through adaptive management of treatment technologies at the CNSC and its licensees, the situation continues to improve in historically-contaminated environments. Modern facilities are now estimating ore selenium content from drilling programs and are modelling the implications to support advanced selenium treatment options (Areva Resources Canada Inc., 2017). This is the best example of how biomarkers have contributed to action on an important emerging issue in environmental protection in the nuclear sector in Canada.

18.4 Environmental Effects Monitoring in Canada

The environmental protection goals of the CNSC overlap with the responsibilities of other jurisdictions and hence joint regulatory activities are common. For example, benthic invertebrate biodiversity monitoring was first implemented as a requirement of the Saskatchewan provincial government. This was done prior to the CNSC's environmental mandate under the *NSCA*. It has also been a core tool for regulatory inference at the CNSC. Monitoring of benthos has continued and is currently a prescriptive element of the federal *MMER* EEM toolbox applicable to all mines in Canada, including uranium mines. Under the *Fisheries Act*, the *MMER* prescribes authorized release limits for a limited number of substances common to mining effluents. The *MMER* also includes effluent and receiving environment monitoring requirements. These have been enforced by the CNSC in licensing. However, independent actions, both effluent treatment and environmental monitoring, have also been taken by the CNSC as outlined earlier (U/Mo/Se).

An important transition in the logic of scientific triggers for regulatory action occurred in 2002 with the implementation of EEM in the metal mining sector under the *MMER* (biological, effluent and water quality monitoring). Similar requirements were first implemented in Canada in 1992 at pulp and paper mills. Lessons learned were summarized in a special issue of the Water Quality Research Journal of Canada (Munkittrick et al., 2002). The EEM components of the *MMER*'s were informed by a large national program evaluating the efficacy of a wide range of monitoring approaches and biomarkers: the Aquatic Effects Technology Evaluation Program or AETE (ESG Interternational Inc., 1999). Monitoring tools considered by AETE addressed at least one of the following questions: 1) are contaminants getting into the system? 2) are contaminants bioavailable? 3) is there a measurable biological response? 4) are the contaminants in the system causing the observed response? Over 100 potential monitoring tools were considered through initial screening, literature (technical) reviews, laboratory and field testing. The biological monitoring component evaluated tools to determine if contaminants were bioavailable, and to measure biological responses in receiving waters. A wide range of tools were considered including metal levels and biochemical indices in plant and animal tissues, metallothionein levels in fish tissues, various measures of benthic invertebrates and surveys of fish including abundance, growth, histopathology, reproduction and organ size.

MMER requirements that evolved out of the first EEM program for pulp and paper and the AETE were applied to the uranium mining sector from 2002 onwards. They included an iterative or phased set of prescribed monitoring and investigative protocols for which there is now more than a decade of experience in their application and interpretation. These activities targeted the detection of statistically-meaningful changes (not necessarily harm) in aquatic ecosystems attributable to releases of contaminants. Protocols included triggers for additional studies of the magnitude and extent of effects, as well "investigation of cause" studies to link presumed causes to effects. Biomarkers were not implemented in the 2002 EEM mining requirements after many years of debate (McCarty & Munkittrick, 1996), but bioindicators were included (McCarty et al., 2002). These targeted ecologically-relevant parameters that were practical to measure in the field and were amenable to hypothesis testing (individual fish health and population indicators, benthic invertebrate biodiversity). Guidance for monitoring fish in particular has evolved through time with many practical insights now available on how to conduct more robust and informative monitoring (Munkittrick et al., 2010; Arciszewski et al., 2017). A short list of factors to be considered includes: the adequacy of reference sites and the need for multiple reference sites, the interpretation of cause and effect in the presence of confounding contaminants, the ecological relevance of changes in endpoints versus proof of actual harm, natural variability limiting statistical inference, selection of representative species and timing of sampling, adequacy of sample sizes in terms of effect sizes and power, many other statistical design considerations; and lastly, quality assurance and quality control for field/laboratory methods. These and other points have all been elegantly discussed in a thorough critical review of the role of "omics" in ecosystem health from the unique perspective of the Canadian EEM program (Bahamonde et al., 2016). In CNSC's experience, an additional lesson learned is the importance of keeping any system simple enough to be understood by diverse stakeholders.

QA/QC issues arising from using biomarker data for inference in a regulatory context are of particular concern above and beyond ecological relevance (Kroon et al., 2017). Use of sophisticated technologies is still often framed in the context of providing practical monitoring tools for regulatory purposes. Regulators such as the CNSC typically require independent confirmation and expert interpretation/peer review of methods and results before acting on sophisticated scientific evidence. This should always be considered when proposing the use of sophisticated techniques in either an assessment or a compliance context, i.e. biomarkers (Feswick et al., 2017). Academic expertise and standards may not be readily-available (and are rarely commercially-available) across government, industry, and consulting. Standardized procedures and access to the necessary technology and expertise by the environmental consulting industry is important for large national industries. It is not economically feasible for government to do the monitoring itself. Similar problems arise in hypothesis testing for cause and effect, even when data quality is good. Specifically, the definition of normal in a statistical sense is not straightforward in an environmental context (Kilgour et al., 2017). Similar issues arise in the biomedical field (Friedrichs et al., 2012).

Lastly, communication of the potential for environmental "harm" to diverse stakeholders based on complex lines of scientific evidence is not trivial. Information dissemination for the public needs to be proactive to achieve transparency and consensus for any regulatory decision that is not prescribed by law. The CNSC has implemented a very public assessment and licensing process for these reasons. This process has been effective in building confidence in decisions to require adaptive management for U/Mo/Se. In recent years, an additional need to facilitate more regulatory transparency was identified. In 2011, the CNSC created a participant funding program, partly as a result of requests from the public for engaging "independent experts". This program provides grants to individuals, not-for-profit organizations and Indigenous groups to participate in the CNSC's environmental assessment and licensing processes.

It is beyond the scope of this paper to do justice to the many lessons learned from the massive EEM program and similar efforts elsewhere (Sanchez & Porcher, 2009). Canadian lessons learned are documented in the third national assessment (ECCC, 2016). Information generated in this program in the nuclear sector has been and continues to be used in the regulatory oversight of the uranium mining sector in Canada. Monitoring protocols have improved with time and a few formal investigations of cause have now been conducted. These have helped to narrow down the scope of potential effects of individual contaminants for which regulations and discharge limits are not prescribed. Although continuing to be researched by the academic community, biomarkers have not yet entered the mainstream of practical biological monitoring in terms of EEM under the *MMER*. The latest proposal to amend the *MMER* since its last revision in 2002, does not include any specific mention of biomarkers (Government of Canada, 2017).

18.5 Nuclear Substances

The previous case histories have focused on the uranium mining sector and on hazardous rather than nuclear substances. After 17 years of experience in ecological risk assessment, there are very few practical environmental issues that are specific to risks from operational releases of nuclear substances (Mihok & Thompson, 2012). Outstanding issues relate mainly to legacy uranium mining and milling rather than to current releases (IAEA, 2014). For example, the persistence of Ra-226 in sediments at legacy sites is a long-term issue of regulatory and scientific interest (Beaugelin-Seiller, 2014). Potential effects on biota from Ra-226 persistence also include the significant contributions of Po-210 (Skipperud et al., 2013) and other uranium decay chain radionuclides to total dose (Beaugelin-Seiller et al., 2016). The CNSC has supported independent research on the effects of alpha emitters such as Ra-226 at environmentally-relevant concentrations to encourage practical research. Studies have included laboratory experiments on ecologically-relevant endpoints such as survival, growth and reproduction. Researchers have explored biomarkers in these laboratory animals in parallel (Walsh et al., 2015; Smith et al., 2015; Shi et al., 2017). Similar field studies in Canada are few (Audette-Stuart et al., 2011), although biomarkers have been used for biodosimetry elsewhere (Ulsh et al., 2003). Biomarkers have also be used as an early warning mechanism (Lourenço et al., 2016). While biomarker research initiatives mature, Canadian regulatory effort remains focused on more practical remediation efforts, including the verification of success through environmental rather than biological monitoring.

At nuclear power reactor sites in Canada, releases of nuclear substances are very low as a result of a long-standing regulatory requirement for ALARA. Public and regulatory interest is therefore mainly focused on low-dose exposure and effects on human health (CNSC 2017a). The health effects of tritium are of particular interest, as well as its environmental behaviour (Mihok et al., 2016). The wider research community also continues to research tritium in multiple contexts, including a recent biomarker study of fish exposed at current drinking water standards (Stuart et al., 2016). Clearly, there are still some nuclear issues where biomarkers can play a role to inform future regulatory oversight of nuclear facilities. The lessons learned from nuclear accidents will also be useful in developing tools for assessing any long-term impacts on the environment (Beresford et al., 2016). However, for biomarker studies to enter the mainstream of routine regulatory oversight of nuclear facilities; they must target relevant levels of exposure and meaningful biological endpoints. Consensus within the scientific community on thresholds of significant effects is also essential (Garnier-Laplace et al., 2015b). These points are a common theme in the literature on hazardous substances, and remain valid today (Hook et al., 2014), as in the past (Forbes et al., 2006).

18.6 Physical Effects at Nuclear Power Plants

A final aspect of the environmental impacts of the nuclear sector deserves special mention: the physical effects (impingement and entrainment of aquatic organisms and thermal releases) on fish as a result of once-through condenser cooling water intakes and discharges at nuclear power reactors. These issues were first addressed in comprehensive ecological risk assessments conducted under the *NSCA* in the early 2000s. They have received the scrutiny of Indigenous communities and stakeholders at the Bruce Nuclear Generating Station (BNGS). The BNGS is one of the largest reactor complexes in the world. It is currently (2017) applying for a ten-year license, which includes the refurbishment of six of the eight reactors. If the reactors are refurbished successfully, the generating station could have competent reactors operating until 2064.

BNGS has two intakes located outside the littoral zone, each with a velocity cap which reduces the velocity of the water around the intake to limit fish impingement.

This is considered industry best practice. One of the caps also has a chain rope barrier that limits the impingement of schooling fish. The BNGS was built before diffusers were included in more modern reactor designs to mitigate thermal effects. Heated water entering Lake Huron has been of particular concern for the protection of fish that spawn nearshore and are sensitive to small thermal changes (lake whitefish, round whitefish). Their eggs normally incubate over winter at temperatures just a few degrees above freezing.

There is an Indigenous commercial fishery for lake whitefish in Lake Huron that is a source of income for the Saugeen Ojibway Nation (SON). Following a public debate about potential impacts at a Commission hearing in 2003, Bruce Power and the SON have participated in research initiatives with academia to study impacts on whitefish. Large collaborative research projects have been undertaken with significant funding provided by Bruce Power. Evidence from DNA biomarkers and stable isotopes suggests that locally-captured whitefish are part of a larger basin-wide or lake-wide population, although separate ecological niches near the site have also been noted (Overdyk et al., 2016; Overdyk et al., 2015; Eberts et al., 2017). The use of the biomarkers was helpful in determining whether the loss of fish due to impingement and entrainment could have an effect on the whitefish population in the vicinity of the site. The absence of a distinct local population helped put the losses into an ecological context.

Laboratory incubation of lake whitefish eggs has indicated that they are sensitive to the increases in temperature that occur in nearshore spawning areas from the operations of the BNGS. This may have an impact on post-hatch survival and results in increased/advanced growth rates, but not increased embryonic mortality (Thome et al., 2016). Laboratory research has included studies on how biomarkers respond to different increments and frequencies of thermal stress, e.g. how heat shock proteins are induced as reflected in mRNA (Whitehouse et al., 2017). This research has shown that thermal stress can initiate a heat shock response at an early age and throughout embryogenesis. However, repeated thermal stress can either result in reduced responsiveness or a downregulation of inducible proteins. As with biomarkers in other contexts (Kroon et al., 2017), short- and long-term effects are quantifiable, but links to ecologically-relevant endpoints require further research to facilitate their application. In the context of thermal effects, heat shock proteins have potential as regulatory tools. However, the details of how this information could be used in a compliance or risk assessment context are still far from clear.

18.7 Conclusions

This retrospective review was intended as an introduction to the use of biological monitoring data in the regulatory oversight of the nuclear sector in Canada. It reflects the experiences and opinions of two scientists who have participated in the development of the environmental protection framework at the CNSC since 2000.

Although few examples could be provided for the direct use of biomarkers in regulation, considerable thought was given to reviewing their potential.

The overall conclusion of this exercise is similar to a repeating theme in reviews of biomarker use by the academic community; i.e. their uptake outside of the research community still requires proof of a direct link between biomarkers and measureable, ecologically-relevant effects. In a nuclear regulatory context, the types of effects to be monitored must also align with fairly broad environmental protection objectives (e.g. prevention of unreasonable risk as in the protection of populations of animals and plants). Evidence of harm must also be clear enough to be understood by diverse stakeholders, sometimes lacking a scientific background. Complex arguments linking causes and effects are unlikely to build consensus. Unnecessary pubic concern can also be kindled if biomarkers or bioindicators simply reflect equivocal changes in non-human biota/the environment, but are not linked to individual or ecological harm. This last issue is especially important for potential radiological effects as there is a heightened level of public perception of risk for nuclear versus hazardous substances.

These points pose very restrictive conditions on the practical application of biomarkers for regulation of the nuclear sector. Presently, the routine application of ALARA and BATEA has nearly eliminated the need for sophisticated biological monitoring. That said, we have shown that biomarkers have played important roles in two key regulatory decisions on effluent controls at uranium mines and mills in Canada. Specifically, biomarkers have provided unequivocal direct evidence of actual harm from selenium through the monitoring of teratogenic deformities in fish in the field. Similarly, biomarkers have provided indirect supporting evidence of potential harm in riparian wildlife from molybdenum through the monitoring of animal physiology and trace element metabolism. Lastly, biomarkers are presently contributing to understanding the physical impacts on fish arising from nuclear power reactor operations, an ongoing issue under scientific scrutiny.

Although operational issues are few at modern facilities, biomarkers can still contribute to scoping future regulatory priorities in environmental remediation at legacy sites. An example would be the use of biomarkers to address uncertainties in the health impacts of alpha-emitters such as Ra-226, or uranium itself, on long-lived fish at historic uranium mines. Integrated tools for monitoring ecosystem health may also prove to be useful for managing long-term consequences at legacy sites. This topic continues to be actively researched and debated in the context of nuclear accidents (Strand et al., 2017).

Acknowledgments The authors thank numerous colleagues at the CNSC who have participated in the development of the environmental protection framework at the CNSC, in particular Patsy Thompson and Mike Rinker. Barbara Dowsley kindly contributed insights on physical effects at nuclear power plants.

References

Arciszewski, T. J., Munkittrick, K. R., Scrimgeour, G. J., Dubé, M. G., Wrona, F. J., & Hazewinkel, R. R. (2017). Using adaptive processes and adverse outcome pathways to develop meaningful, robust, and actionable environmental monitoring programs. *Integrated Environmental Assessment and Management, 13*(5), 877–891. https://doi.org/10.1002/ieam.1938

Areva Resources Canada Inc. (2011). *Midwest project environmental impact statement*. Saskatoon.

Areva Resources Canada Inc. (2017). Written submission from AREVA resources Canada Inc. in the matter of McClean Lake Operation application for the renewal of the uranium mine operating licence for the McClean Lake Operation. Canadian Nuclear Safety Commission Member Document CMD 17-H91, June 7, 2017.

Audette-Stuart, M., Kim, S. B., McMullin, D., Festarini, A., Yankovich, T. L., Carr, J., & Mulpuru, S. (2011). Adaptive response in frogs chronically exposed to low doses of ionizing radiation in the environment. *Journal of Environmental Radioactivity, 102*(6), 566–573. https://doi.org/10.1016/j.jenvrad.2011.03.001

Bahamonde, P. A., Feswick, A., Isaacs, M. A., Munkittrick, K. R., & Martyniuk, C. J. (2016). Defining the role of omics in assessing ecosystem health: Perspectives from the Canadian environmental monitoring program. *Environmental Toxicology and Chemistry, 35*(1), 20–35. https://doi.org/10.1002/etc.3218

Beaugelin-Seiller, K. (2014). The assumption of heterogeneous or homogeneous radioactive contamination in soil/sediment: Does it matter in terms of the external exposure of fauna? *Journal of Environmental Radioactivity, 138*, 60–67. https://doi.org/10.1016/j.jenvrad.2014.07.027

Beaugelin-Seiller, K., Goulet, R., Mihok, S., & Beresford, N. A. (2016). Should we ignore U-235 series contribution to dose? *Journal of Environmental Radioactivity, 151*(Part 1), 114–125. https://doi.org/10.1016/j.jenvrad.2015.09.019

Beresford, N. A., Fesenko, S., Konoplev, A., Skuterud, L., Smith, J. T., & Voigt, G. (2016). Thirty years after the Chernobyl accident: What lessons have we learnt? *Journal of Environmental Radioactivity, 157*, 77–89. https://doi.org/10.1016/j.jenvrad.2016.02.003

Biron, P. A., Massarin, S., Alonzo, F., Garcia-Sanchez, L., Charles, S., & Billoir, E. (2011). Population-level modeling to account for multigenerational effects of uranium in *Daphnia magna*. *Environmental Science & Technology, 46*(2), 1136–1143. https://doi.org/10.1021/es202658b

Bréchignac, F., Oughton, D., Mays, C., Barnthouse, L., Beasley, J. C., Bonisoli-Alquati, A., Bradshaw, C., Brown, J., Dray, S., Geras'kin, S., Glenn, T., Higley, K., Ishida, K., Kapustka, L., Kautsky, U., Kuhne, W., Lynch, M., Mappes, T., Mihok, S., Møller, A. P., Mothersill, C., Mousseau, T. A., Otaki, J., Pryakhin, E., Rhodes, O. E., Jr., Salbu, B., Strand, P., & Tsukada, H. (2016). Addressing ecological effects of radiation on populations and ecosystems to improve protection of the environment against radiation: Agreed statements from a Consensus Symposium. *Journal of Environmental Radioactivity, 158*, 21–29. https://doi.org/10.1016/j.jenvrad.2016.03.021

CCME. (2011). *Canadian water quality guidelines for the protection of aquatic life: Uranium*. Canadian Council of Ministers of the Environment.

Chambers, D. B., Osborne, R. V., & Garva, A. L. (2006). Choosing an alpha radiation weighting factor for doses to non-human biota. *Journal of Environmental Radioactivity, 87*(1), 1–14. https://doi.org/10.1016/j.jenvrad.2005.10.009

Chaudhry, M. A. (2008). Biomarkers for human radiation exposure. *Journal of Biomedical Science, 15*(5), 557–563. https://doi.org/10.1007/s11373-008-9253-z

CNSC. (2017a). *The biological mechanisms acting at low doses of radiation*. Canadian Nuclear Safety Comnission, Ottawa, Canada, CMD 17-M46, November 9, 2017.

CNSC. (2017b). Environmental protection: Policies, programs and procedures, regulatory document REGDOC-2.9.1, version 1.1. Canadian Nuclear Safety Commission, Ottawa, Canada.

CNSC (2017c) *Regulatory oversight report for uranium mines and mills in Canada: 2016*, Canadian Nuclear Safety Commission, Ottawa, Canada, CMD: 17-M47, October 13, 2017.

CNSC, Environment Canada. (2009). *Risk management of uranium releases from uranium mines and mills, 2007 annual report*, Canadian Nuclear Safety Commision, Ottawa, Canada, Report INFO-0771.

CNSC, Environment Canada. (2012). *2010 Annual report on uranium management activities*, Canadian Nuclear Safety Commision, Ottawa, Canada, Report INFO-0830.

Colin, N., Porte, C., Fernandes, D., Barata, C., Padrós, F., Carrassón, M., Monroy, M., Cano-Rocabayera, O., de Sostoa, A., Piña, B., & Maceda-Veiga, A. (2016). Ecological relevance of biomarkers in monitoring studies of macro-invertebrates and fish in Mediterranean rivers. *Science of the Total Environment, 540*, 307–323. https://doi.org/10.1016/j.scitotenv.2015.06.099

DeForest, D. K., Brix, K. V., Elphick, J. R., Rickwood, C. J., deBruyn, A. M. H., Tear, L. M., Gilron, G., Hughes, S. A., & Adams, W. J. (2017). Lentic, lotic, and sulfate-dependent water-borne selenium screening guidelines for freshwater systems. *Environmental Toxicology and Chemistry, 36*(9), 2503–2513. https://doi.org/10.1002/etc.3793

Eberts, R. L., Wissel, B., Simpson, G. L., Crawford, S. S., Stott, W., Hanner, R. H., Manzon, R. G., Wilson, J. Y., Boreham, D. R., & Somers, C. M. (2017). Isotopic structure of lake ehitefish in Lake Huron: Evidence for tegional and local populations based on resource use. *North American Journal of Fisheries Management, 37*(1), 133–148. https://doi.org/10.1080/02755947.2016.1245225

ECCC. (2016). *Third national assessment of environmental effects monitoring information from metal mines subjected to the metal mining effluent regulations*, Environment and Climate Change Canada, Ottawa, Canada, Report En14-64/2016, December 2015.

Ecometrix Inc. (2008). *A review of environmental management criteria for selenium and molybdenum.* Mine Environmental Neutral Drainage (MEND) Program, Ottawa, Ontario, Report 1011, November 2007.

Environment and Climate Change Canada. (2017). *Fisheries act, proposed regulatory initiative: Coal mining effluent regulations*. Canada Gazette (pending).

Environment Canada, Health Canada. (2004). Canadian Environmental Protection Act, 1999. Priority substances list assessment report. Releases of radionuclides from nuclear facilities (Impact on non-human biota). Report En40-215/67E. Canadian Environmental Protection Act, 1999, Prioirty substances list assessment report, vol En40-215/67E. Government of Canada, Ottawa, Canada,

ESG Interternational Inc. (1999). *Aquatic Effects Technology Evaluation (AETE) program synthesis report of selected technologies for cost-effective environmental monitoring of mine effluent impacts in Canada*, AETE Report 4.1.4. Canada Center for Mineral & Energy Technology and Mining Association of Canada, Guelph, Ontario.

Feswick, A., Isaacs, M., Biales, A., Flick, R. W., Bencic, D. C., Wang, R.-L., Vulpe, C., Brown-Augustine, M., Loguinov, A., Falciani, F., Antczak, P., Herbert, J., Brown, L., Denslow, N. D., Kroll, K. J., Lavelle, C., Dang, V., Escalon, L., Garcia-Reyero, N., Martyniuk, C. J., & Munkittrick, K. R. (2017). How consistent are we? Interlaboratory comparison study in fathead minnows using the model estrogen 17α-ethinylestradiol to develop recommendations for environmental transcriptomics. *Environmental Toxicology and Chemistry, 36*(10), 2614–2623. https://doi.org/10.1002/etc.3799

Forbes, V. E., Palmqvist, A., & Bach, L. (2006). The use and misuse of biomarkers in ecotoxicology. *Environmental Toxicology and Chemistry, 25*(1), 272–280. https://doi.org/10.1897/05-257R.1

Frank, A. (2004). A review of the "mysterious" wasting disease in Swedish moose (*Alces alces* L.) related to molybdenosis and disturbances in copper metabolism. *Biological Trace Element Research, 102*(1), 143–159. https://doi.org/10.1385/BTER:102:1-3:143

Frank, A., Galgan, V., & Petersson, L. R. (1994). Secondary copper deficiency, chromium deficiency and trace element imbalance in the moose (*Alces alces* L.): Effect of anthropogenic activity. *Ambio, 23*(4/5), 315–317.

Friedrichs, K. R., Harr, K. E., Freeman, K. P., Szladovits, B., Walton, R. M., Barnhart, K. F., & Blanco-Chavez, J. (2012). ASVCP reference interval guidelines: Determination of de novo reference intervals in veterinary species and other related topics. *Veterinary Clinical Pathology, 41*, 441–453. https://doi.org/10.1111/vcp.12006

Froehlich, M. B., Steier, P., Wallner, G., & Fifield, L. K. (2016). European roe deer antlers as an environmental archive for fallout ^{236}U and ^{239}Pu. *Journal of Environmental Radioactivity, 151*(Part 3), 587–592. https://doi.org/10.1016/j.jenvrad.2015.06.008

Gagnaire, B., Adam-Guillermin, C., Festarini, A., Cavalié, I., Della-Vedova, C., Shultz, C., Kim, S. B., Ikert, H., Dubois, C., Walsh, S., Farrow, F., Beaton, D., Tan, E., Wen, K., & Stuart, M. (2017). Effects of in situ exposure to tritiated natural environments: A multi-biomarker approach using the fathead minnow, *Pimephales promelas*. *Science of the Total Environment, 599*, 597–611. https://doi.org/10.1016/j.scitotenv.2017.04.210

Garnier-Laplace, J., Alonzo, F., & Adam-Guillermin, C. (2015a). Establishing relationships between environmental exposures to radionuclides and the consequences for wildlife: Inferences and weight of evidence. *Annals of the ICRP, 44*(1 suppl), 295–303. https://doi.org/10.1177/0146645315572311

Garnier-Laplace, J., Beaugelin-Seiller, K., Della-Vedova, C., Métivier, J. M., Ritz, C., Mousseau, T. A., & Møller, A. P. (2015b). Radiological dose reconstruction for birds reconciles outcomes of Fukushima with knowledge of dose-effect relationships. *Scientific Reports, 5*, 16594. https://doi.org/10.1038/srep16594

Gombeau, K., Bourdineaud, J.-P., Ravanat, J.-L., Armant, O., Camilleri, V., Cavalie, I., Floriani, M., & Adam-Guillermin, C. (2017). Epigenetic, histopathological and transcriptomic effects following exposure to depleted uranium in adult zebrafish and their progeny. *Aquatic Toxicology, 184*(Supplement C), 14–25. https://doi.org/10.1016/j.aquatox.2016.12.004

Goulet, R. R., Thompson, P. A., Serben, K. C., & Eickhoff, C. V. (2015). Impact of environmentally based chemical hardness on uranium speciation and toxicity in six aquatic species. *Environmental Toxicology and Chemistry, 34*(3), 562–574. https://doi.org/10.1002/etc.2834

Government of Canada. (2017). Regulations amending the metal mining effluent regulations. *Canada Gazette, 151*(19) May 13, 2017.

Government of Saskatchewan. (2017). Water quality guidelines for the protection of aquatic life – Molybdenum (Mo) [DRAFT].

Guéguen, Y., Roy, L., Hornhardt, S., Badie, C., Hall, J., Baatout, S., Pernot, E., Tomasek, L., Laurent, O., Ebrahimian, T., Ibanez, C., Grison, S., Kabacik, S., Laurier, D., & Gomolka, M. (2016). Biomarkers for uranium risk assessment for the development of the CURE (Concerted Uranium Research in Europe) molecular epidemiological protocol. *Radiation Research, 187*(1), 107–127. https://doi.org/10.1667/RR14505.1

Hall, J., Jeggo, P. A., West, C., Gomolka, M., Quintens, R., Badie, C., Laurent, O., Aerts, A., Anastasov, N., Azimzadeh, O., Azizova, T., Baatout, S., Baselet, B., Benotmane, M. A., Blanchardon, E., Guéguen, Y., Haghdoost, S., Harms-Ringhdahl, M., Hess, J., Kreuzer, M., Laurier, D., Macaeva, E., Manning, G., Pernot, E., Ravanat, J.-L., Sabatier, L., Tack, K., Tapio, S., Zitzelsberger, H., & Cardis, E. (2017). Ionizing radiation biomarkers in epidemiological studies – An update. *Mutation Research, Reviews in Mutation Research, 771*, 59–84. https://doi.org/10.1016/j.mrrev.2017.01.001

Heijerick, D. G., Regoli, L., & Carey, S. (2012). The toxicity of molybdate to freshwater and marine organisms. II. Effects assessment of molybdate in the aquatic environment under REACH. *Science of the Total Environment, 435-436*(Supplement C), 179–187. https://doi.org/10.1016/j.scitotenv.2012.05.075

Hinton, T. G., Bedford, J. S., Congdon, J. C., & Whicker, F. W. (2004). Effects of radiation on the environment: A need to question old paradigms and enhance collaboration among radiation biologists and radiation ecologists. *Radiation Research, 162*(3), 332–338. https://doi.org/10.1667/rr3222

Hook, S. E., Gallagher, E. P., & Batley, G. E. (2014). The role of biomarkers in the assessment of aquatic ecosystem health. *Integrated Environmental Assessment and Management, 10*(3), 327–341. https://doi.org/10.1002/ieam.1530

IAEA. (2014). *Modelling of biota dose effects: Report of Working Group 6 Biota Dose Effects Modelling of EMRAS II Topical Heading Reference approaches for biota dose assessment, Environmental Modelling for Radiation Safety (EMRAS II) programme*. Dose thresholds of effect for benthic invertebrates associated with uranium mining and milling in Canada, IAEA TECDOC 1737, International Atomic Energy Agency. Vienna, Austria.

ICRP. (2003). *A framework for assessing the impact of ionising radiation on non-human species. Annals of the ICRP, ICRP Publication 91*. International Commission on Radiological Protection. https://doi.org/10.1016/S0146-6453(03)00022-8

Janz, D. M. (2012). Selenium. *Fish Physiology: Homeostasis and Toxicology of Essential Metals, 31A*, 327–374. https://doi.org/10.1016/S1546-5098(11)31007-2

Kilgour, B. W., Somers, K. M., Barrett, T. J., Munkittrick, K. R., & Francis, A. P. (2017). Testing against "normal" with environmental data. *Integrated Environmental Assessment and Management, 13*(1), 188–197. https://doi.org/10.1002/ieam.1775

Kilgour, B. W., Dowsley, B., McKee, M., & Mihok, S. (2018). Effects of uranium mining and milling on benthic invertebrate communities in the Athabasca Basin of northern Saskatchewan. *Canadian Water Resources Journal*. (in review).

Kroon, F., Streten, C., & Harries, S. (2017). A protocol for identifying suitable biomarkers to assess fish health: A systematic review. *PLoS One, 12*(4), e0174762. https://doi.org/10.1371/journal.pone.0174762

Lourenço, J., Mendo, S., & Pereira, R. (2016). Radioactively contaminated areas: Bioindicator species and biomarkers of effect in an early warning scheme for a preliminary risk assessment. *Journal of Hazardous Materials, 317*(Supplement C), 503–542. https://doi.org/10.1016/j.jhazmat.2016.06.020

Massarin, S., Alonzo, F., Garcia-Sanchez, L., Gilbin, R., Garnier-Laplace, J., & Poggiale, J. C. (2010). Effects of chronic uranium exposure on life history and physiology of *Daphnia magna* over three successive generations. *Aquatic Toxicology, 99*(3), 309–319. https://doi.org/10.1016/j.aquatox.2010.05.006

Mathews, T., Beaugelin-Seiller, K., Garnier-Laplace, J., Gilbin, R., Adam, C., & Della-Vedova, C. (2009). A probabilistic assessment of the chemical and radiological risks of chronic exposure to uranium in freshwater ecosystems. *Environmental Science & Technology, 43*(17), 6684–6690. https://doi.org/10.1021/es9005288

McCarty, L. S., & Munkittrick, K. R. (1996). Environmental biomarkers in aquatic toxicology: Fiction, fantasy, or functional? *Human and Ecological Risk Assessment: An International Journal, 2*(2), 268–274. https://doi.org/10.1080/10807039609383607

McCarty, L. S., Power, M., & Munkittrick, K. R. (2002). Bioindicators versus biomarkers in ecological risk assessment. *Human and Ecological Risk Assessment: An International Journal, 8*(1), 159–164. https://doi.org/10.1080/20028091056791

McKee, M., & Phaneuf, M. (2014). Risk based environmental assessment for uranium mines – The Canadian experience. Paper presented at the International symposium on uranium raw material for the nuclear fuel cycle: Exploration, mining, production, supply and demand, economics and environmental issues, International Atomic Energy Agency, IAEA-CN-216, Vienna, Austria (Proceedings in preparation),

Mihok, S., & Thompson, P. (2012). Regulatory experience in applying a radiological environmental protection framework for existing and planned nuclear facilities. *Annals of the ICRP, 41*, 256–262. https://doi.org/10.1016/j.icrp.2012.06.024

Mihok, S., Wilk, M., Lapp, A., St-Amant, N., Kwamena, N. O. A., & Clark, I. D. (2016). Tritium dynamics in soils and plants grown under three irrigation regimes at a tritium processing facility in Canada. *Journal of Environmental Radioactivity, 153*, 176–187. https://doi.org/10.1016/j.jenvrad.2015.12.025

Munkittrick, K. R., McMaster, M. E., & Courtenay, S. C. (2002). Introductory remarks – Scientific concepts associated with the development of the Canadian environmental effects monitoring program. *Water Quality Research Journal of Canada, 37*(1), 3–6.

Munkittrick, K. R., Barrett, T. J., & McMaster, M. E. (2010). Guidance for site-specifically assessing the health of fish populations with emphasis on Canada's Environmental Effects Monitoring program. *Water Quality Research Journal of Canada, 45*(2), 209–221.

Muscatello, J. R., Bennett, P. M., Himbeault, K. T., Belknap, A. M., & Janz, D. M. (2006). Larval deformities associated with selenium accumulation in northern pike (*Esox lucius*) exposed to metal mining effluent. *Environmental Science & Technology, 40*(20), 6506–6512. https://doi.org/10.1021/es060661h

Muscatello, J. R., Belknap, A. M., & Janz, D. M. (2008). Accumulation of selenium in aquatic systems downstream of a uranium mining operation in northern Saskatchewan, Canada. *Environmental Pollution, 156*(2), 387–393. https://doi.org/10.1016/j.envpol.2008.01.039

Oughton, D. H. (2016). Ethical foundations of environmental radiological protection. *Annals of the ICRP, 45*(1 suppl), 345–357. https://doi.org/10.1177/0146645316639836

Overdyk, L. M., Braid, H. E., Crawford, S. S., & Hanner, R. H. (2015). Extending DNA barcoding coverage for lake whitefish (*Coregonus clupeaformis*) across the three major basins of Lake Huron. *DNA Barcodes, 3*(1), 59. https://doi.org/10.1515/dna-2015-0007

Overdyk, L. M., Braid, H. E., Naaum, A. M., Crawford, S. S., & Hanner, R. H. (2016). Real-time PCR identification of lake whitefish *Coregonus clupeaformis* in the Laurentian Great Lakes. *Journal of Fish Biology, 88*(4), 1460–1474. https://doi.org/10.1111/jfb.12922

Robertson, E. L., & Liber, K. (2007). Bioassays with caged *Hyalella azteca* to determine in situ toxicity downstream of two Saskatchewan, Canada, uranium operations. *Environmental Toxicology and Chemistry, 26*(11), 2345–2355. https://doi.org/10.1897/06-489R.1

Sanchez, W., & Porcher, J.-M. (2009). Fish biomarkers for environmental monitoring within the Water Framework Directive of the European Union. *TrAC Trends in Analytical Chemistry, 28*(2), 150–158. https://doi.org/10.1016/j.trac.2008.10.012

SENES. (2006). *Uranium in effluent treatment process*. Prepared for the Canadian Nuclear Safety Commission, Ottawa, Canada by SENES Consulting Ltd. Research Contract RSP-0204.

Shi, X., Smith, R., Seymour, C., & Mothersill, C. (2017). Ra-226 bioaccumulation and growth indices in fish. *International Journal of Radiation Biology, 93*(6), 617–626. https://doi.org/10.1080/09553002.2017.1297900

Skipperud, L., Jorgensen, A. G., Heier, L. S., Salbu, B., & Rosseland, B. O. (2013). Po-210 and Pb-210 in water and fish from Taboshar uranium mining Pit Lake, Tajikistan. *Journal of Environmental Radioactivity, 123*, 82–89. https://doi.org/10.1016/j.jenvrad.2012.03.013

Skorupa, J. P. (1998). Selenium poisoning of fish and wildlife in nature: Lessons learned from twelve real-world examples. In W. T. Frankenberger & R. A. Engberg (Eds.), *Environmental chemistry of selenium*. CRC Press.

Smith, R. W., Wang, J., Mothersill, C. E., Lee, L. E. J., & Seymour, C. B. (2015). Proteomic responses in the gills of fathead minnows (*Pimephales promelas*, Rafinesque, 1820) after 6 months and 2 years of continuous exposure to environmentally relevant dietary ^{226}Ra. *International Journal of Radiation Biology, 91*(3), 248–256. https://doi.org/10.3109/09553002.2014.988894

Snape, J. R., Maund, S. J., Pickford, D. B., & Hutchinson, T. H. (2004). Ecotoxicogenomics: The challenge of integrating genomics into aquatic and terrestrial ecotoxicology. *Aquatic Toxicology, 67*(2), 143–154. https://doi.org/10.1016/j.aquatox.2003.11.011

Stark, K., Goméz-Ros, J. M., Vives i Batlle, J., Lindbo Hansen, E., Beaugelin-Seiller, K., Kapustka, L. A., Wood, M. D., Bradshaw, C., Real, A., McGuire, C., & Hinton, T. G. (2017). Dose assessment in environmental radiological protection: State of the art and perspectives. *Journal of Environmental Radioactivity, 175–176*, 105–114. https://doi.org/10.1016/j.jenvrad.2017.05.001

Strand, P., Sundell-Bergman, S., Brown, J. E., & Dowdall, M. (2017). On the divergences in assessment of environmental impacts from ionising radiation following the Fukushima accident. *Journal of Environmental Radioactivity, 169–170*, 159–173. https://doi.org/10.1016/j.jenvrad.2016.12.005

Stuart, M., Festarini, A., Schleicher, K., Tan, E., Kim, S. B., Wen, K., Gawlik, J., & Ulsh, B. (2016). Biological effects of tritium on fish cells in the concentration range of international drinking water standards. *International Journal of Radiation Biology, 92*(10), 563–571. https://doi.org/10.1080/09553002.2016.1222090

Thome, C., Mitz, C., Somers, C. M., Manzon, R. G., Boreham, D. R., & Wilson, J. Y. (2016). Incubation of lake whitefish (*Coregonus clupeaformis*) embryos in cooling water discharge and the impacts of fluctuating thermal regimes on development. *Canadian Journal of Fisheries and Aquatic Sciences, 73*(8), 1213–1221. https://doi.org/10.1139/cjfas-2015-0286

Tse, J. J., Gallego-Gallegos, M., Franz, E. D., Liber, K., & Pickering, I. J. (2012). Selenium speciation and localization in chironomids from lakes receiving treated metal mine effluent. *Chemosphere, 89*(3), 274–279. https://doi.org/10.1016/j.chemosphere.2012.04.036

Ulsh, B., Hinton, T. G., Congdon, J. D., Dugan, L. C., Whicker, F. W., & Bedford, J. S. (2003). Environmental biodosimetry: A biologically relevant tool for ecological risk assessment and biomonitoring. *Journal of Environmental Radioactivity, 66*(1), 121–139. https://doi.org/10.1016/S0265-931X(02)00119-4

UNSCEAR. (2011). Effects of ionizing radiation on non-human biota. Sources and effects of ionizing radiation. United Nations Scientific Committee on the Effects of Atomic Radiation. UNSCEAR 2008, Report to the General Assembly with Scientific Annexes, Volume II, Annex E.

US EPA. (2016). *Aquatic life ambient water quality criterion for selenium in freshwater 2016.* United States Environmental Protection Agency Factsheet.

Vanhoudt, N., Vandenhove, H., Real, A., Bradshaw, C., & Stark, K. (2012). A review of multiple stressor studies that include ionising radiation. *Environmental Pollution, 168*, 177–192. https://doi.org/10.1016/j.envpol.2012.04.023

Walsh, S., Satkunam, M., Su, B., Festarini, A., Bugden, M., Peery, H., Mothersill, C., & Stuart, M. (2015). Health, growth and reproductive success of mice exposed to environmentally relevant levels of Ra-226 via drinking water over multiple generations. *International Journal of Radiation Biology, 91*(7), 576–584. https://doi.org/10.3109/09553002.2015.1028598

Whitehouse, L. M., McDougall, C. S., Stefanovic, D. I., Boreham, D. R., Somers, C. M., Wilson, J. Y., & Manzon, R. G. (2017). Development of the embryonic heat shock response and the impact of repeated thermal stress in early stage lake whitefish (*Coregonus clupeaformis*) embryos. *Journal of Thermal Biology, 69*(Supplement C), 294–301. https://doi.org/10.1016/j.jtherbio.2017.08.013

Wiramenden, C. I. E., Forster, E. K., & Liber, K. (2010). Selenium distribution in a lake system receiving effluent from a metal mining and milling operation in Northern Saskatchewan, Canada. *Environmental Toxicology and Chemistry, 29*(3), 606–616. https://doi.org/10.1002/etc.63

Wismer, D.A., & McKee, M. (2007). *Key Lake present operations cumulative effects: CNSC staff determination of environmental risk* (210 pp., August 2006). Referenced in Canadian Nuclear Safety Commission CMD: 07-H5, January 25, 2007. Canadian Nuclear Safety Commission.

Index

M. D. Wood et al. (eds.), *Biomarkers of Radiation in the Environment*, NATO Science for Peace and Security Series A: Chemistry and Biology,
https://doi.org/10.1007/978-94-024-2101-9